ML

WI

Cosmic Legacy

Space, Time, and the Human Mind

Greg F. Reinking

VANTAGE PRESS
New York

FIRST EDITION

Copyright © 2003 by Greg F. Reinking

Published by Vantage Press, Inc.
516 West 34th Street, New York, New York 10001

Manufactured in the United States of America
ISBN: 0-533-14027-7

Library of Congress Catalog Card No.: 01-126753

0 9 8 7 6 5 4 3 2 1

To my wife, Trang,
our families,
and friends.

Contents

Preface

The ancient Greek philosopher Aristotle found humans infinitely curious about their own nature. As sentient intelligent beings, we have the ability to pose questions, pursue solutions, comprehend analyses, and share knowledge among ourselves. There is an innate interest in understanding what we are, how we came to exist, what makes us human, and what the future may hold. The equally intriguing question of *why* is perhaps more in the realm of philosophy, metaphysics, and religion—however, understanding the nature of things does provide insight.

The questions investigated in this book are not new; they have been asked for countless generations. It is fortunate for us that progress in the sciences has raised our understanding of the world to ever higher levels. In fact, it has been estimated that about 80 percent of all scientific discoveries have been made in just the last 100 years. Although I have made attempts to highlight the latest developments in many of the topics covered, continuing investigations will undoubtedly modify or even radically change many of the current theories.

It is interesting to follow the change in perspective man has had of the world and his place in it. The domain of our environment has been repeatedly redefined throughout history. To prehistoric people, the world was limited to their local habitat and hunting grounds. The picture was broadened to include the planet Earth by the philosophers of ancient Greece. Eventually, the Universe became a place that contained many stars but was limited to what we now call the Milky Way. In the twentieth century, as astronomy and cosmology progressed, the Milky Way was found to be just one galaxy within a vast Universe that contains up to 100 billion galaxies. Furthermore, some cosmologists propose that our Universe may be one of many (perhaps *infinitely* many). Humbling.

As the theories of physics and cosmology converge on the origin of the Universe, it is becoming increasingly apparent that there are direct relations with its structure in the ultramicroscopic realm. In fact, some of the different theories of the Universe's beginning suggest a common underlying theme where the fabric of space and time on the smallest scales is unlike a smooth continuum, as had been expected, but more akin to a discontinuous foam with an ultimate lower limit to its divisibility (including both spatial and time dimensions). It may be the structure of this foam that holds the answer to how the Universe began and how its properties from the smallest to grandest scales are determined (matter, energy, forces, and perhaps much more). It was from the material and properties of our curious Universe that the ingredients of terrestrial life were formed and came together.

All terrestrial life, including ourselves, is made of star dust collected from the cosmos billions of years ago. Is it actually possible to explain the existence of life from the materials that the Universe has provided? The scientific disciplines of biogenesis and evolution seek the answer. In the case of animal life, the cognitive properties of sentience and intelligence are a challenge to define and rigorously attribute to neuroanatomy and biochemistry. Throughout written history, man has questioned his origin, capabilities, and destiny. It is these questions, and many more, that are explored in the pages of *Cosmic Legacy*.

Cosmic Legacy represents a compilation of years of research, presenting its extensive thesis in a roughly chronological sequence, from the beginning of the Universe, the origin of life, the arrival of man, to the end of everything. It is intended to address basic questions of our existence and humanity, using understandable, succinct, and thought-provoking style, supported by accepted theories and the latest discoveries. *Cosmic Legacy* is not meant to be the final word or even a comprehensive source but a framework to give the reader an exposure to the vast complexity of what we are.

I must acknowledge the work of the countless scientists, philosophers, and historians who have dedicated their time to improving our knowledge base.

Cosmic Legacy

Part I

The Beginning

The die is cast.

—Gaius Julius Caesar, 49 B.C.

1

Origin of the Universe

Comprehending the origins of life, matter, and the Universe itself has been an elusive subject of human curiosity since time immemorial. As the tools and methods of the quest to explain the Universe's birth have become more sophisticated, the dense veil of mystery that cloaks the events of that nascent period is slowly lifting. Recent endeavors to decipher this monumental instant are focusing the attention of investigators on the fundamental principles governing the architecture of space, matter, force, and time. Indeed, these presumably discrete elements of nature that we experience in our existence, alternatively, may ultimately represent permutations of a single medium through a multidimensional construct.

It is generally believed that the nature of our Universe, including its eventual fate, was determined at a fiery birth billions of years ago. The formation of the matter of which everything is composed occurred within just an infinitesimal part of the first second of that colossal inferno. In fact, the very laws of physics that govern our natural environment were established in those early moments. Predestinarians argue that all events—including our thoughts and intentions—are the direct and inevitable consequence of the unfolding of creation. Such believers contend that once matter and its forces (which can be mathematically described) came into existence, they were destined to dutifully follow predictable patterns throughout time, rendering the future purely determined by events of the past and present. However, the elegance of that philosophy is shaken when quantum effects—properties on the smallest scales—are considered.

The desire to uncover the mysteries of the birth of our Universe has challenged humanity for millennia. In ancient times theories were rooted in forms of philosophical, religious, or quasi-scientific speculation. Ancient Greek philosophers proposed the Universe began in a formless state of elements, called *chaos,* which somehow achieved order to form the cosmos. The transition was a role delegated to various characters in mythology. However, centuries of philosophers attempted vainly to explain the process through material principles rather than the whims of deities. In the mid–seventeenth century, James Ussher, Archbishop of Armagh, Ireland, consulted biblical references to pinpoint the Earth's creation at 6:00 P.M. on October 22, 4004 B.C. Thousands of years of philosophical reasoning and, more recently, scientific and mathematical investigations have not been able to satisfactorily describe the exact nature of the Universe's origin. In fact, the moment the Universe began is considered indefinable by some investigators as they argue that time itself may not have existed in the conventional sense.

The task of retracing the footsteps to the Universe's earliest moments was the realm of pure speculation until only recently. How is it possible to even consider such a mystical journey? Advances in physics and applications of new technologies are bringing us ever closer to the beginning. Strangely enough, the capability to actually "see" the early Universe, and watch it age, exists.

Unveiling the Past

By the middle of the twentieth century the ability to look back in time at our Universe was recognized. The "time machine" had been used by mankind for centuries and is known as the astronomical telescope. Telescopes collect the photons (the particle waves that comprise light) emitted from stars and focus them into an image, analogous to the focus of light from a camera's lens to produce a portrait. Larger telescopes collect more light and therefore can detect fainter (and more distant) sources of photons than telescopes with relatively smaller apertures. The light of the stars originating in other galaxies travels great distances to reach the Earth. The light emitted by these distant stars, although traveling very fast through intergalactic space (300,000,000, or 3×10^8, meters each second), does not arrive at our telescopes instantaneously; it actually takes a great deal of time to make that journey. Thus the images of these distant stars that we see in our telescopes today are how the stars appeared when their light was emitted at some time in the past. How far in the past a particular star's image dates depends upon the distance between us and that star.

This critical concept in astronomy and cosmology, of "looking into the past," deserves particular attention. This "distance versus time" phe-

4

nomenon is related to the speed at which information travels. In the case of astronomy, information is transmitted by photons of electromagnetic radiation in the form of light. A beam of light may be thought of as a chain of photons; the typical 100-watt bulb in a desktop lamp emits about 10^{20} photons every second. The wavelength (or frequency)—related to color—of the photons of electromagnetic radiation may be visible to the eye or reside at wavelengths that lie outside the visible spectrum, but all photons travel at the same speed. It is from the finite speed at which light travels that a delay develops that allows us to look into the past. This time-delay phenomenon not only applies to the celestial objects of the cosmos but also can be related to everyday life.

Returning to the analogy of the photographer composing a portrait, the light from the subject must traverse a distance before it enters the camera lens to the detector (the film or photographer's eye). Light travels about one meter every three nanoseconds (a nanosecond is a billionth of a second). Thus what the detector sees is an image of the subject at some point in the past. Furthermore, let's assume both the photographer and subject have highly accurate watches that are synchronized. If the subject of the portrait is seated about three meters from the camera, the light that forms the subject's image takes about nine-billionths of a second to reach the camera; the subject's watch in the camera viewfinder will register a time that is about nine-billionths of a second behind the time of the photographer's watch. In taking a portrait, as with the detection of an image by the human eye, the exposure time necessary to create the image is much longer than the transit period of light over such a short distance, rendering the time delay phenomenon inconsequential. However, at increased subject distances, the effect of the finite transit time of light becomes noticeable. During the *Apollo* moon explorations, the one-way transit time of the communications between the Earth and Moon was a little over one second. Light from the Sun takes about five hundred seconds, a little over eight minutes, to reach the Earth. Transit periods of communications with interplanetary space probes at more distant locations are correspondingly increased, which requires designing such probes with capabilities to function with some degree of independence, as simply relying on continuous commands sent from the Earth can be disastrous (the signal transit time can be too prolonged for practical operation).

The distances in astronomy are so vast that conventional units to measure distances become cumbersome and inadequate. A conceptually simple and convenient measurement unit to define astronomical distances is

5

the light-year, which is the distance that light travels in one year; 9,500,000,000,000 kilometers (9.5×10^{12} kilometers). Thus if we were to measure the distance to the nearest star (aside from our Sun), Proxima Centauri, we would find that it is located about 40,000,000,000,000 (4×10^{13}) kilometers away. At that distance, the light originating from Proxima Centauri takes about 4.3 years to reach us—Proxima Centauri lies 4.3 light-years away. The distances to other galaxies range from millions to billions of light-years! If we are focused on the faint image of a galaxy that is 10 billion light-years distant, the photons have been traveling through space for 10 billion years and the image they form in our telescope is of how that galaxy appeared 10 billion years ago. Likewise, by the time the light emitted today by our Sun reaches an observer in a galaxy 10 billion light-years away, the Sun will have been burned out for about 5 billion years (despite what the observer sees).

Technological advances are now enabling astronomers to peer much deeper into space than before, and they are detecting galaxies that had formed within the first billion years of our Universe's existence. These distant early galaxies (proto-galaxies) typically have forms that are different from the nearby contemporary galaxies that lie in closer proximity to our galaxy (Figure 1.1). If we continue along this line of reasoning, shouldn't there be something at about 14 billion light-years away that *is* the beginning? It isn't quite that simple, as most cosmologists believe that the Universe began in a highly compacted state where the point of origin has been stretched out over the aeons so that it actually exists "everywhere" in our expanded Universe of today. However, astronomers have actually detected a residual glow from the nascent Universe, which indeed exists in all directions of the cosmos.

Thus far we have been limiting our discussion to the photons of visible light. The electromagnetic spectrum provides many other forms of radiation at shorter and longer wavelengths than visible light. Through a window of cosmic light at wavelengths beyond the visible spectrum, photons beginning their journey in the distant past are revealed. This very ancient source of photons at longer wavelengths forms the "cosmic microwave background radiation," which is a representation of the Universe as it appeared as early as 379,000 years after its birth (Figure 1.2). That is very close to the beginning; roughly 0.003 percent of its estimated current age. If we use the analogy of a seventy-five-year-old man, going back to 0.003 percent of his age (including gestation), he would be a single cell (zygote) just over eighteen hours old. The fertilized egg, a zygote, is a

6

Figure 1.1 NASA Hubble Space Telescope image of the galaxy cluster CL1358+62 revealed a gravitationally-lensed image of one of the most distant galaxies known (arrow). [Upper Right] Enlargement of the distant galaxy clearly shows irregular structure with bright areas that may represent regions of active star formation. [Lower Right] Computer enhancement of the galaxy image to illustrate its actual appearance by removing the artifacts of the gravitational lens. *Credit: Marijn Franx (University of Groningen, the Netherlands), Garth Illingworth (University of California, Santa Cruz), and NASA.*

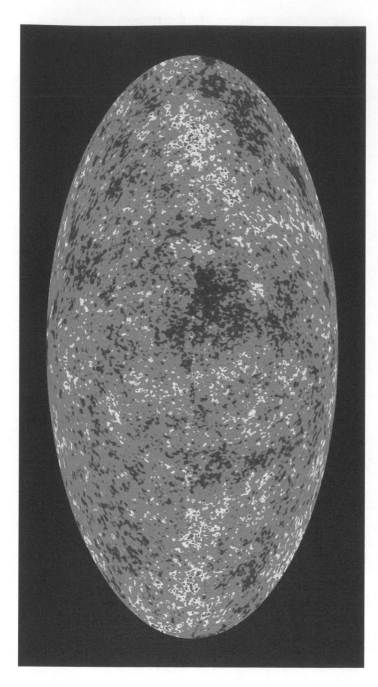

Figure 1.2 The cosmic microwave background radiation is a remnant of the early Universe, which has an underlying anisotropy representing fluctuations in density. Although very slight, these early density variations led to the complex structure we see in our Universe today. In February 2003, the WMAP satellite team released this image from 1989. Analysis of the WMAP image data suggests the microwave background radiation was released 379,000 years after universal expansion began and that the Universe is 13.7 billion years old (with a one percent margin of error). Credit: NASA and WMAP Science Team.

8

cell that makes its first division (to become two cells) about thirty hours after fertilization. This cycle of cell division and specialization continues until there are billions of cells forming the different tissues of the human body. As with the early Universe, a zygote has a much different appearance from what it eventually becomes.

Unfortunately, there are no other readily detectable lingering phenomena known to provide a peek at the Universe prior to the formation of the microwave radiation background. In theory, gravitational waves would have been produced during the earliest instants of the Universe's birth and may have left their fingerprints as gentle ripples on the cosmic microwave background radiation, which formed hundreds of thousands of years later. However, the presence of the primordial gravitational waves remains unproven. To turn the cosmic clock back to the earlier times, before it reached 379,000 years of age, we must rely on theoretical models that make predictions as to what the Universe was like. The field of "computational cosmology" is used to create virtual universes that are based on theoretical models. The simplest of these inherently complex models have about fifteen parameters that may be varied. These models utilize the principles of physics and must satisfy certain boundary conditions such that the early Universe they are designed to characterize will develop into a virtual universe that is similar to ours of today. The investigator chooses the input parameters and allows the computer to calculate the characteristics of the virtual universe from its program. Cosmological models that fail to produce a universe that is consistent with the cosmic microwave radiation background and other observable phenomena (e.g., distribution of matter) of our Universe must have their validity questioned.

Although it remains a topic of heated controversy, most estimates of the Universe's age fall between 10 and 16 billion years, with the latest estimates suggesting 13 to 15 billion years. Despite the high technology and many teams of astronomers trying to solve the ancient question, the debate continues. In May 1999 the most recent and "accurate" estimate was announced at a press conference by Wendy Freedman and her team at Carnegie Observatories using observational data from the Hubble Space Telescope. Their investigations suggest the Universe has existed between 12 and 13.5 billion years. Almost immediately, Allan Sandage and his team, also of Carnegie Observatories, also using data based on Hubble Space Telescope observations, suggested an older Universe of between 11 and 18 billion years. Subsequent WMAP data imply both groups were off.

The desire to understand our Universe and its origin has generated

9

much attention through the centuries and formed the central thesis of the science of cosmology. Only recently, through advances in science and technology, it has become possible for us to develop theories capable of describing the conditions that existed within an infinitesimal fraction of a second of the moment of origin. Carried by the wings of cosmology, based on theoretical, particle, and quantum physics, our journey will take us to the beginning of our Universe: initial singularity.

History of Cosmology

We will begin with a brief history of cosmology, the field of science that provides the most plausible theories of our Universe's birth. The creatures of Earth live in a world of complex and mysterious natural forces, which humbled early man. To those living in ancient times, tales of an original creation were handed down as answers to a fundamental question for which they otherwise had no plausible explanation. Such stories, often told through a cast of characters including man, beast, or gods, cleverly explained the mysterious origins of everything and defined a place for humans in the curious natural world. These stories were taught and accepted without what we would consider scientific scrutiny and verification. In ancient times, scientific evaluation didn't exist, as contemporary understanding of the physical world was in its infancy and largely guided by superstition. Although the stories did provide satisfactory explanations for countless generations, these beliefs were eventually questioned. It must be remembered that investigations of the Universe by the ancients were limited to what one could see by simply looking at the sky. It wasn't until the sixteenth century that Galileo Galilei fashioned the first telescope to visually bring these distant mysterious places a little closer for inspection.

Since the times of antiquity, philosophers pondered the nature of the Universe in the pursuit of intellectual solace. Ahead of his time, Aristotle, in the fourth century B.C., went as far as to postulate that the Earth was a sphere located at the center of the Universe. Aristotle's model described the cosmos as a series of concentric transparent spherical shells spinning about the Earth, and on these shells were the familiar celestial bodies. However, this model was not able to accurately predict important celestial events such as eclipses or planetary motions. By the third century B.C., Eratosthenes was able to prove the world was indeed round by measuring the shadows cast by stakes placed at different latitudes on the Earth's surface. The passage of the centuries brought other models, but it was Claudius Ptolemaeus (Ptolemy), an astronomer at Alexandria in the second century A.D., who proposed a more sophisticated model of the Universe, which became widely accepted. The Ptolemaic model, which also had the Earth at the Universe's center, described the motions of the celestial bodies as following complex circular paths (epicycles). This much more mathematically complex model improved the ability to predict celestial events. The church adopted and supported his theory, which held for over a millennium. In 1514, Nicholas Copernicus, a clergyman, placed the Sun at the center of the Universe to make sense of the otherwise extremely complicated paths the planets followed through the sky. He is given credit as the founder of modern astronomy for dethroning the Earth from center stage. Johannes Kepler, a German astronomer and mathematician, continued the work a century later by changing the celestial orbits from "perfect circles" to ellipses. In the late seventeenth century, Sir Isaac Newton published his classical theories of solid body motion and gravitation in *Principia,* which further refined the accuracy of calculating orbits and interactions of celestial bodies through a set of mathematical equations. The success of the improved cosmological models gradually brought their acceptance, which consequently displaced some of the mysticism lingering from previous theories.

Until the twentieth century, it was generally held that the Universe was a homogeneous place (looks the same in all directions) in a state of constancy across the ages (static). Thus the Universe would have looked essentially the same in the distant past as it would far into the future, composed of the same celestial bodies that had been observed throughout recorded time. The Universe was considered a boring, stagnant place where any sort of dynamic change was unthinkable. The concept of a static Universe came into question early in the twentieth century.

Profound cosmological insight may be gained simply by looking to the cosmos: the sky at night is dark. In an infinite Universe, one that has an infinite number of stars that have been shining for an infinite amount of time, starlight would reach us from all directions. No matter which direction an observer peered, there would be a star. Even if some stars were very far away, their light would still reach the observer, as the photons had an infinite amount of time to arrive. This hypothetical universe would produce a bright sky in all directions, which is clearly in contradiction to the dark night sky of our cosmos. This phenomenon is known as Olbers' paradox, credited to the nineteenth-century German astronomer Wilhelm Olbers. The paradox implies the Universe is not necessarily infinite in content or age but may have had a beginning.

Serious doubt of a static universe was cast when Albert Einstein published his General Theory of Relativity in 1915. Einstein was able to mathematically describe physical properties of the entire Universe, which was not possible with the classical physics models of times prior. Also known as Einstein's Theory of Gravity, the General Theory of Relativity may be written in a simplified form as $G_{\mu\upsilon} = 8\pi T_{\mu\upsilon}$, where $G_{\mu\upsilon}$ is related to space and time while $T_{\mu\upsilon}$ is related to mass and energy. He proposed a relation between objects and their inertial frame (location) creating a four-dimensional continuum composed of the three dimensions of space combined with one dimension of time. Einstein's theory describes the Universe as a smooth space-time continuum where the presence of matter geometrically distorts the continuum, producing the effect of gravity. Experimental verification of Einstein's theory was confirmed at the total solar eclipse of 1919, when astronomers were able to detect a shift in position of stars near the Sun consistent with a gravitational distortion of space by the Sun's mass. However, Einstein's equations are complicated and to accurately describe the Universe in detail becomes exceedingly difficult. Simplifications of these equations can be made that lead to approximate solutions. Furthermore, the predictions made by the simplified equations can be correlated with astronomical observations to ascertain their validity. The approximations must produce solutions that correlate with the characteristics of our Universe; otherwise, the simplifications are assumed invalid.

In order to simplify the equations of General Relativity, Einstein assumed that the Universe has a uniform density. This assumption may, at first, seem inaccurate since we clearly know the Universe has an extremely heterogeneous density locally; the Earth is solid, the atmosphere is gaseous, and interplanetary space is basically a vacuum. If one considers the Universe on a vast cosmological scale, in terms of billions of light-years, it actually approaches a pattern of uniform structure and density. On such vast scales, the Universe looks nearly the same in all directions (isotropic) and matter is uniformly spread out (homogenous). Upon applying this simple assumption, to his surprise he found that his mathematical model predicted a dynamic Universe, one in either expansion or contraction. He felt the solution was flawed, as he believed the Universe was static, the prevalent contemporary dogma.

In an attempt to conform to contemporary bias, Einstein introduced *lambda*—later renamed the Cosmological Constant—to alter the unexpected solution so that his model described a static, rather than a dynamic, universe. Translating this theoretical mathematical constant to the physical world, it would represent a force of opposition to gravity. When this factor was included, the solution was a static universe, but one that was highly unstable. In 1922 Alexander Friedmann, a Soviet mathematician, contested the cosmological constant and solved the equations of General Relativity without it. He was later credited as the first individual describing the currently accepted idea of an expanding universe. Einstein later referred to his cosmological constant as his "greatest blunder." He evidently was a man given to self-criticism, as he later described a secret letter he wrote to President Roosevelt in 1939 urging the development of the atomic bomb, which eventually led to the Manhattan Project, as the "single greatest mistake" of his entire life.

In 1927, Edwin Hubble and Milton Humason, working with the 100-inch Mount Wilson telescope (the largest at the time), were able to offer convincing observational data to support the expanding universe model. In the early twentieth century, the Universe was thought to be much smaller than we now recognize, basically confined to the dimensions of our own galaxy. Astronomers had long known the existence of many peculiar small patches of faint light (nebulae) scattered throughout the sky, which were assumed to be within our galaxy, as was everything else in the Universe. Hubble was intrigued with these nebulae and carefully studied the light from them. Eventually he was able to demonstrate that many of these "nebulae" were in fact other galaxies far outside our Milky Way galaxy. Furthermore, Hubble examined the properties of the light emitted by these galaxies to determine their relative velocities (velocity of either approach or recession). He applied the principle of the "Doppler Shift" of light to the observed emission spectra (the component colors) of the light of these galaxies to determine their motion relative to the Earth.

The Doppler Shift of light refers to the phenomenon by which the color spectrum of light is skewed when its source is in motion relative to the observer. The magnitude of color shift is proportional to the relative velocity of approach or recession to the observer. The wavelength (color) of light is compressed when its source approaches the observer, and conversely, the wavelength is stretched with a receding source. Blue light has a shorter wavelength than red light. As a result, an approaching light source appears blueshifted while a receding source appears redshifted. In cosmology, the redshift effect is related to the expansion of space through which the light travels rather than a true velocity of recession between the light source and observer. Hubble's observations revealed redshifted spectra indicating the galaxies were spreading away (with very few exceptions) from us in all directions. Analyzing the amount of spectral shift, which corresponds to the relative velocity of galactic recession, he also found that more distant galaxies were moving away faster than those located closer to us. A pattern quickly developed where the rate of galactic recession increased in direct proportion to the distance of a galaxy, the proportionality being Hubble's Constant (h). The principle of the increasing recession velocity with increasing galactic distance is Hubble's Law. The combination of these observational findings leaves us with a Universe that is expanding in all directions.

The concept of an expanding universe has powerful ramifications. A major implication is that the Universe has a finite age. If all of the pieces of the Universe are moving away from one another, it is logical to assume that in the past they were closer together. If we continue back, there must be a time when all the matter was collected and compressed into one very dense hot point when the expansion started; this is the basis for the standard Big Bang theory of the Universe's origin. Since Hubble's Law is basically a linear relationship between galactic recession velocity and distance (from Earth), with Hubble's Constant as the ratio, a simple reinterpretation can theoretically predict the age of the Universe. If we assume galaxies are moving away from a point of origin and know the relative distances to the galaxies, it is a simple calculation to determine how long they have been traveling (assuming a constant rate of expansion), which gives us the age of the Universe. Under Hubble's Law, the age of a universe undergoing a constant rate of expansion is the reciprocal of Hubble's Constant.

Hubble's initial estimation of the Constant was approximately 500 km/sec per mega-parsec (a mega-parsec is 1 million parsecs or 3.26 million light-years). According to Hubble's first prediction, the Universe would be about 2 billion years old, less than half the age of our Sun. However, the most recent estimates place Hubble's Constant from 70 to 73 km/sec per mega-parsec with a range between 65 and 81 km/sec per mega-parsec. Under such a rate of expansion, two points in space one light-year apart would separate at a rate of a couple centimeters each second. Calculations using the revised estimates of Hubble's Constant place the maximum age of the Universe at about 10 billion years. Such results are inconsistent with observational findings, as many celestial objects are

14

believed to be more than 10 billion years old, making nonsense of the result: how can a Universe be younger than the stars it contains? The paradox lies with the assumption of a constant rate of expansion. Recent evidence suggests the rate of cosmological expansion has not been constant, and therefore the Universe's age is not directly related to the inverse of Hubble's Constant.

Models of the Universe based on Einstein's gravity suggest the total energy density (which determines the expansion rate) is dependent upon two factors: the mean mass density (Ω_m) and the Cosmological Constant (Ω_Λ). The Universe contains mass in many forms—the majority being "dark matter"—which contributes to the mean mass density. Mass tends to slow expansion through the collective gravitational tug of all matter between its particles. However, observations of distant supernovae and the cosmic microwave background suggest the cosmic expansion rate is actually increasing, which has renewed interest in Einstein's Cosmological Constant. The Cosmological Constant, a product of the General Theory of Relativity, represents a uniform energy density of the vacuum, independent of mass density. If the Cosmological Constant has a positive value, it is analogous to an antigravity force, which would accelerate the rate of expansion of the Universe. The contribution of these factors complicates attempts at estimating the age of our Universe, as each new factor has its own range of uncertainty. Combining recent estimates of these factors (h, Ω_m, and Ω_Λ) into a modified Big Bang model suggests an age of 13.4 billion years, with a range of 1.6 billion years above and below the result. The exact value of Hubble's Constant and the other factors, their relative contributions, and the appropriate theoretical model remain subjects of debate in cosmological circles, which will continue to raise skepticism of any reports claiming an "accurate" age of the Universe.

Undoubtedly the early Universe was much hotter and more tightly packed (dense) than it is today. As the constituent particles of the Universe (or any matter) are compressed, they are more likely to interact or "bump" into their neighbors, which is analogous to a rise in temperature (an increase in the average kinetic energy of the particles). Conversely, as a substance is allowed to expand (as the Universe does), its overall temperature decreases. Thus, assuming that the early Universe was bathed in heat, it was proposed that a background radiation would exist today, a lingering warm residue from its steamy birth. The conflagration of creation would be cooled (redshifted) during the intervening billions of years of expansion. Furthermore, it would exist in all directions as a nearly uniform energy. The cosmic microwave background radiation was accidentally discovered in 1965 by Arno Penzias and Robert Wilson, two radio engineers at the Bell Telephone Lab in New Jersey. It had been predicted by Big Bang cosmological models over fifteen years earlier that there should be a low-temperature (microwave) background radiation. In fact, every cubic meter of space contains about 400 million of these microwave background photons from the early Universe. More recent observations by NASA's Cosmic Background Explorer (COBE) satellite in 1989 refined the spectrum of this energy, without the filtering effects of Earth's atmosphere, verifying its "black body" temperature spectrum of 2.73 K. Furthermore, the shape of the thermal curve indicates that 99.97 percent of the energy of the early Universe was released in its first year (implying a rapid process of energy release, consistent with a Big Bang).

The discovery of the microwave background, with its excellent correlation to the predictions of previous theoretical models, essentially eliminated the possibility of a static universe. In addition to its existence, the recent high-resolution observations have detected tiny variations in the microwave background intensity, suggesting a primordial density inhomogeneity in its distribution (cooler areas being more dense than warmer ones) (Figure 1.2). The distribution of energy of the microwave background corresponds to that of our early Universe almost four hundred thousand years after its birth. These density fluctuations, called *defects* in the background radiation distribution, have significant repercussions in the eventual distribution of matter. The defect cosmological models describe an expanding (and cooling) Universe that contains these early inhomogeneities, which provide the seeds about which matter coalesced to form the stars, galaxies, and their clusters.

Initial Singularity

In retracing the steps of the formation and evolution of our Universe, we will follow the celebrated prototype model: the Big Bang. It is conceptually simple as far as cosmological models are considered, and it was the first plausible model that fit the early astronomical observations and correlated with the astrophysical calculations. Although over the years some of its fundamental parameters have become suspect, it generally provides an excellent framework from which to proceed.

If we travel back through time to the Universe's beginning, we arrive at what some cosmologists call initial singularity, a term borrowed from mathematicians, which designates an undefined (infinite) solution to a mathematical problem. Initial singularity represents a condition of infinite density and temperature where all of the Universe's mass is concentrated into an infinitesimal ("zero-sized") volume. Under such extreme conditions, the properties of matter, space, and time are inexplicable by known physical laws. The extremous implied by a singularity, which counters natural phenomena, raises question that such a condition could ever exist. However, it is clear that the early conditions of the Universe approached those predicted by a theoretical singularity. As one compresses material into a state approaching the conditions of a singularity, the component particles of matter (the building blocks of atoms) are so tightly packed together, the properties of matter and its interactions are far different from what we observe in our experience of everyday life. Under these extreme conditions, the forces of matter with which we are familiar (gravity, electromagnetism, the strong and weak nuclear forces) take on a new form and theoretically merge into a single force (Unified Field). Description of the properties of matter and energy on these ultramicroscopic scales of a highly compressed tiny (quantum) Universe requires the principles of quantum physics.

Energy and matter are directly related ($E = mc^2$) and therefore can be described by characteristics common between them. According to quantum physics, matter can be thought of as having a characteristic wavelength. On the scale of the environment in which we interact, the wavelengths associated with familiar objects (e.g., a rock, a building, a person) are much smaller than the object itself. In this domain, the equations of classical physics adequately describe the physical properties because the quantum effects are negligible. However, when considering very small objects, such as the components of an atom, the wavelength relative to the object size becomes substantial. Under such ultramicroscopic conditions the laws of classical physics fail to accurately describe physical behavior because on such small scales the quantum effects dominate. Furthermore, in the ultramicroscopic realm, pairs of related properties such as particle position and velocity or energy and time cannot be independently determined with precision (Heisenberg Uncertainty Principle). To describe the characteristics of matter and energy under these conditions, quantum mechanics was developed in the early twentieth century. From initial singularity until about 10^{-43} seconds (the Planck Epoch) the

Universe was smaller than its quantum wavelength; therefore, we must invoke the uncertainties of quantum mechanics to describe its properties. Unfortunately, we are not yet able to clearly define the properties that existed at the earliest times, when the unified field existed—in fact, even time itself may not have existed as we know it!

The relationships among mass, gravity, and the curvature of space were defined by the four-dimensional space-time continuum of Einstein's General Theory of Relativity. In regions where the force of gravity is extreme, the effects on space and time are equally extreme. This situation arises as we approach the conditions of a singularity, as in the center of a black hole or at the beginning of our Universe. The conditions affecting space-time found at initial singularity certainly would have been the most extreme at any moment in the history of our Universe.

Some solutions to Einstein's equations suggest that as we approach singularity, space and time may be so distorted that one can travel not only through space but through the dimension of time as well. In fact, according to theory, if one enters a gravitational distortion of space-time in the vicinity of a spinning black hole, the traveler may exit in a different place and time (assuming he survives the trip). However, quantum distortion of space and time around such objects likely precludes successful entry in the first place, making it an impractical method of travel. According to certain models of quantum space-time, the very fabric of space and time is constantly forming and breaking on subatomic scales. Theories invoking the existence of parallel universes suggest that these fluctuations, as well as implications of time travel, are interactions with an ubiquitous set of coexistent universes. The proponents of parallel universe theories can conveniently avoid theoretical paradoxes of time travel (e.g., a traveler preventing his own birth) by claiming the traveler actually enters a parallel universe where any changes he may instigate do not affect the universe from which the traveler originated—by virtue of changing the past, the traveler has entered a different universe. However, the creation of an entire new universe every time a coin is tossed, a photon emitted,or a decision made, may be considered a violation of the medieval scientific and philosophical precept Occam's Razor: *pluralitas non est ponenda sine necessitate*, "entities should not be multiplied unnecessarily."

In 1983, Stephen Hawking and James Hartle suggested that at initial singularity the fourth space-time dimension (time) was united with the three spatial dimensions, merging all four into a common form. In their approach, they applied the principle of a quantum mechanical wave function to the cosmos to determine the initial conditions of the Universe. In effect, the direction of time breaks down as it assumes the properties of a spatial dimension. Conveniently, this model actually avoids a singularity since there is no real time of creation: time wouldn't exist in the conventional sense. In essence, the space-time of a compacted universe would curve upon itself without becoming a true singularity. Mathematically the concept of *imaginary time* is introduced, under which the Universe has no boundaries, no beginning or end. Through Einstein's General Theory of Relativity, space and time are not independent of each other but are related, a consequence of the speed of light (c) being absolute from all observer references. Furthermore, time itself is affected by gravity; time passes slower in high gravitational fields relative to that in a lower gravitational-strength environment. Indeed, the extreme density of the early quantum Universe would produce indefinable "gravitational" fields strongly affecting both space and time.

Peculiar interrelationships of space and time in the General Theory of Relativity are imposed by the constancy of the speed of light being independent of an observer's reference. In our familiar (nonrelativistic) situations, the speed of an object is measured by the difference between the "absolute" speed of the object and that of the observer: if an observer is running after a rolling ball, the ball's speed (relative to the observer) is reduced by the observer's speed. If the ball is moving slower than the pursuing observer, she will eventually catch up to the ball. In the case of the speed of light, its speed is the same irrespective of the observer's speed. Furthermore, the speed of light is the absolute limit as to how fast material or information may travel through space. The restraints imposed by these restrictions lead to situations where time passes more slowly for a relativistic traveler (time dilation), object size asymptotically approaches zero in the direction of motion, and object mass asymptotically approaches infinity as speed approaches that of light (Lorentz Transformations). These effects are negligible in the nonrelativistic events of our daily lives: the passage of time as measured by a passenger after a commercial transatlantic flight is slowed by about forty billionths of a second relative to those on the ground. In the case of light, its composite photons must be massless and are not subject to the passage of time from our reference, because any hint of a photon mass would be infinitely amplified by relativistic effects, making the photons' existence impossible. Einstein further assumed that the motion of material in space is not limited to the three spatial dimensions but also includes a fourth dimension (time). He envisioned the ma-

19

jority of "travel" of an object moving at nonrelativistic speeds was in the dimension of time whereas objects moving at relativistic speeds have a more significant component of their motion in the spatial dimensions. Therefore, light which lies at the relativistic extreme confines its motion in the spatial dimensions and is "stationary" in time.

The properties of space and time on a quantum level may preclude the ability to define the instant of the Universe's beginning. Extension of quantum theory to describe the fundamental forces, including Quantum Gravity models, points to a discrete nature for time itself. Quantum theories of time suggest that time cannot be continuously divided into infinitesimally smaller and smaller intervals but actually has an absolute limit where shorter intervals are nonexistent. Recent investigations indicate the minimum interval into which time may be divided is about 10^{-95} seconds. Thus the earliest moments of the Universe may be nonexistent in the conventional sense, because the fragmented nature of space and time obscures the initial quantum epoch.

A corollary of initial singularity is a condition of symmetry at the beginning. An original symmetry is an aesthetically appealing initial condition in cosmological theory, as many cosmologists favor a beginning when the Universe was in a perfect symmetrical state of both space and time. Symmetrical space implies that all locations in the Universe were identical. A "degenerate" solution (which leads to a condition of symmetrical space) of spatial geometry occurs when all space is concentrated into a single point: a singularity. A condition of symmetrical time implies an ambiguity in direction in the time domain. However, when the Universe began, an origin of time was defined from which it moves forward, resulting in a broken symmetry. Thus under a theoretical initial condition of symmetry, an asymmetry must have developed to create a universe that corresponds to the one in which we live. If a symmetrical condition persisted, the Universe could have remained in the form of an expanding ball of energy without solid matter to form the stars and galaxies. An early asymmetry (not necessarily at the same instant expansion began but very soon after) is required to explain the presence of matter and its distribution throughout the Universe of today. An early asymmetry that created ripples leading to a heterogeneous distribution of matter would allow it to coalesce into scattered collections to eventually form the celestial bodies. A singularity also implies a condition of uniformity, which may be applied to a variety of parameters: temperature, density, energy, particles, and forces. However, the concept of a singularity (i.e., a Universe of zero size, infinite density, etc.)

predicted by the smooth space-time continuum of General Relativity raises suspicion of the theory's validity at the Universe's beginning: "unrealistic" solutions, such as singularities, of mathematical equations describing physical models, generally imply the principles of the underlying theory are invalid under the conditions being investigated.

Although the theoretical initial singularity represents an infinitesimal early Universe, the problem remains as to where all of the mass originated. The "vacuum fluctuation" principle of quantum theory supports the spontaneous formation of particle pairs in a vacuum, i.e., "something from nothing." Strangely enough, such a theory does not necessarily violate such fundamental principles as conservation of energy. If we consider the entire energy of a star, including its mass, heat, and light, it is balanced by the *negative* energy of its gravity—an overall energy of zero! Thus the spontaneous formation of an entire star through a chance event of quantum statistics is not impossible. The same argument can be applied to the Universe, suggesting a net energy of zero. To produce an entire universe by chance events may seem pretentious, but the possibility exists as a nonzero quantity. Understandably, critics dislike the assumption of the Universe coming into existence from nothing. By the anthropic principle, one can argue (on a teleological premise) that if such an unlikely event had never occurred, then we wouldn't be here to ponder it—in other words, we wouldn't exist if the Universe didn't fall into place exactly the way it did. Theoretically, such an event is required to occur only once in an eternity, which has some statistical support as, in theory, if given enough time anything can happen, including the spontaneous creation of an entire universe! Alternatively, the singularity could be the remnant of a collapsed predecessor universe. If so, the question still remains: where did *that* Universe come from? In an almost circular argument, some theoretical models of the early expanding Universe that incorporate quantum effects postulate scenarios of simultaneous creation of multiple Universes. In fact, through an extreme distortion of space-time, a closed time curve can be formed in which one of the simultaneously created Universes would in fact be the original Universe! In any event, the origin of the initial singularity remains a mystery.

It seems natural to assume that the Universe had a particular "location in space" where it came into existence. From theory and observational data, we know that the galaxies and their clusters are diverging in an expanding Universe. Are they diverging from a definable point of origin? Actually, the diverging pattern is a consequence of the expansion of space

21

around us, rather than the cosmos expanding into a larger "empty" volume. The Universe is its own boundary and is expanding throughout itself; space itself is driving the expansion. One way to visualize the expansion concept is to consider the Universe as the two-dimensional surface of a balloon with the galaxies drawn on its surface. As the balloon is filled with air, each "galaxy" on the surface moves away from its neighbors. The balloon's two-dimensional surface has no true center about which all of the points diverge. In addition, if all of the balloon's air was removed, the points on the surface would be drawn closer and eventually merge into one point (assuming the balloon is made of infinitely thin rubber), placing the "origin" at every position on the balloon's surface. Another example, a somewhat comical three-dimensional analogy, is that of a cosmic blueberry muffin. As the muffin is cooked, it expands (we'll assume in three dimensions), so that the blueberries (galaxies) separate from one another as the intergalactic dough rises. If an observer stands on any blueberry and watches neighboring blueberries, all of them will be moving away (receding). However, it is not necessary to stand on the "central" blueberry of the muffin to achieve the same observational result. Thus the expansion does not necessarily stem from a single point in the "middle" of the Universe despite the inclination of the observer to assume such a reference.

A consequence of a Universe originating as a symmetric singularity is that its site of origin is "everywhere." The expansion of space itself "stretches" out the origin, leaving everything in the Universe at the origin despite increasing dimensions of space and time. Furthermore, according to General Relativity, travel between points in space approaching relativistic velocities (i.e., near the speed of light) requires a complimentary dilation of time. In the extreme of relativistic travel, photons of light traveling through a vacuum experience no passage of time on their journey, indicating that the moment they leave their point of origin is identical to the time of their arrival at their destination, no matter how far they travel. From the perspective of the photon, the two points (its site of origin and destination) are identical in time. Another intriguing property of space and time is that of entanglement, where two quantum particles may be interlinked. Entanglement, which has been experimentally verified, is a strange phenomenon where distant quantum particles can instantaneously influence each other. As an example, consider two photons traveling in different directions with opposite polarizations (i.e., opposite directions of their oscillating electromagnetic fields). The photon polarizations are indeterminate until measured. As soon as one photon's polarization is measured, the polarization of the other photon becomes simultaneously fixed. The physical separation of the photons is irrelevant, as if all points in space are connected, thereby permitting entanglement effects to instantaneously involve particles separated by any distance—a phenomenon also known as nonlocality. Thus one can envision a supreme cohesive nature to space and time that permeates the entire Universe.

The Expanding Universe

At the beginning of the Big Bang we have an incomprehensibly dense, hot, and dimensionless particle representing the entire Universe. During the first 10^{-43} second (Planck Epoch), the Universe was smaller than its quantum wavelength and therefore governed by principles of quantum mechanics. During this period, the Universe was analogous to a single particle associated with its single force. The single force represents a unification of the four forces with which we are familiar: gravity and the electromagnetic, strong, and weak nuclear forces. As expansion proceeded from initial singularity until 10^{-43} second, when the visible Universe was less than 10^{-32} cm across (over a billion billion times smaller than a proton), the previously dominant single force began to lose its identity as gravity was taking form as a separate and independent force. Under these extreme conditions Einstein's General Theory of Relativity is not valid; instead the forces at play are described by a Quantum Gravity model. Quantum Gravity Theory describes the force of gravity as a consequence of graviton carrier particles, combining the principles of General Relativity and quantum mechanics; a completely satisfactory form of the Quantum Gravity Theory has yet to be established.

During the early moments, the evolving Universe went through a series of phase transitions (symmetry breaking) during which various particles came into existence and new forces took form. Following the Planck Epoch until about 10^{-35} second, the energy density of the Universe was so great that the electromagnetic, strong, and weak nuclear forces remained unified, leaving only two forces; gravity and the "electronuclear" (Grand Unified) force. During this Grand Unification Epoch, the Universe was an extremely hot ball of energy where particles, antiparticles, and photons were in abundance. At such high temperatures the particles and antiparticles were in a relativistic state and as such behaved like photons. Particle-antiparticle pairs were continuously being formed and annihilated. As expansion continued to cool the environment, the photon energies began to fall short of those required to form the higher mass particles (and antiparticles). The heavier particles, which were being formed less frequently as a consequence of the decreasing photon energies, "froze out" to establish their cosmic abundance that still persists.

As the Universe expanded until the Grand Unification symmetry was broken, irreversible decay of some of the exotic particles that froze out took effect. As these exotic particles decayed, an excess of quarks over antiquarks developed. The conditions that allow an excess of matter over antimatter come from CP (charge conjugation parity) violation. The principle of charge conjugation implies that particles undergo the same interactions as their oppositely charged antiparticle counterparts. *Parity* refers to mirror symmetries of elementary particle spin as it relates to particle interactions. It has been experimentally shown that the combination of charge conjugation and parity may be violated as a natural property of some elementary particle reactions. In 1964 violations of charge parity were demonstrated in certain K meson decay reactions, and more recent experiments, in 1999, suggest a CP violation in B meson decay. K and B mesons are formed from quarks, a reaction that is thought to be the source of the asymmetry. Stemming from a CP violation in the early Universe, (probably during the breaking of the Grand Unification Epoch around 10^{-35} second), the excess of particles over antiparticles was established. If such an asymmetry did not occur, recombination of matter and antimatter would leave the universe as pure energy.

As the early Universe continued to expand, the Grand Unification Epoch gave way to the Electroweak Epoch, which lasted until about one ten-billionth (10^{-10}) of a second. At that time there were three forces, because the electromagnetic and weak nuclear forces remained combined ("electroweak") and independent of the strong nuclear force and gravity. The electroweak force separated into the electromagnetic and weak forces by 10^{-9} second, establishing the four independent forces that remain in effect to this day. As expansion continued, by 10^{-5} second, particles and antiparticles began to combine. The excess of particles over antiparticles was about one in 30 billion; for every 60 billion particles that combined (into energy) there was only one particle of matter left (30,000,000,001 particles + 30,000,000,000 antiparticles = energy + 1 particle). In the aftermath, it is predicted that about 10^{80} particles were left to populate the Universe. Thus all the matter in our Universe represents only a minuscule fraction of what was present before the period of particle-antiparticle annihilation. During those few microseconds, at a temperature of a trillion degrees, the Universe was a sea of free quarks swimming in a hot quark-gluon plasma (the "quark era"), as the gluons were beginning to bring the quarks into stable triad configurations to form protons and neutrons. When the Universe was about one second old, neutrinos were no longer subjected to particle interactions and began to travel freely through space. The moment when the Universe became transparent to the neutrinos is analogous to the much-later release of the photons that formed the cosmic microwave background radiation. Over the next few minutes, the components of ordinary matter began to appear with formation of free electrons and nuclei of light elements with scant amounts of heavier nuclei.

It is interesting to consider the size of the early Universe at these monumental instants (according to the Big Bang model). At the beginning, during the condition of initial singularity, the Universe had a theoretical radius of zero. At 10^{-43} second, the quantum Universe (horizon distance) had a radius of only 3×10^{-33} cm—in other words, light, traveling at 3×10^{10} cm/sec, would have traveled only 3×10^{-33} cm to define the size of the visible Universe. By 10^{-35} second, when gravity was becoming an independent force, the Universe was only 6×10^{-25} cm in diameter (still much smaller than an electron)! At that rate of expansion, the visible Universe would have been the size of a beach ball by one-billionth of a second, when the four forces were beginning to take form.

From about one second following initial singularity onward, we no longer rely as heavily on theoretical models of particle physics but enter the realm of experimental particle physics. Despite the extremely high energies and particle densities of the early Universe at these times, such conditions can be reliably described by both theoretical and experimental models on Earth—however, it is not possible to experimentally simulate the governing properties in effect during the extreme particle densities and temperatures of the earlier moments. As spatial expansion and cooling of the particles ensued, eventually the strong nuclear force could overcome the kinetic (thermal) energy of the particles. By 100 seconds following the initiation of expansion, the Universe enters the period of nucleosynthesis when the milieu of particles was able to combine through nuclear particle (protons and neutrons) fusion reactions to form stable nuclei of the lightest elements (hydrogen and helium isotopes and lithium). After the first fifteen minutes, these reactions would cease, leaving a fixed relative abundance of the protons, neutrons, and light elements. Calculations based on these proposed particle interactions predict an abundance of about 75 percent hydrogen to 25 percent helium. The observed abundance of these elements in the Universe today and experimental results from particle accelerators confirm these predictions, adding credence to the Big Bang cosmological theory.

At this point, the early Universe became a plasma (a sea of electrons and nuclei) bathed in radiation. In fact, until 10^4 years after expansion began, radiation was the dominant component of the Universe, the "radiation era." After about three hundred thousand years of expansion, matter began to dominate with the formation of stable atoms from the plasma: free electrons combined with ions (atomic nuclei). By this time the photons emitted from various reactions were no longer being continually scattered by (the dwindling population of) free electrons and were able to travel freely through space. The Universe was a thousand times smaller than it is today and was at a temperature of 4,000 K when it became transparent to these photons. The cosmos would have been akin to a dense fog that cleared as it cooled with the simultaneous release of the photons that form the cosmic microwave background radiation that we detect today. Over the intervening billions of years of expansion, this radiation has continued to cool (redshift) to a temperature of 2.73 K—just a few degrees above absolute zero.

As the photons we detect from the radiation era continue to cool, they prelude a period in our Universe's history when the foundations of large-scale cosmic structure were laid. For hundreds of millions of years, the expanding mass of the early Universe did not produce the light-emitting stars that currently fill our sky. In fact, there was a dark period following the radiation era that lasted for about 250 million years before the first stars ignited. The most distant images telescopes have thus far obtained suggest the early Universe was much more complicated in its structure following the dark period than in times prior. During the dark period, the relatively formless mass (with some heterogeneity evidenced in the microwave background) of "dark matter," hydrogen and helium, evolved into a Universe containing relatively small scattered proto-galaxies filled with regions of active star formation. The lack of radiation from the dark period hinders astronomical study. However, the significantly complicated structural changes of the Universe during this relatively brief period make it a topic of intense investigation.

Massa Incognitus

Analogous to the condition of a third of adults in industrialized nations: the Universe has a weight problem. Complicating our cosmological model, observational data of the visible Universe fail to detect enough mass to account for some of its properties. There is abundant indirect evidence of the existence of this matter throughout the cosmos, but it is invisible to conventional methods of detection. The presence of "dark matter" (matter invisible to telescopes) is perhaps best-known through its effect on the gravitational dynamics within and between galaxies; to explain the dynamic properties that we detect, there must be more mass present within these systems than meets the eye. In fact, if the small group of galaxies of which we are a part (the Local Group) were composed of only the matter we see, it would have dissipated billions of years ago—our small galactic cluster must be held in check by the gravitational pull of unseen matter that is dispersed throughout our galactic neighborhood. Furthermore, studies of spiral galaxies indicate that the orbital speed of material as one moves away from the galaxy center remains relatively constant or even increases in some cases—if all of the mass was contained in the visible disk, the rotation speed should decrease with distance just as the planets in our solar system orbit the Sun at slower speeds farther out. The stability of galactic and larger-scale structure implies the presence of dark matter.

Estimates indicate that stars account for less than 1 percent of all the matter detected in the Universe, leaving the majority in the form of dark matter. In fact, the mean density of stars in the Universe is approximately 7.5 percent of the predicted mean density of baryonic material, suggesting that most of the baryons created at the beginning of the cosmos are in the form of dark matter. It is estimated that dark matter halos, spheres of material measuring up to 3 million light-years across with masses of at least 5 trillion suns, encase each visible galaxy. Using an analogy where our Milky Way galaxy is scaled down to the size of a twenty-five-centimeter dinner plate, its dark matter halo would measure 7.5 meters across (enough to fill an average room)—on this scale, the Andromeda galaxy would be a slightly larger dinner plate lying about 5.5 meters away from the Milky Way, with overlap of the margins of their respective galactic dark matter halos. In fact, some studies suggest the galactic halos may be up to ten times larger than these estimates! Although astronomers are convinced of the existence of dark matter, they are at a loss for its composition. In addition to its cohesive principles that bind the galaxies and their clusters, the presence of dark matter has a significant role in the ultimate fate of our Universe.

One plausible source of dark matter is in the form of cool objects such as brown dwarf stars, neutron stars, black holes, or even large planets. Some of these bodies would have radiation spectrums at longer wavelengths than visible light, making them difficult to detect by ground-based or even orbiting telescopes (unless such telescopes are sufficiently cooled and shielded from the Sun's heat)—ambient radiation obscures the infrared light of these objects. Thus far, studies from the Hubble Space Telescope have failed to find dwarf stars in sufficient quantity to account for any significant role. A newly described class of celestial objects, dubbed L dwarfs by the California Institute of Technology astronomical team that discovered them, represents accumulations of gas and dust into objects about the size of Jupiter. These objects are too small to initiate the nuclear fusion reactions of stars but do emit radiation from the heat of the gravitational collapse that formed them. Initial surveys suggest they are in a numerical abundance similar to that of conventional stars. In late 2003, NASA plans to launch the SIRTF satellite into a solar orbit trailing behind the Earth, where it will be able to scan the Universe at infrared wavelengths, with hopes of discovering these and other difficult to detect low-temperature objects.

The early Universe provides a theoretical source of vast amounts of material that may pose as dark matter candidates. Antimatter was abundant in the early Universe (as was conventional matter), at least up until the period of annihilation. In 1997, after compiling years of data from the Compton Gamma Ray Observatory satellite, Dr. William Purcell announced the discovery of a large source of antimatter within our own galaxy, but much of its properties and the significance of this newly discovered source remain unclear. Generally, studies of the abundance of antimatter come up far too short to account for a significant contribution to the dark matter ledger. Another class of dark matter candidates that has received much attention is the exotic particles, WIMPs (weakly interacting massive particles). They represent particles left over from the early moments of the Universe, and their presence would have significant impact on cosmological theories. Neutrinos are weakly interacting particles, a characteristic that enabled them to avoid detection for decades despite active attempts to find them. Their existence was predicted in order to satisfy the laws of conservation of mass and energy, which were otherwise violated in certain experimental radioactive decay reactions. Original predictions of neutrino properties included a neutral charge and a rest mass that was assumed to be zero. Although in great cosmic abundance, as massless particles neutrinos would not make satisfactory dark matter candidates; however, experimental investigations have placed their massless status in question.

Recent investigations of electron neutrinos at the Super-Kamiokande neutrino observatory suggest a small mass for at least one of the three known types of these particles (electron, muon, and tau neutrinos). The observatory is about a kilometer beneath a mountain in an old zinc mine near Takayama, Japan, and features tanks containing nearly 50 million liters of water. Interactions of neutrinos with the water produce photons, which provide a detectable signature of the neutrino encounter. Experimental results indicate that at least one of the types of neutrino particles has a mass through its interaction with matter, but the exact amount of mass remains uncertain. The suggested mass of this neutrino particle is probably less than 0.001 percent the mass of an electron. Although such a particle mass may be infinitesimal, the abundance of neutrinos makes them potential dark matter candidates. It is predicted that billions, or even trillions, of neutrinos harmlessly pass through each of us every second, some of which were released during the beginning of our Universe. Some physicists suggest that in addition to the three familiar neutrinos, two additional neutrinos may also exist: a superheavy neutrino and an extremely elusive "sterile" neutrino.

Dark matter may be ordinary matter that is emitting radiation at undetected wavelengths, thereby masking its presence. Computer models of the expanding Universe by Jeremiah Ostriker and Renyue Cen of Princeton suggest half of the ordinary matter of the Universe is in the form of a gas at a temperature where emission is in the extreme ultraviolet wavelengths. Such radiation may be easily overlooked because absorption by dust clouds or overwhelming emission from other sources can interfere with detection. Preliminary investigations using ultraviolet spectroscopy satellites suggest that there are vast clouds of hydrogen gas in intergalactic space that may account for up to half of the missing baryonic material.

In order to explain some of the observations of the amount of matter in the Universe and the rate of universal expansion, some proponents of the Inflationary Theory cosmological model have proposed the existence of a form of dark matter that produces an antigravitational effect. The Superstring Theory cosmological model proposes its own form of dark

29

matter. This theory requires a particle symmetry such that particles (electrons, neutrinos, photons, and quarks) have corresponding "superpartners." Superpartners refer to pairs of particles, which differ by a half-unit of quantum mechanical spin, that must exist in a supersymmetric universe. As the particles of matter have a quantum spin of one-half and the force carriers have a quantum spin of one, there is suggestion of a partnership between matter and force. Although these superpartners have not been experimentally detected, if they are found to exist they could account for a significant amount of the dark matter in the Universe. An alternative theory, MOND (Modified Newtonian Dynamics), proposes that the observed affects attrtibuted to dark matter actually represent an incomplete understanding of the physical laws of gravitation and motion.

Will cosmic expansion continue indefinitely? The answer lies in the balance between the energy causing expansion versus the force of gravity, which tries to pull everything back together. The pull of gravity is related to the amount of mass in the Universe. On the one hand, should the energy of expansion overcome the collective gravitational attraction of the particles in space, the Universe will expand forever (open universe), making it infinite in extent. On the other hand, if there is enough mass so that gravity wins, the Universe will eventually collapse upon itself (closed universe) and will therefore be finite. In a closed universe, space is curved, so that if one travels in a particular direction he will eventually arrive at the starting point, never reaching a boundary. A third option lies between the first two where gravitational attraction exactly balances the expansion, which ultimately results in an infinite flat universe (Figure 1.3).

The geometry of the universe is related to its fate as well. A flat universe has Euclidean geometry, which is consistent with space locally. A closed universe has a spherical geometry where its extent is finite, while an open universe has a hyperbolic geometry, which is infinite. Observations from within a universe should be able to determine the geometry. In a flat universe, as one peers deeper into intergalactic space the apparent sizes of objects are inversely proportional to their distance, congruent with our sense of perspective. In a universe with a spherical geometry, distant objects would appear correspondingly larger than in a flat geometry. Distant objects in an open universe would appear smaller than in a flat universe. The observational data from the COBE and WMAP satellites combined with data from more recent high-resolution studies from high-altitude balloons suggest the Universe has a flat geometry. In the next few years, additional data from NASA's Wilkinson Microwave Anisotropy Probe

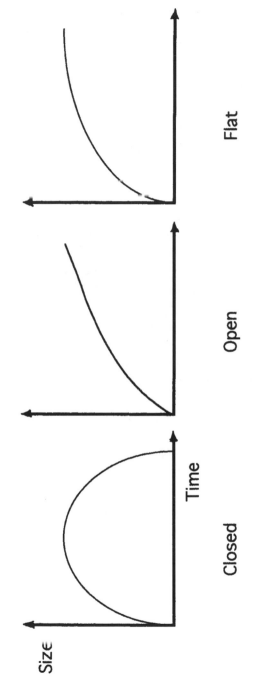

Figure 1.3 The time course evolution of Universe size according to the three popular models of cosmic expansior; closed, open, and flat.

(WMAP) and the European Planck Probe will provide higher-resolution measurements of the cosmic microwave background. These data may further refine the geometry of our Universe and perhaps clarify its fate.

The rate of our Universe's expansion is close to the balanced expansion rate, but uncertainties of the data cannot entirely exclude a closed or open system. The fine-tuning of the rate of expansion is so critical to produce a universe such as ours that only an infinitesimal increase or decrease in the initial rate would either spread out matter too quickly to form any structures (i.e., stars or galaxies) or have collapsed long ago, before any galaxies would have had time to form. In fact, by one second after the Universe began a difference of one part in 10^{10} in the expansion rate would mean the difference between recollapse in a few million years and dispersal of matter so rapidly that stars would have never formed. The ability to discriminate between the three models of expansion is difficult, as it requires extremely precise measurements of the expansion rate. The Standard Big Bang model is unable to explain why our expansion rate is so close to the balanced value—the "flatness problem."

The ultimate fate of our Universe, at least in part, relies on how much mass is present to create the gravitational force necessary to halt expansion or on its lack of mass to allow continued expansion. One line of investigation involved the solar wind project of the *Apollo 11* Moon landing. The project's goal was to determine the amount of deuterium ("heavy hydrogen") in the cosmos. Deuterium could have been created only in the early moments following the Big Bang, and its abundance provides a means to estimate the total mass of the Universe. Analyses concluded there was insufficient deuterium detected to account for enough mass within the Universe to halt its expansion. Furthermore, observations of visible matter fall short of sufficient mass to cease the expansion; at best only about 5 percent of the required mass has been detected. If we include the inferred quantity of dark matter from observational data, it still falls short of the critical mass necessary to prevent the continued expansion. The recent observational evidence suggesting an accelerating expansion rate with lack of sufficient matter to produce a flat geometry favors either an open Universe or one that is flat but contains a form of "dark energy" that accelerates the rate of expansion. The balance of data and theory favors a flat Universe with an accelerating expansion.

Cosmological Revision

Thus far, we have followed the birth and evolution of the Universe as it is described by the classic Big Bang cosmological model. Although this model is elegant with its conceptual simplicity, there are constraints of the standard Big Bang model that do not parallel astronomical observations and lose credibility under closer inspection. Modifications have been proposed over the years that can account for some of the inconsistencies encountered by the Big Bang model; however, even the latest cosmological models are unable to completely explain all of the observational and experimental data.

If we assume the Universe's expansion was constant and simply put it into reverse, we find that at time zero the radius of the visible Universe is much too large; it would not have collapsed into a zero volume by time zero to satisfy the initial condition of singularity. To explain the nearly homogenous density (to one part in 100,000) of the microwave background, it is assumed that the different regions of the early cosmos were in close-enough proximity to develop a near-equilibrium state. The original Big Bang theory assumes the homogeneity as a fortuitous initial condition, which has become known as the "horizon problem." The horizon distance describes the radius of a sphere that began at time zero (singularity) and expands at the speed of light. The matter in the Universe would be contained within this relativistically expanding sphere, as matter cannot travel faster than light (to get outside the horizon sphere). Interestingly, the expansion of space itself is not limited to nonrelativistic speeds (as are phenomena that travel through space). In fact, the expansion of the early Universe was likely not constant and it must have expanded more rapidly at some point to account for the size discrepancy. The Inflationary Theory cosmological model conveniently solves this problem by proposing an early transient period of increased rate of expansion so that the size of the expanding shell of initial singularity overlies (and easily surpasses) that of the visible Universe (Figure 1.4).

The Inflationary Theory assumes there were regions within the primordial Universe of differing temperatures such that some would *supercool* as they expanded. Under the quantum conditions of the early Universe, these regions would develop a *negative* pressure leading to a *repulsive* gravitational force (reminiscent of Einstein's Cosmological Constant). This early repulsive force would produce an exponentially

33

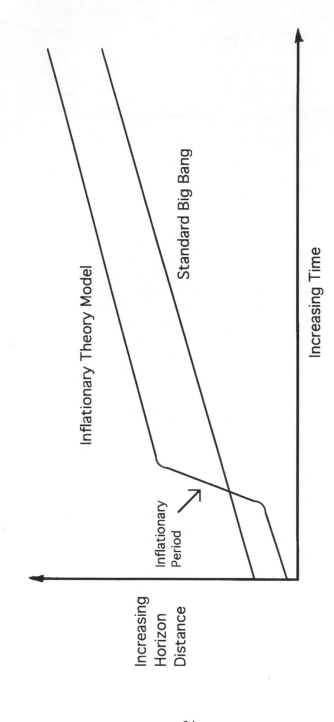

Figure 1.4 The relation of the size of the Universe between the Inflationary Theory and the Standard Big Bang models as it varies with time.

34

accelerated (inflationary) expansion of space that was very brief, continuing for about 10^{-32} second. From about 10^{-35} second to 10^{-32} second, the inflationary process could expand the Universe by a factor of up to 10^{50}, from smaller than an electron to the size of a grapefruit (or even much larger). The inflationary period coincides with the breaking of the force symmetry of the Grand Unification Epoch as the Electroweak Epoch arrived. The transition between the epochs may have simultaneously released tremendous energy into the continuum, which then drove the accelerated expansion. Under conditions of an accelerated expansion, the entire Universe would rapidly expand to proportions vastly larger than the visible Universe with which we are familiar. Although the rate of spatial expansion was effectively greater than the speed of light, the expansion was actually acting on the space around matter, such that no particles necessarily travel faster than light within their inertial reference frames (avoiding violation of Einstein's relativistic restriction that "nothing" travels faster than light). In the early quantum universe, random spatial variations in energy (quantum noise) would have been rapidly expanded to macroscopic proportions by the inflationary process to form scattered regions of differing matter-density. Regions of relatively increased density would eventually develop into stars and galaxies, while regions of relatively decreased matter-density would leave large cosmic voids. As the inflationary field dissipated, its energy was left in an inhomogeneous distribution throughout space to form the matter and radiation pattern detected in the cosmic microwave background. Following the inflationary period, by about 10^{-32} second, a phase transition took place slowing the rate of expansion to that described by the Big Bang model. Thus the currently observable distributions of matter and energy in our Universe (general uniformity with density variations), unexplained by the Big Bang model, are predicted by Inflationary Theory. Under the conditions of a transient rapid accelerated expansion, the Inflationary Theory favors a flat universe (any primordial curvature of space would have been "stretched out" by the transient inflation) with expansion continuing indefinitely. However, there are observational data that suggest the rate of cosmic expansion is not constant—it is accelerating.

Recent observations of dozens of distant (type Ia) supernovae were made to determine the shape of the cosmos and its rate of expansion. The intrinsic brightness characteristics of type Ia supernovae are well understood in theory and consistent to within about 12 percent, making them a useful tool to determine their actual distance ("standard candle"). Re-

35

peated surveys of thousands of distant (about 4 to 7 billion light-years) galaxies occasionally reveal these particular supernovae, enabling astronomers to estimate the distance to the host galaxy of a detected supernova. It was found that the light of these distant supernovae was about 20 percent dimmer than expected based on their distance according to the redshift data; the distance to these supernovae may be greater than that predicted by the redshift data. The data may be explained by an accelerating rate of expansion of space, which would increase the distance the light of these distant supernovae must travel. Alternatively, an overall negative curvature (hyperbolic geometry) of space could produce the same effect by spreading out the light to make it appear fainter than it would in a flat geometry continuum. However, even assuming the maximal theoretical negative curvature allowed by observational data cannot completely explain the distance discrepancy.

A variable rate of cosmic expansion permits the geometry of space to remain flat, which is favored by the majority of cosmologists and supported by the microwave background radiation data. Assuming the Universe is flat, it must have a total energy and mass density equivalent to the theoretical critical density. The critical density is nearly a vacuum as calculations place it at only about six hydrogen atoms per cubic meter. Surveys of the Universe have found only about 40 percent of the matter required to achieve the critical density. The implications of an underlying dark energy—a mysterious energy that accelerates cosmological expansion—may be sufficient to bring the total energy density to the critical value. This dark energy would be distributed throughout the Universe with a higher density than matter—its (negative) influence on gravity overcomes that exerted by matter leading to an accelerating expansion. Sometime between four and eight billion years ago, the decelerating effect of gravity gave way to an acceleration of cosmological expansion. This dark energy may be thought of as an energy with properties echoing Einstein's Cosmological Constant.

The proposed cosmological dark energy, which exists everywhere, has antigravity properties in the present epoch, but behaved very differently in the past. In the early universe, a time when all matter was tightly packed together, gravity was a dominant force as it tried to pull the expanding material back together which tended to slow cosmological expansion. With time, as the universe expanded, the decelerating effect of gravity waned as the matter spread further apart. However, as the volume of space expanded, the strength of the dark energy increased which has accelerated cosmological expansion. Two popular theories have surfaced as candidates for the dark energy: vacuum energy and quintessence. The existence of vacuum energy has been established experimentally by its subtle effects on atomic energy levels and its force on closely apposed metallic plates. Theoretical calculations of the magnitude of the vacuum energy suggest it is 120 orders of magnitude greater than the cosmological dark energy. However, uncertainties of the properties of the vacuum energy leave the accuracy of such calculations in doubt. Quintessence, a theoretical quantum field, may exhibit properties which vary with space and time. In the early universe, quintessence may have behaved similar to gravity, but about 50,000 years after the beginning of cosmological expansion, it began to exert an antigravitational force which has been growing in strength and thereby accelerating expansion.

The standard Big Bang model assumes that the Universe began in a very hot and dense state and has been cooling through continued expansion ever since. It assumes the original mass consistency of the Universe was nearly (but not completely) uniform, to allow the eventual collection of material into stars and galaxies. It further assumes that the rate of expansion lies close to that yielding the open, closed, and flat universes, to match the observational data. Inflationary Theory places fewer restrictions on the initial boundary conditions prior to the inflationary period such that the Universe was not necessarily nearly uniform originally and could have been in a chaotic state that created a phase transition that led to the transient inflationary state. Most cosmologists favor an inflationary model, as it depicts a universe that is more consistent with astronomical observations than the Standard Big Bang model. The merging of these principles is the basis of the popular Inflationary Big Bang model.

As the cosmological theories permute and multiply, the support of conclusive observational evidence is hotly pursued. Recent observational evidence supporting a flat Universe was obtained using highly sensitive measurements of small regions of the microwave background with instruments carried aloft by high-altitude balloons above Antarctica and the southwestern United States. Small sections of the sky were examined at microwave wavelengths with a resolution about forty times higher than the COBE satellite data. Although results have been promising, cosmologists await more complete high-resolution sky surveys. There is optimism that high-resolution imaging of the cosmic microwave background radiation and other observational data will clarify the underlying boundary conditions to narrow the range of plausible cosmological models.

Unification

In the current epoch, we live in an expanding universe of (at least) four-dimensional space-time containing matter and energy (which are interchangeable) and four forces. It seems reasonable that these components are somehow related. Analogous to the spatial dimensions, time may exist as an extended quantity—block time (what we call "past, present, and future" simultaneously exist just as distances throughout the Universe exist)—with its passage an illusion created by conscious interpretation of our senses and memory. Experimental investigations have uncovered some of the interrelations between the forces, and Einstein's famous equation, $E = mc^2$, implies the equivalence of energy and matter. The objective of understanding these relationships is to determine a fundamental theory that describes and correlates the natural phenomena of matter, force, space, and time—the definition of the fabric of our Universe.

Recent developments have included revisions in the concept of the vacuum of space. The definition of a vacuum implies an empty region void of content or structure. In fact, what we perceive as solid matter is actually almost entirely empty. A solid object, such as a piece of wood or steel, is about 99.9999999999999 percent empty, with the infinitesimal remaining volume composed of the nuclear particles and electrons. The atoms and molecules maintain their integrity through the electromagnetic force of their bonds, which lock them into a rigid structure. Thus our concept of a solid material seems more like an illusion, as such materials are mostly "nothing." However, the emptiness in which the particles of matter are suspended has properties of its own. In fact, the particles and their forces of interaction may be manifestations of a multidimensional (space-time) microstructure within the vacuum.

According to Heisenberg's Uncertainty Principle of quantum mechanics, atomic particles maintain a veil of mystery, which precludes complete determination of simultaneous particle location and momentum. In fact, there is a nonzero probability that one can pass through a solid wall to materialize on the other side, but it would take an interval about as long as the Universe has existed before such an event would be expected to occur. The premise of quantum uncertainty also includes the properties of energy and time, implying that empty space (a vacuum) actually has intrinsic energy. Einstein's famous formula points out the equivalence of energy and matter, leading to the conclusion that the vacuum of space can create particles. These "spontaneous" particles appear and decay over very brief intervals. In fact, these virtual particles produce measurable forces. Some theorists suggest these quantum fluctuations that permeate the fabric of space's vacuum may have been responsible for the initiation of the expansion of the Universe. Although not widely accepted by the scientific community, the energy of the vacuum, called the zero-point energy, is under investigation as an exploitable source of energy. Proponents argue that it is a form of electromagnetic radiation that permeates all space and provides stability to matter. It prevents electrons from colliding into the nucleus, and on larger scales it contributes to the gravitational stability of the matter of the Universe. The theoretical amount of energy within a vacuum is not precisely known (some suggest it could be infinite), but experimental work has thus far yielded very little extractable quantity.

By the early 1930s, only a few subatomic particles were known to exist: the neutrons and protons of nuclei, the electron, and the photon. These four particles have both striking similarities and vast differences, which hinted at an elementary particle picture that was somehow incomplete. It was known that the proton and neutron have similar mass, but one is charged and the other neutral, yet the electron, which is almost two thousand times less massive, has a negative charge equivalent in magnitude to the proton. In 1933, positrons were discovered in cosmic rays, establishing the existence of antimatter (e.g., a positively charged "antielectron"), which had been predicted a few years earlier by the British physicist Paul Dirac. In the years that followed, more and more particles were discovered until the "elementary" particles numbered in the hundreds! Fortunately, patterns of similarity began to emerge, enabling classification of particles into groups. A major division of elementary particle types separates them into matter and antimatter, where each particle has its corresponding "antipartner." The physical combination of a particle with its antipartner results in an annihilation event with release of energy by conversion of their mass ($E = mc^2$). Thus nature has revealed a form of elementary particle symmetry, as they may be divided into matter and antimatter pairs. Some neutral particles are actually their own antipartner, but they tend to have very brief lifetimes of stable existence. An interesting case is the photon, which is its own antipartner; since the photon has no mass and travels at the speed of light, it isn't truly a particle and through relativistic effects its lifespan is undefined. Physicists divide the basic particles into hadrons (hundreds of types) and leptons (only six types), where the division relies on the ability to affect a particle by the strong nuclear force (color force)—hadrons are affected by the strong nuclear force, while leptons are not. The leptons appear to be fundamental, while hadrons have underlying structure, as they are composed of quark triads. The third category of fundamental particles is the carrier particles (gauge bosons) of the forces (Figure 1.5).

Fundamental Particles

Leptons: electron, muon, tau, and neutrinos (e, μ, τ)
Quarks: up, down, strange, charmed, bottom, top
Carriers: gluons (8 types), photon, weak gauge bosons (3 types), graviton

Force	Relative Strength	Carrier Particle
Strong Nuclear	1	gluon
Electromagnetic	10^{-2}	photon
Weak Nuclear	10^{-5}	W^+, W^-, Z^0
Gravity	10^{-44}	graviton (theoretical)

Figure 1.5 The fundamental particles and their relation with the four forces. The graviton is the proposed carrier particle of gravity from Quantum Gravity Theory.

In an attempt to organize the properties of matter and its forces, the Standard Model was developed, and over the years it gradually expanded as new discoveries were made. The Standard Model, which is the prevailing theory of particle physics, divides the substance of nature into particles that interact by various forces through an exchange of carrier particles. On the ultramicroscopic level, the interaction of the basic forces between particles is envisioned as volleys of carriers, which cause the particles to behave according to which force is in effect. The theory includes electromagnetism, which had been well described for decades prior to the concept of the Standard Model. In the 1960s the electroweak theory was developed, which revealed a relationship between the electroweak force and electromagnetism. The field of Quantum Chromodynamics describes the properties of the quarks. These point particles are held together by gluons and combine to form protons and neutrons. A list of eighteen different quarks (each of the six may come in three different "colors") has been compiled, and each of these quarks has an antimatter correlate. The merging of the electroweak theory and Quantum Chromodynamics provides the framework of the Standard Model. The Standard Model is cumbersome with its complex assortment of various particles and force carriers: it is a conglomerate of experimental and theoretical results. Furthermore, the Standard Model is entangled with the *hierarchy problem* where it cannot satisfactorily explain the extreme range of differences in the various particle masses and forces. Proponents of the model suggest that scalar fields—uniform energy fields that exist everywhere in the vacuum, i.e., Higgs fields—interact with particles to confer inertia upon most of the particles, effectively giving them mass. Although powerful, the Standard Model lacks the simple organization that many physicists envision elementary matter and its forces must ultimately hold.

The Grand Unified Theory (GUT) was developed in attempts to define interrelationships of three of the major forces: the unification of electromagnetism and the strong and weak nuclear forces. An interesting consequence of the GUT is that matter must have a finite lifetime, after which it will eventually disintegrate into its constituents. The theory requires that all nuclei are radioactive and will eventually decay (including protons, where the lifetime is estimated at about 10^{31} to 10^{34} years). The theory provides an avenue for experimental verification: detection of the proton decay (using huge underground storage tanks of water), but no decay event has thus far been detected. The lack of a detected decay event suggests a lower limit of the half life, pushing the proton decay time to perhaps greater than 5×10^{32} years. Although some may argue that the failure to detect a proton decay event invalidates GUT, proponents of the theory claim the lack of decay detection neither proves nor disproves the theory. Another consequence of the GUT is that neutrinos must have a tiny mass—recent investigations suggest neutrinos may indeed have mass. However, GUT falls short of a completely unifying theory by its failure to relate gravity with the other three forces.

Early in the twentieth century, the revolutionary theories of relativity and quantum mechanics were introduced. Relativity elegantly describes the properties of matter and force on a cosmological scale, while quantum mechanics accurately characterizes particle properties in the ultramicroscopic realm. The two theories are extremely accurate, but on vastly different scales. To those seeking unification laws in nature, it was disturbing that the properties of matter were so vastly different at these extremes, requiring completely different sets of physical analyses to describe its behavior. Physicists were searching for a unified theory that would be useful at all levels and furthermore be able to incorporate the elusive force of gravity. Ill-conceived early attempts at directly combining the two concepts into a useful Quantum Field Theory were plagued with nonsense solutions, leading to years of frustration and disillusionment. The two theories are incompatible, as the General Theory of Relativity assumes space is a smooth continuum, whereas the basic premise of Quantum Mechanics requires that on ultramicroscopic scales—on the order of Planck's length (10^{-33} cm) or less—the fabric of space is filled with random fluctuations of spatial geometry and time (sometimes referred to as a *quantum foam*).

Despite the instabilities of Quantum Field Theory, continued attempts at solving the complicated relativity equations paid off in the mid-1980s. Abhay Ashtekar, professor of physics at Pennsylvania State University, was able to greatly simplify the equations of the General Theory of Relativity through substitution of a pair of variables that represented a four-dimensional space-time geometry. These simplified expressions eventually allowed physicists Carlo Rovelli and Lee Smolin to produce a set of solutions of Quantum Field Theory where lines of gravitational force are interpreted as closed loops. In Loop Quantum Gravity, a Quantum Gravity model, the space-time continuum of Einstein's General Theory of Relativity is replaced by a quantized microstructure in which regions less than about 10^{-35} meter are unobservable. They propose that space itself is composed of tiny discrete building blocks defining an absolute minimum volume (volumes less than the minimum do not exist). In this model, the vacuum of space has discrete states, just as the electrons of an atom occupy specific energy states. The quantization of space suggests that it consists of closed loops of gravitational force that are linked and knotted with one another. The finite size of the building blocks of space and the loops of gravitational force lines avoid the undefined solutions that frustrated previous attempts at describing quantum fields. The properties of the infinitesimal pieces of space (called spin networks) of Loop Quantum Gravity are more useful in describing three-dimensional space than four-dimensional space-time. Attempts at extending the model to add the fourth dimension include the incorporation of spin foams to describe the properties of spin networks per interval of time.

As work continues to iron out the complexities of Loop Quantum Gravity, its major rival for a unified theory has an altogether different approach. String Theory redefines matter by stretching the point particles of relativity and quantum mechanics into strings. Its proponents argue that the most promising unifying theory of General Relativity and Quantum Mechanics thus far is this complicated mathematical model based on strings: one-dimensional vibrating objects that form the elementary particles and forces. In addition, String Theory requires a higher-dimensional space-time to explain the presence of all four known forces. In Einstein's General Theory of Relativity, the force of gravity is thought of as a consequence, or distortion, of four-dimensional space-time. Thus the attractive force between two masses is considered a distortion of space-time geometry such that the paths of motion taken by moving objects follow the distortion contours produced by the presence of a mass (rather than being "pulled" by a force). Extending the concept to include all four natural forces as distortions of space-time suggests the existence of more dimensions than the four proposed by General Relativity. String Theory is a relatively recent development in theoretical physics, and its full potential is not yet clear.

String Theory is a mathematical model composed of differential equations with a complex set of conditions that describes the fabric of space, its particles, forces and time. The theory broadly describes physical interactions of matter and energy, seems to bear resemblance to reality, and in addition has an aesthetically appealing underlying condition of ultimate physical simplicity. The treacherous singularities encountered when quantum effects are introduced into the geometric continuum of General Relativity, which stymied Quantum Field Theory, are avoided by redefining the particles as tiny one-dimensional strings. The oscillating string loops have dimensions on the order of Planck's length (smaller than 10^{-32} cm, less than a billionth of a billionth of the size of a proton) or smaller. In fact, as they are the fundamental elements of nature, their size defines the smallest dimension into which space may be divided (volumes of space smaller than this have no physical meaning). The dimensionless point particles of the Standard Model are replaced by infinitesimal (but not dimensionless) strings.

In essence, String Theory proposes that matter is composed of tiny one-dimensional structures. The strings may be "open"-ended or a "closed" loop and may combine or come apart during particle and force interactions. The vacuum of space is thought to be an active multidimensional foam through which quantum forces are transmitted and where particle pairs are continuously created and recombined. The interactions cause vibrations of the mesh, and through relations between the frequency energy and its equivalence with mass, the vibrations produce equivalent mass quanta. According to Einstein's energy-mass relation, strings that vibrate at higher frequencies correspond to higher energies that represent particles of higher mass. Furthermore, these strings may vibrate and rotate with various specific modes. The frequency of vibration distorts the geometry of space around it, producing the effects predicted by our conventional laws of physics: the properties of matter. The lowest vibration mode of an open string corresponds to electromagnetism (photons), while the lowest mode of vibration of a closed loop produces the force of gravity (gravitons). Other vibration modes of strings produce the other forces (carrier particles) and subatomic particles: all formed from the same basic ingredient (i.e., a string). In the case of a closed loop superstring with a length of zero, "supergravity" is obtained, which is an extension of supersymmetry to include gravity. The geometry of space-time defines gravity, as was shown by Einstein's General Theory of Relativity, and through the geometry of the other dimensions, the other forces of nature are transmitted as permutations of an ultimate form of gravity.

Superstring Theory—String Theory under the conditions of supersymmetry—accounts for the matter particles and all four forces but requires the inclusion of superpartner particles. The notion of supersymmetry was hinted at as the Standard Model was being pieced together where symmetries were suggested between the particles and force carrier particles. However, theoretical study revealed that simply pairing of the known particles would not satisfy superpartner pairing characteristics. Thus the number of fundamental particles was doubled to include the (theoretical) superpartners. From quantum mechanics, all particles originate from two types: fermions (e.g., matter) and bosons (e.g., photons, gravity). In supersymmetry, each particle has its dual particle with a different spin (angular momentum). Supersymmetry requires a symmetry between fermions and bosons, just like there is a symmetry between electricity and magnetism. These modifications led to the Supersymmetric Standard Model.

At first it may seem undesirable to expect a supersymmetric condition: it further complicates things by doubling the number of fundamental particles with a host of undiscovered superpartners. However, there are several reasons to anticipate supersymmetry. Perhaps the simplest reason is the expectation of symmetry out of aesthetic principles. There are many dualities and symmetries in nature, and it would be disconcerting to find it lacking on a fundamental level. The incorporation of supersymmetry into the Standard Model allows much less constraint on various quantum effects, which otherwise require exceedingly precise properties to explain their behavior. In addition, the force unification of the GUT is actually a near miss without adding in the contribution of superpartner particles. In support of the lack of verification of the superpartner particles, they are expected to be extremely heavy (at least 1,000 times the mass of a proton), and the particle accelerators of today are unable to produce the energies necessary to detect them. The Large Hadron Collider (in Geneva, Switzerland) may be able to detect the existence of superpartner particles in the coming decades.

Unlike the other models, Superstring Theory incorporates all of the matter particles (including superpartners) and the four forces into a unified form. An interesting constraint of this model is that it limits the total number of space-time dimensions to eleven. Although initially an attractive solution, in 1984 the eleven-dimensional supergravity model fell out of favor, as it was unable to explain some of the lack of symmetry of forces in the real world. It was later found that some of these properties may be explained through at least five string model permutations in ten-dimensional space-time (Type I, Type IIA, Type IIIB, Heterotic O[32] and Heterotic E_8XE_8). These five models include the three spatial and single time dimensions (the familiar four-dimensional space-time) with an additional six spatial dimensions curled up into a Calabi-Yau space. The three dimensions of space we are familiar with are vast and may even extend to the edge of the Universe to curl upon themselves (i.e., a traveler moving in a particular direction would eventually return to his starting point), with the other six dimensions existing everywhere curled into an infinitesimally small Calabi-Yau geometry. The most popular ten-dimensional superstring model is the Heterotic E_8XE_8, which is compatible with the experimentally tested Standard Model when applied at low energies. Investigations into the geometric forms and energies of the five models have shown dualities between them, indicating that they are in fact related to each other.

In an attempt to unify the popular ten-dimensional solutions with a supersymmetry, a sixth solution was derived. The unifying model, M-theory (Membrane theory), is an eleven-dimensional space-time where strings are replaced by membranes: strings are one-dimensional solutions while membranes are higher-dimensional. In M-theory, the curling of two-dimensional (membrane) space can reduce an eleven-dimensional supersymmetry model into the five string solutions of ten dimensions. In addition, the eleven-dimensional M-theory is linked to an eleven-dimensional supergravity model, which is a string model that combines General Relativity with supersymmetric Quantum Field Theory.

Superstring Theory predicts an original symmetry reminiscent of the symmetry suggested by an "initial singularity" of the Big Bang models. At the beginning of our Universe, one can envision a time of perfect symmetry with one superforce. However, the spatial dimension of the Universe at its origin would not have been zero but on the order of a Planck length: strings have a nonzero size. In addition, the original multidimensional universe may not have been stable, just as natural phenomena favor a minimum energy state. In the pursuit of a minimum energy state, the original multidimensional universe with its single force collapsed into two universes where one universe degenerated into infinitesimal proportions (Planck size), leaving the other (the four dimensions with which we are familiar) to expand to large proportions. The collapsed dimensions are everywhere, curled up into infinitesimal size (as strings in Calabi-Yau space), and exist at all points in our four-dimensional space-time. The geometry of the extra dimensions produces the four forces from the original supergravity. Furthermore, the cataclysmic event of the collapsing dimensions may have set the expansion of our Universe into motion. In fact, both Superstring Theory and the Inflationary Theory model suggest that the creation and dynamics of the Universe lie in the fundamental structure and properties of space itself.

Thus there are basically two contenders for a unified theory (popularly referred to as the Theory of Everything) of the Universe: the Quantum Gravity Theory (e.g., Loop Quantum Gravity) and Superstring Theory. Both of these theories lack significant experimental verification. Interestingly, these mathematical models share certain qualities, but they do have a fundamental difference: the strings of Superstring Theory permeate space, while the loops of Loop Quantum Gravity actually comprise space. In order for the forces to be unified, particle separations must be in the range of 10^{-32} cm, which places an upper limit on the size of the one-dimensional strings of Superstring Theory. Quantum Gravity Theory, through the discrete nature of quantum mechanics, also implies a lower limit to the dimension of size, which may be in the same range as that for superstrings. As both Superstring Theory and Quantum Gravity Theory continue to evolve, it has been suggested that their similarities may be a hint that the two different theories may actually be descriptions of the same theory.

The models of the Universe discussed are all plausible candidates currently. The Inflationary Big Bang is the most widely accepted model, but as new theories are proposed popular opinion may change. Experimental verification of the mathematical models of Quantum Gravity and Superstring Theory is actively sought to support or disprove their premise. In fact, it is the fine structure of the cosmic microwave background radiation that may be able to define the mechanism of the initial expansion of the Universe to provide credence for one theory over another.

Is there only one universe? The principles of quantum cosmology allow the possibility of innumerable universes. Just as the Heisenberg Uncertainty Principle tells us elementary particles may place themselves at different states through quantum fluctuations, a quantum universe may do the same under the auspices of quantum cosmology: coexistent parallel universes. In theory, if many such universes exist, the majority probably do not have the stability of matter that is necessary to complete the extended and complex pathway required for the development of intelligent life.

Elementary Complexity

Ever since the philosophers of ancient Greece put forth their theories of the origins of the Universe, there has been a desire to seek a beginning where conditions were simple and of a pure symmetrical form. It seems natural to expect the Universe was a place of symmetry in both space and time at its inception, that all positions within it were the same and time did not exist. Perhaps a break in the symmetry set into motion the expansion of space that led to the unilateral progression of time and a heterogeneous distribution of matter, which eventually led to our Universe of today.

The Universe is not only asymmetric in that time moves forward and matter is distributed in clumps, but the forces of nature are also asymmetric. We are familiar with four very different forces: gravity, electromagnetism, and the strong and weak nuclear forces. Gravity and electromagnetism work over infinite distances, whereas the other two are effective over distances within an atomic nucleus. These forces act through different force-carrying particles: gravity uses gravitons, electromagnetism uses photons, the strong nuclear force uses gluons, and the weak nuclear force acts through weak bosons. Furthermore, electromagnetism has both attractive and repulsive forces, while gravity is a force of attraction only. Although these forces behave very differently from one another under normal conditions, similarities in these disparate forces occur if we "turn up the heat."

Under conditions of extreme temperature and pressure, electromagnetism and the weak nuclear force begin to combine into one common force as their particles of interaction (photons and weak bosons) become interchangeable. As conditions of heat and pressure are further increased, the particles of interaction for the strong nuclear force (gluons) become interchangeable with those of electromagnetism and the weak nuclear force. These conditions have been experimentally produced by high-energy physicists. Theoretically, if conditions are elevated to even further thermal extremes, one would presume that all four forces would combine into a single force (Figure 1.6). However, such conditions cannot be experimentally created at this time, leaving only theory to predict these behaviors.

If the disparate forces of nature aren't confusing enough, matter itself is exceedingly complex. Conventional matter consists of atoms, which in turn have nuclei that contain protons. The proton is basically composed of three quarks held together by the strong nuclear force mediated by gluons. The quarks are moving about at relativistic speeds "within" the proton, encased by a shell of fleeting virtual quarks and antiquarks, which briefly come in and out of existence continuously. Furthermore, the mass of the three quarks totals only about 2 percent of the proton mass, with the majority of the proton's energy residing in its gluons. In the least, the fundamental particles required to create matter include the electrons, neutrinos, and a variety of quarks. Why are nuclear particles composed of quark triads held together by gluons? Why are there so many different types of quarks and other particles? Some theoretical physicists actively seek a simple, elegant explanation of the origins of energy, matter, and the forces. Albert Einstein dedicated his final years to this challenge and was unsuccessful. String Theory suggests that the properties of the elementary particles and the forces lie within the multidimensional geometry of space-time, which may be thought of as an extension of Einstein's radical leap in describing gravity as a property of the space-time continuum. Perhaps the day will arrive when an elegantly simple explanation of our Universe will be revealed, to the delight of all.

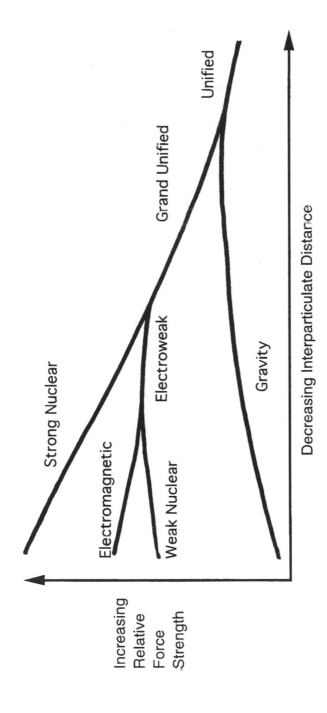

Figure 1.6 A comparison of the relative strengths of the four forces of nature. As the interparticulate distance is decreased (analogous to increasing density, temperature, and energy), the four forces are altered. The GUT applies with the combination of strong nuclear force and the electroweak force (itself a combination of the electromagnetic and weak nuclear forces). At energy levels approximately 10^4 times higher than GUT, theory predicts the force of gravity will merge with the GUT force to create a single Unified Field force.

49

2

Origin of the Galaxies

If the expanding Universe began and remained in a perfectly symmetric and homogeneous state, the formation of galaxies, stars, planets, etc., never would have taken place. An early inhomogeneity comprised of scattered variations in mass density allowed matter to coalesce into clumps that eventually formed galactic systems and the stars they contain. The cosmic microwave background radiation provides clues to the mass density distribution of the early Universe. Beginning in 1989, the COBE satellite mapped the microwave background radiation with much higher precision than previous attempts. Minute fluctuations were found in the microwave pattern, implying density inhomogeneities existed in the Universe when the cosmic microwave background energy was released (almost four hundred thousand years after the Universe's birth) (Figure 1.2). The microwave background is an image of the Universe at a time when whole atoms were being formed from the plasma of light atomic nuclei and electrons.

The density fluctuations imprinted on the microwave background existed before the decoupling of matter and radiation. However, it is not entirely clear how these heterogeneous clumps of matter came to exist in the first place. According to the Inflationary Theory, the initial quantum vacuum contained random density fluctuations that were amplified by the inflation process to form the heterogeneous zones imprinted on the microwave background. In the absence of an inflationary process, other mechanisms have been proposed: the presence of WIMPs could act as seed points by their gravitational interaction with surrounding subatomic particles that gradually coalesced into vast collections of matter. As expansion continued, gravitational effects of dark matter enabled material to collect into sheets and clusters of galaxies leading to its current state of distribution.

50

Galactic Formation

The early Universe evolved from its previous homogenous (smooth) state to one where matter was gathering in clumps (to form galaxies) with interconnecting filamentous streams (Figure 2.1). Galactic formation began about 100 million years after the beginning of the Universe's expansion. Recent observations suggest that there was a peak of galaxy formation when the Universe was less than 2 billion years old. Galactic formation remained active until about 10 billion years and has been on a decline ever since.

In the early Universe, the average mass density was much higher than it is currently, which allowed much more frequent interaction between the collections of matter that existed. Although the matter of the Universe was initially expanding, heterogeneities allowed fluctuations in the local gravitational environments to reduce the rate of matter dispersion locally. These regions of reduced local expansion would approach an equilibrium between the gravity and kinetic energy of their composite matter. Once a state of equilibrium is achieved, a galaxy becomes stable and no longer contracts or expands. The early galaxies, ones that formed in the first billion years, were relatively smaller, with irregular shapes, and harbored higher rates of star formation than those of today. Many of the early forms contained bright, dense areas thousands of light-years across, which are presumed to represent regions of active star formation. Interactions between galaxies were more common in the early Universe, which may simply be a consequence of more crowding when the Universe was much smaller and denser. Thus the abundant pregalactic fragments (collections of stars, gas, and dust) could gravitationally interact with one another to combine and create larger conglomerate proto-galactic forms. These proto-galaxies would continue to evolve to gradually resemble the galactic forms seen in more recent epochs. During the early galactic aggregation process, gravitational interactions and fluctuating densities of the clouds of gas and dust would induce regions of star formation within a coalescing proto-galaxy. It is estimated that star formation in the early galaxies was about one thousand stars per year—star formation is about one new star each year in a typical galaxy, such as the Milky Way, in the current era. Some lines of evidence suggest that star formation actually preceded the formation of galaxies. After an initial rapid rise in the rate of star formation, it has been on a continual decline for at least the last 9 billion years.

51

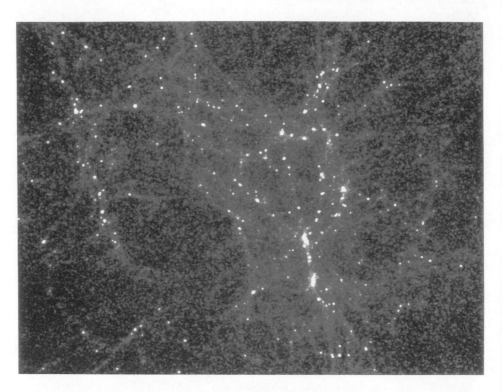

Figure 2.1 A computer model of the early Universe pictorially displays how it may have appeared. As conglomerations of matter were brought together by gravity to form the first galaxies, there were filaments of matter bridging the space between. *Credit: Frank Summers, Donna Cox, Robert Patterson, Erik Wesselak, and Barry Sanders. Rendering program written by Loren Carpenter.*

Some cosmologists are less convinced of a proto-galactic aggregation process and propose that individual galaxies form from a single large star-forming gas cloud. These collections of star-forming clouds of gas, perhaps averaging a few thousand light-years in diameter, would gradually coalesce to form a galaxy. In the early twentieth century, Sir James Jeans, a British applied mathematician, calculated that a mass of about 10^5 times that of our Sun (Jeans mass) is required to initiate such a collapse. In other words, gas collections of Jeans mass or larger would become unstable and begin to internally collapse to form galaxies. Other theories of early galactic formation suggest that local thermal variations within a large gas cloud may induce collapse into a galaxy — warmer regions would favor expansion while cooler ones would tend to contract. As the cooler regions contract under the influence of their own gravity, collapse eventually leads to formation of a galaxy. There is ample evidence to support either theory of galactic formation, so that one may consider galaxy formation as a diverse process that may involve both mechanisms. After a few billion years, galactic forms like those of today became more abundant, with a relative stability of this pattern over the last several billion years.

Observations indicate that the early galaxies were much brighter and active than those of the present. In the past, quasars and radio galaxies were more common. Models suggest that such galaxies contain extremely massive objects at their galactic cores; black holes are one such possibility. As time passed, such active galaxies have become much less common. The characteristics of these early galaxies are the subject of much recent research and have been studied by the Hubble Space telescope in the Hubble Deep Field study (Figure 2.2). In December 1995, the Hubble Space Telescope focused on a small patch of "empty" space in Ursa Major (near the Big Dipper) in a direction that provides an unobstructed view out of our galaxy into intergalactic space. Hundreds of exposures were made over a ten-day period to produce the Deep Field image. This small region of space, which appears as a void in Earthbound telescopes, contains about three thousand galaxies. Computer statistical analysis of the intergalactic background of the Deep Field image failed to reveal any patterns indicative of more distant sources of light. The analysis suggests the limit of the visible Universe may have been reached: there may be little beyond the most distant galaxies captured in the image. In October 1998, another deep field survey was carried out by the Hubble Space Telescope but was directed near the south celestial pole. The Hubble Deep Field South, like its northern counterpart, reveals a similar image containing thousands of dis-

53

Figure 2.2 The Hubble Deep Field image, made by the Hubble Space Telescope in December 1995, reveals a plethora of galactic forms in the distant Universe. Some of the galaxies imaged may have formed less than a billion years after the Big Bang. The image represents a composite of 276 exposures over a ten-day period. *Credit: Robert Williams, the Hubble Deep Field Team, and NASA.*

tant galaxies in various forms. As part of the Deep Field South survey, a ten-day exposure of a tiny region in the southern constellation of Tucana purposely included a quasar to detect the gas and dust of galaxies that lie along the line of sight. The absorption of different frequencies of the quasar's light by intervening gas and dust enables scientists to determine the composition and distance to these materials, providing information to improve cosmological models of galactic evolution.

A survey of the Universe reveals that most galaxies lie within a spectrum of size from between about one-tenth to about one hundred times the diameter of our Milky Way. Our galaxy falls into the mainstream with its 200 billion stars. However, not only do galaxies come in a wide range of sizes; they also have a variety of basic structures. Galaxies are typically categorized according to shape, with three major types: elliptical (E), spiral (S), and irregular (Irr). Elliptical galaxies decrease in brightness from the center to their edge with subclasses according to elliptical shape: E0 for spherical and up to E6 for a highly eccentric elliptical galaxy. Spiral galaxies are flat, disklike galaxies, often with two arms originating from a central bulge, the nucleus. Spiral galaxies are further subclassified according to relative size of the nucleus, compactness of the spiral arms, and continuity of the star mass in the arms. Irregular galaxies typically do not show the ordered structure of the elliptical or spiral galaxies. Generally, elliptical galaxies have older star populations than spirals, while irregular galaxies are often the most active, harboring extensive areas of star formation.

As matter aggregates to form a galaxy, several factors become important in determining the galactic form it will eventually assume. Initially, the temperature will not change much with contraction until the density is high enough for thermal interactions between the particles to develop. Further contraction is checked when a thermal equilibrium is reached where further compression is limited by the expansion force produced by the heating of the matter. Another contraction limit is encountered when the rotational energy balances the cloud's gravitational energy. Rotation of a pregalactic collection of matter may be set in motion by the tug of gravity from another collection of matter drifting nearby. As a cloud contracts, its rate of rotation increases—conservation of angular momentum, as when the spinning skater turns more quickly when her arms are brought closer to her body—until balanced by gravity. Eventually, under such conditions, a flattened, disk-shaped galaxy (spiral) is formed.

According to astrophysical theory, an elliptical galaxy forms when star formation halts further contraction of the cloud before the overall rotation of the proto-galaxy has a significant effect on material distribution. In regions without sufficient rotation or thermal effect to counteract gravitational contraction, continued agglomeration of material can produce massive collections, which tend to be in the central region of the galaxy. If enough material collects into a single body, it may collapse into itself by virtue of its own immense gravitational field, leading to a black hole. However, galactic core black holes are not limited to elliptical galaxies. In fact, at the center of our Milky Way galaxy there is a massive object (nearly 3 million times the mass of our Sun), which has an enormous disk of material spiraling into it at a rate of about fourteen hundred kilometers per second. Observations suggest that the inner region of the disk ends in the form of a ring about ten light-years in diameter, from inside of which no radiation emanates. Perpendicular to the plane of the disk of our galaxy is a stream of antimatter particles shooting out from the galactic core into space to about five thousand light-years, where it terminates through annihilation with matter. Astrophysicists envision that at the center of our galaxy is a black hole that is consuming the matter of an accretion disk.

Galaxies may also be categorized by the degree of radiation activity found at their cores. Despite the apparently active core of our galaxy, it remains relatively sedate and falls within the range of that detected in "typical" galaxies. Some models of galactic evolution propose that galactic core activity may even vary such that a relatively quiescent galaxy, such as our own, may at times have a much more active core. These fluctuations may occur over periods lasting millions of years. Although more common in the early Universe, some galaxy cores contain intense sources of energy at radio wavelengths. Most of these radio galaxies are elliptical, where the radiation source is presumed to lie within the galactic core in the form of a quasar.

Observations from the Hubble Space Telescope support elliptical galaxy formation from the collision of spiral galaxies. As the majority of a galaxy is empty space, a galactic "collision" more closely resembles a "passing through" phenomenon, since very little of the solid matter will actually collide (Figure 2.3). It takes about a half-billion years for colliding spiral galaxies to gradually settle into an elliptical form. Although such collisions were much more frequent in the early Universe, measurements of the motions of some collections of older stars near our galaxy's center indicate circulating remains of a collision with a small (dwarf) galaxy that occurred about 10 billion years ago. Furthermore, Omega Centauri, the most massive globular star cluster in the Milky Way, may actually be the

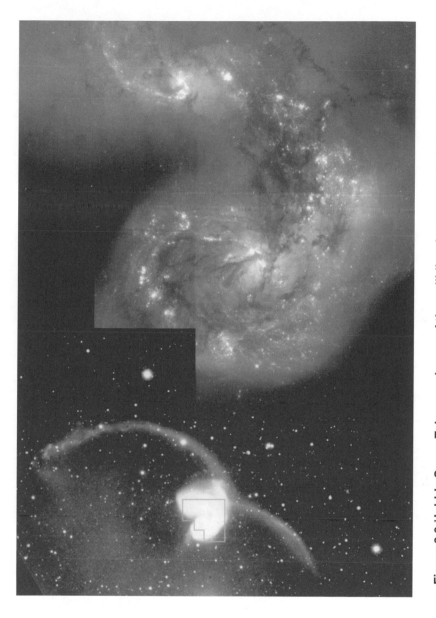

Figure 2.3 Hubble Space Telescope image of the colliding Antennae galaxies (NGC 4038/4039) with a ground-based image on the left. The galaxies lie 63 million light-years away. Many of the bright patches are large star clusters; vast areas of active star formation fueled by the merging of gas clouds from both galaxies. *Credit: Brad Whitmore and NASA.*

cannibalized nucleus of a captured small galaxy. On the far side of the Milky Way, in the direction of the constellation Sagittarius, is a dwarf galaxy that is being tidally pulled apart as it is drawn into and absorbed by our galaxy. The gravitational dynamics of these collisions with relatively small galaxies will leave the overall spiral structure of the Milky Way intact. In the case of the early Universe, observations of earliest galaxies indicate that on average galaxies were typically irregular and smaller than those of later periods, which suggests subsequent galactic growth was an aggregation of neighboring galaxies to form the larger ones that would follow. Studies of galaxy clusters indicate that about four billion years ago spiral forms made up about 30 percent of the total, but they represent only about 5 percent in galaxy clusters in more recent times. It has been proposed that the decline in the relative number of spiral galaxies reflects a tendency for the spiral galaxies to combine with one another or otherwise be re-formed into elliptical types by their own internal dynamics.

As clouds of gas and dust were aggregating to create the first galaxies, star formation was simultaneously taking place. When the accumulation of material within a moderate-sized cloud of gas forms a mass comparable to that of our Sun, a typical main sequence star may result. At the core of the young proto-star, the pressure from the overlying layers of hydrogen compresses and contains the hydrogen nuclei, allowing the reactions of nuclear fusion to ensue. The energy released from the fusion reactions at the core balances the compressive gravitational forces, giving the star a stable spherical form. In cases where the condensing material is of insufficient mass to produce the pressures required for nuclear fusion, the proto-star never ignites; depending on the amount of mass collected, a red dwarf, brown dwarf, or gaseous planet may result. The object is warmed initially by the kinetic energy liberated through its aggregation and contraction, but over time this heat energy is radiated into space as the object cools. However, when larger clouds of gas condense, forming much more massive objects, the results can be spectacular.

In stars such as our Sun, the energy released from nuclear fusion at its core will prevent continued collapse of surrounding layers, but with an excessive gravitational force this balance can be overcome. When a star (such as our Sun in its distant future) can no longer maintain the nuclear reactions to generate the thermal energy to combat the contracting force of gravity, it will be crushed by its own weight into a dense white-hot body (white dwarf) that will slowly cool over many billions of years. During this process, the outer layers of gas of a large star may explode as a supernova, leaving the dense core behind (Figure 2.4). When these stars exhaust their nuclear fuel, collapse of the core is so intense that free electrons are compressed into neighboring protons, forming a solid body of neutrons (a neutron star), which is as dense as the nucleus of an atom. Further collapse under the intense gravity is prevented by support of the physical structure by the strong nuclear force. The density of the neutron star is so great that one teaspoon would weigh about 10^8 tons. It would require compressing our Sun into a sphere with a ten-kilometer radius to transform it into a neutron star. In the formation of a neutron star, as the collapsing core shrinks, its rate of rotation increases by conservation of angular momentum—some neutron stars rotate at rates of 700 times per second. Radio telescopes detect the extremely strong magnetic fields that are emitted as regularly pulsing radio waves from the spinning star. These amazingly regular beacons of radio wave energy pulses are known as pulsars—when first detected, these mysterious interstellar beacons raised suspicions of signals from extraterrestrial civilizations.

If sufficient mass is collected at the core of a condensing gas cloud, its collective force of gravity may be strong enough (from the greater mass) such that the collapse may continue beyond those stages of a typical star. If the mass of the condensing core is enough to overcome the degenerate neutron pressure, the collapse continues until the object disappears from the Universe as a black hole. From classical physics, the force of gravity between objects is directly proportional to the product of their masses and inversely related to the square of the distance between them. Thus more massive objects have higher gravitational fields than less massive objects, and as one gets closer to an object, the pull of gravity increases. In addition, if a mass is compressed, decreasing its radius, the force of gravity at its surface will increase. However, the gravitational force in the vicinity is not altered as an object is transformed into a black hole. If our Sun were somehow squeezed into a black hole (about three kilometers in diameter), the gravitational attraction between the Sun and Earth would remain unchanged—the Earth would remain in the same orbit and not be pulled into the black hole (but the Earth would be a very dark place). Even Mercury, the closest planet to the Sun, would not experience a change in the force of gravity; the amount of mass producing the gravitational field is the same, and consequently the magnitude of the gravitational field it produces remains unchanged. It is when objects get too close to a black hole that bad things happen. As one draws closer to the center of a black hole, ever-higher velocities are necessary to break free of its gravitational grasp.

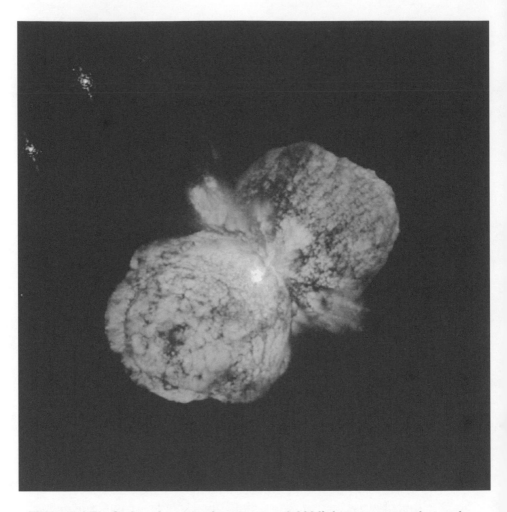

Figure 2.4 Eta Carinae is a massive star over 8,000 light-years away that under-went a tremendous explosion, seen during the mid-nineteenth century, when it became one of the brightest stars in the southern sky. Surprisingly, the star survived the explosion and is flanked by two expanding lobes of gas and dust. *Credit: Jon Morse (University of Colorado) and NASA.*

Alternatively, black holes may never exist in a "mature form" (i.e., never condense into a singularity). As material collapses to form a black hole, the distortion of the space-time continuum—by the increasing gravitational field—locally slows the forward progression of time. In fact, theoretically, it would take an eternity to form a fully mature black hole. Thus, the condensing matter would resemble a collapsing sphere of dense hot material that is continually shrinking to enter within its own event horizon as it becomes frozen in time. Material falling into an eternally forming black hole would become plastered on its surface, which lies just outside the event horizon. However, some experts argue further that black holes can not be formed at all. The strong nuclear force, which increases as particles are brought closer together, may prevent the complete collapse of matter (into a singularity), thereby eliminating the formation of mature black holes altogether.

Related to the gravity and mass of an object is its escape velocity. The escape velocity is that speed one must acquire directed away from an object's surface to balance the gravitational field. On the Earth, the escape velocity is about eleven kilometers per second (km/sec); if a projectile were directed away from the Earth's surface with an initial velocity of eleven km/sec, it would break free to come to rest in outer space. Initial velocities above the escape velocity are sufficient to allow a projectile to break free of the Earth's gravity, while projectiles with initial velocities less than that needed for escape will return to the surface by the pull of the Earth's gravity. When an object has just enough mass so that its escape velocity is equal to the speed of light, it becomes a black hole. With slightly less mass than necessary to create a black hole, light can escape the object, but it will be redshifted. Once the mass is sufficient to raise the escape velocity beyond that of light, nothing escapes; nothing can travel faster than light. Depending on the black hole's mass, at a certain radial distance (Schwarzschild radius) from its center lies a region where the escape velocity is equivalent to the speed of light. This region is in the form of an imaginary spherical surface and may be loosely thought of as the black hole's surface (there is no true surface, as the center is a singularity). In a nonrotating black hole, the spherical surface delineated by the Schwarzschild radius is called the event horizon: the region separating that which lies within our Universe to that which lies outside. Anything that lies within the Schwarzschild radius is excluded from the Universe, as it cannot escape.

Theoretically, black holes can be formed when the mass of an object lies within its event horizon; should the mass of the Earth be squeezed to about two centimeters, it would become a black hole. In theory, black holes may eventually disintegrate through loss of energy of the gravitational field by formation of particle-antiparticle pairs outside the event horizon. In this case, one particle can escape while its partner enters the black hole by virtue of oppositely directed velocities through conservation of momentum. The black hole thus loses the energy equivalent of the escaped particle. As the black hole decreases in size, the rate of pair productions increases, which in turn accelerates the decay process. The finale would be a flash of particles and radiation as the residual of the black hole returns to the Universe. The process is slow, taking perhaps 10^{67} years for an average-sized black hole to disintegrate.

The intense gravitational fields produced by a black hole wreak havoc with Einstein's space-time continuum. Space-time curves upon itself within a black hole, and at its center lies a singularity (analogous to the origin of the Universe). Theoretically, if the collapse of the black hole's core converges into a singularity, the gravitational field at the center will become infinite as the dimension of time freezes to a halt. To an observer, an unfortunate traveler entering a black hole would slow as he approached the event horizon with the light of his image becoming increasingly redshifted by the intense gravity. In theory, the observer would never see the traveler enter the black hole, since it would take an infinite time to approach and cross the event horizon. As for the traveler, he would be heated to extreme temperatures and ripped to pieces by the black hole's gravity, rendering such a journey extremely unpleasant. Should the hypothetical traveler somehow survive a trip into a black hole, he would watch an eternity pass in the Universe around him during the microsecond that he crosses the event horizon. Once inside the black hole, the fabric of space-time breaks down, as does reality as we know it.

When the collapsing matter that forms the black hole has a rotation, it may produce an infinitesimally small spinning ring at the center in the plane of the accretion disk (Kerr black hole). By approaching the center of a spinning black hole from one of its poles, in theory a traveler would encounter an intense, but not infinite, gravitational field. The extreme and distorted gravity at the center creates an interdimensional conduit, called an Einstein-Rosen bridge (wormhole), which could theoretically release the traveler elsewhere in the Universe. The stability of such theoretical phenomena is tenuous, and it is entirely possible that quantum effects make them altogether impossible.

As the greatest amount of concentrated material lies at the center of most galaxies, massive black holes at their cores are not uncommon. A survey of twenty-seven of the nearest galaxies to the Milky Way suggested that at least eleven had convincing evidence of massive objects, which are probably black holes within their cores. Galactic core black holes can be immense, having accumulated the mass of millions or even billions of stars. It is estimated that there are billions of black holes throughout the Universe.

Although black holes cannot be seen, their effect on other objects in their vicinity (i.e., nearby stars) can be detected. Quasars (Quasi-Stellar Radio Sources) are strong sources of light (radio and visible) that lie at great distances (billions of light-years) and have been attributed to massive black holes at the centers of early galaxies, which are consuming the matter of nearby stars. The matter is heated to extreme temperatures as it is consumed by the black hole, thereby emitting the enormous radiant energy that is detected by telescopes. Quasars are more common at distances of about 10 billion light-years and become less abundant at distances greater or less than that. This distribution pattern suggests an epoch when quasars were more common and that they have a lifespan (perhaps 10 million years). Since most galaxies may contain massive black holes at their centers, the presumed mechanism to generate a quasar is prevalent and the brief lifespan may be related to a temporary source of fuel. Studies suggest that most quasars (about 75 percent) are found in galaxies that are undergoing galactic collisions. Thus a colliding galaxy can be cannibalized of some of its matter to fuel a core black hole, producing a quasar. The quasar continues to shine until its fuel supply is exhausted. Quasars were most abundant when the Universe was between 2 and 4 billion years old. At that time in the early Universe, galaxies were forming and relatively close together, permitting increased frequency of collisions. Such collisions have decreased as the continued expansion disperses the matter of the Universe, with consequent decrease in quasar incidence.

After the first few billion years, the galactic forms began to resemble those of the present. Thus many of today's galaxies have remained unchanged for several billion years. The Milky Way galaxy formed at least 12 billion years ago through a succession of mergers of smaller precursor galaxies until it developed into the spiral form we have today. The 200 billion stars that form the visible spiral structure of the Milky Way, approximately one hundred thousand light-years across, lie within a spherical dark matter halo extending out at least a half-million light-years in all directions. The dark matter halo, which occupies more than a thousand times the volume of the spiral disk, contains approximately 95 percent of our galaxy's mass. Some of this dark matter may be in the form of massive compact halo objects (MACHOs), composed of baryonic material such as planets, brown dwarfs, neutron stars, and black holes. The Milky Way has two companion irregular galaxies, the Large and Small Magellanic Clouds, which are slowly being pulled apart as they are drawn closer through mutual gravitational attraction—the merger process may continue for another 10 billion years. The central part of our galaxy is in the form of a luminous barlike bulge from which the spiral arms emanate. The central core of the Milky Way, about twenty-six thousand light-years from the Earth (in the direction of the constellation Sagittarius), harbors a black hole designated Sagittarius A*, with a mass of approximately 2.6 million suns. The first stars in the Universe were likely massive (hundreds of times larger than our Sun), composed of the hydrogen and helium from creation, burning brightly for just a few million years before exploding as supernovae. They left black hole remnants within their proto-galaxy hosts; gravitational seeds around which the first galaxies arose (and providing nidi of galactic core black holes).

Our galaxy has been in existence for billions of years, with relatively little change in its structure during the recent epoch. It is unlikely that any large mergers have occurred in the last 5 billion years; mergers involving more than a few percent of the Milky Way's mass would significantly distort the disk. The Milky Way has a rotation about its center that carries our Sun (and planets) around once every 250 million years. Thus our solar system has made it around the galaxy about twenty times since it formed. The arms of spiral galaxies have compression zones between them where areas of active star formation are located. In these zones the gas densities are increased, favoring aggregation of some of the material to form stars. Our Sun has passed through these zones many times, recently exited a compression zone, and is currently passing through a low-density dust cloud. It has been proposed that passing through relatively dense interstellar clouds may have been the cause for the Earth's ice ages—the Earth could experience a global decrease in temperature, as sunlight would be partially obscured by the intervening dust. However, other ice age mechanisms have received more support, including the Earth's carbon cycle with its associated atmospheric carbon dioxide level fluctuations that affect global temperature.

Large-Scale Structure

Most galaxies are found in groups or clusters, and our Milky Way is no exception. We are part of the *Local Group,* which is composed of about three dozen galaxies and is just a few million light-years across—some galactic clusters contain tens of thousands of galaxies, which tend to have an overall elliptical shape. The next level of organization is the galactic supercluster, representing a conglomeration of galactic clusters. Such large structures become apparent on scales approaching hundreds of millions of light-years. We are at the edge of a supercluster, which is centered in the constellation Virgo, about 60 million light-years away. The Virgo supercluster is about 100 million light-years across. Superclusters are not dynamically stable, as the Universe has not been in existence long enough to allow the mutual gravitational attractions to form a stable configuration. Therefore, the overall structure of galactic superclusters is irregular. The largest-scale galaxy surveys and theoretical models suggest superclusters are actually concentrations of galaxies at the nodes (intersections) of an overall filamentary distribution of galactic material.

The developing pattern of the large-scale distribution of matter is that of a gigantic web or "foam," with vast voids several hundred million light-years across lined by sheets or filaments of galaxies and their clusters. In addition to the filamentous regions and voids are vast clumps of galactic material that include the Local Supercluster (of which we are part), the Great Attractor, and the Shapley Concentration. Gravitational dynamics among the galaxies within cluster concentrations can be predicted by computer models and compared with observational data of their relative velocities. In fact, the presence of dark matter in the early Universe and its temperature has implications on its own distribution and that of the visible matter in cosmological models. Some of these models have demonstrated striking similarities to the observational evidence, including the pattern of galactic distribution.

The weblike pattern of galactic distribution is seen at scales in the range of several hundred million light-years; but over much larger distances the distribution begins to take on a more uniform pattern (Figure 2.5). Previously, astronomers had considered galactic distribution random, with some galaxies residing in clusters. Beginning in 1986, Margaret Geller and John Huchra, astronomers at the Harvard-Smithsonian Center for Astrophysics, produced a map of a thin slice of the northern sky com-

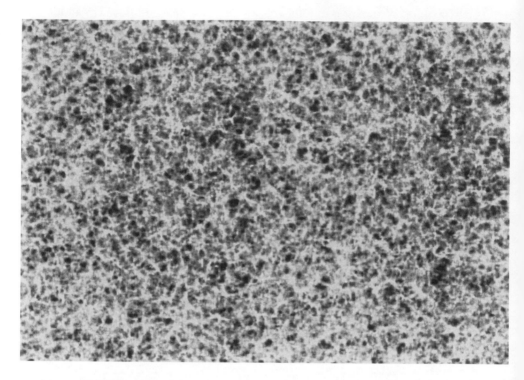

Figure 2.5 An enhanced image from a computer model of the expanding universe clearly demonstrates the weblike pattern of the distribution of matter. The image represents a region of almost 10 billion light-years across. The filaments are composed of thousands of galaxies, which are separated by dark, empty voids. On this scale, our Milky Way galaxy would be a nearly imperceptible dot along one of the filaments. *Credit: The "Virgo Consortium for cosmological simulations," a collaboration of astrophysicists based at universities in the United Kingdom, Germany and Canada.*

posed of 1,057 galaxies; the project continued with the addition of more galaxies to the census and further expansion of the map's boundaries, which clearly brought the weblike distribution pattern into focus. Another study, by Enzo Branchini and his group at the Kapteyn Institute in Groningen, the Netherlands, using infrared satellite image data of 15,500 galaxies covering 84 percent of the sky, were able to construct a three-dimensional density map of the local universe, which confirms the clumping of matter into clusters that are separated by vast voids (Figure 2.6). The Sloan Digital Sky Survey, using a 2.5-meter ground-based optical telescope, will record 100 million celestial objects, including the spectra of 1 million of the brightest galaxies. This ambitious project will produce a three-dimensional map of about half of the northern celestial hemisphere. When completed, the Sloan survey will provide a vast data bank of information for astronomers, which may enable many to spend time reviewing the archive rather than competing for observatory time. Although estimates of the size of the visible Universe predict it contains about 50 to 100 billion galaxies, there is optimism that the entire visible Universe will be mapped before the end of the twenty-first century. The data of these surveys may further refine understanding of the rate of universal expansion and the fate of the cosmos.

In the late 1980s study of the motion of our Local Group and other nearby galaxies indicated a collective motion of 630 km/sec toward a largely unseen mass, the Great Attractor. Direct visualization of the Great Attractor is obscured by intervening dust within our own galaxy. It is predicted to lie about 200 million light-years away and presumed to be in the direction of the Centaurus constellation in the southern sky. The Virgo and Hydra-Centaurus clusters are also affected, suggesting the Great Attractor has a collective mass of about fifty thousand galaxies. More recent maps of matter distribution and relative galactic velocities, in part, dispute some of these motions, suggesting a general flow toward the Shapley Concentration, which lies beyond the Great Attractor. The Shapley Concentration is a supercluster in the southern hemisphere with three galaxy clusters at its core and lies about 700 million light-years away. Although the Shapley Concentration is in a state of dynamic gravitational disequilibrium, it may be the largest gravitationally bound structure in the observable universe. It is estimated that up to 25 percent of the motion of our galaxy through space is through the gravitational influence of the Shapley Concentration. The accurate determination of the relative velocities of galaxies and their clusters is fraught with pitfalls, making nearly any claim subject to debate.

67

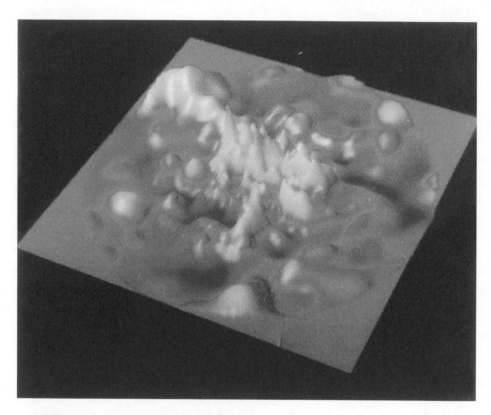

Figure 2.6 A plane of the density field of the local Universe out to about 500 million light-years, with the Milky Way located at the center. The density distribution reveals areas of high density separated by vast voids. A large density structure extends from the Perseus-Pisces cluster, just to the left of the map center, which extends to the upper left to reach the Shapley Concentration at the map edge. *Credit: "A Non-Parametric Model for the Cosmic Velocity Field," by E. Branchini, et al., Blackwell Science.*

However, one may accept that the motions of the galaxies and clusters are more complicated than simply the relative velocity of Universal expansion or gravitational effects within the clusters themselves; e.g., there are also relative motions between the clusters and superclusters.

3

Origin of the Solar System and Earth

Under the influence of gravity, local inhomogeneities of relatively increased matter density in interstellar clouds of hydrogen and helium gas will begin to coalesce. As the density of hydrogen atoms at the core of a collapsing region of a gas cloud increases, so will the temperature. As the hydrogen atoms are brought closer and closer together by gravity, eventually they are forced close enough to undergo fusion reactions. The first-line nuclear fusion reaction (nucleosynthesis) of a star involves the formation of helium from hydrogen. The star releases radiant energy from the proton-proton reaction whereby four hydrogen nuclei (each a proton) are combined into an α-particle (a helium nucleus), with the subsequent emission of gamma radiation and neutrinos. Our Sun currently fuses 600 million tons of hydrogen nuclei every second. In the fusion of hydrogen into helium, there is a "loss" of 0.66 percent of the particle mass in the reaction: the α-particle has less mass than the four protons used to make it. This missing mass changes form into energy by Einstein's famous relation: $E = mc^2$. In the equation, c^2 (the square of the speed of light) is a very large number, which indicates the massive amount of energy released in the conversion from matter to energy. In the case of our Sun, approximately 4 million tons of matter is converted into energy every second. Photons created by the fusion reactions near the Sun's center undergo numerous absorption and reemission events (in random directions) within the dense surrounding plasma before they reach the Sun's surface, where they are free to escape into space. The photon journey to the surface takes about thirty thousand years—an average velocity that is less than one millimeter per second!

The proton-proton nuclear fusion reaction yields intense gamma radiation and extreme heat, which balances the gravitational contraction of the surrounding layers of gas, stabilizing the star into a spherelike shape. As a star depletes its hydrogen at the core, there is a consequent reduction in the nucleosynthesis reaction. This decrease in nucleosynthesis tips the balance of force in favor of gravity, which leads to further core contraction. The increasing pressure and temperature of the core enable the fusion of helium nuclei. The fusion of three helium nuclei to form a single carbon nucleus goes through an unstable beryllium isotope. The mass of the carbon nucleus is less than the sum of the helium nuclei used in its synthesis, allowing the release of energy from the nucleosynthesis reaction. The conversion to carbon nucleosynthesis at a star's core with its increased temperatures eventually causes the star's outer layers to expand about one hundred times and cool (redden) Thus during the phase of carbon nucleosynthesis a star becomes a red giant. Our Sun will become a red giant in about 5 billion years.

In stars with sufficient mass, nucleosynthesis progresses to heavier elements as previous ones decrease in quantity in the core. Nucleosynthesis proceeds from carbon to other heavier nuclei until the core becomes saturated with iron. Iron cannot be used as a nuclear fusion fuel because it weighs less than its corresponding fusion products (i.e., fusion of iron requires energy). At this juncture, the core can no longer release energy through further fusion reactions and gravity becomes dominant: the core collapses. A star that has a mass similar to our Sun (or even up to about six times as much) will ultimately collapse into a white dwarf. A white dwarf supports itself against further contraction (under gravity) by the degenerate electron pressure produced by crowding of the electron quantum energy levels of its atoms. Our Sun will contract to about 1 percent of its current size when it becomes a white dwarf. White dwarfs no longer undergo nuclear fusion; they are white-hot from the thermal energy derived by the tremendous pressures of the collapse and then slowly cool over billions of years.

Massive stars typically have relatively short lifespans and dramatic finales. Stars about ten to fifty times as massive as the Sun will eventually undergo a final collapse but will shed their outer layers in a violent release of X rays, gamma rays, and neutrinos as a supernova. The outer layers would have been active in nucleosynthesis, such that the supernova explosion releases the heavy molecules into interstellar space for use in future-generation stars and their systems. These heavier elements were present in the cloud from which our solar system formed—about 2 percent of our Sun's mass is composed of these heavier elements. The Earth and the other rocky inner planets are composed mostly of heavier elements, including silicon, oxygen, carbon, and lead. In fact, the carbon in our bodies

71

was formed in the centers of other stars billions of years ago and subsequently collected in the solar nebula during the formation of our solar system.

The fate of the heavier stars, determined primarily by their mass, is to become either a neutron star or a black hole. If massive enough, the core remnant following a supernova will collapse beyond the degenerate electron pressure barrier until a balance is reached between gravity and the degenerate neutron pressure (free electrons and protons are forced together to create neutrons)—a neutron star is formed. Stars that begin with a mass at least fifty times that of our Sun may form black holes because their cores have sufficient mass to overcome the degenerate neutron pressure. As such a star core rapidly collapses to lie within its event horizon, its light becomes ever increasingly redshifted until it disappears. Some black holes may form with an intermediate neutron star stage. In this situation, the preliminary core remnant of the supernova collapses into a neutron star, but as some of the material initially blown into space from the supernova explosion returns, sufficient mass may accumulate on the neutron star to increase the gravitational well enough to convert the remnant into a black hole. Recent observations have found unusual properties of a few neutron stars, suggesting these cases are actually quark stars—gravitational collapse has overcome the color force whereby the neutrons have degenerated into their constituent quarks. The findings are preliminary and controversial. Further observations and analyses are required to confirm the existence of this new form of matter.

In some cases, insufficient mass is collected to form a star. Coalescing clouds of gas that contain less than a tenth of a solar mass, have insufficient material to initiate hydrogen fusion. If about 10 percent of the Sun's mass is collected, a brown dwarf results. These objects are heated through the energy of their contraction but do not undergo nucleosynthesis and will cool over time. Even less massive objects may form from smaller clouds and are basically gaseous planets. Such objects may be abundant in the Universe and are not easily detected by conventional methods (since they don't emit the high-energy photons of stars), placing them on the list of dark matter objects.

About 4.5 billion years ago the Sun ignited at the center of a collapsing cloud of interstellar gas and dust within a spiral arm of the Milky Way galaxy. Most likely the Sun formed with several other sibling stars from the same cloud of gas (Figure 3.1). These stellar nurseries dissipate after several million years through the dispersal of the sibling stars under their

Figure 3.1 The dark pillars of the Eagle Nebula (M16) are a star-forming cloud of hydrogen gas and dust 7,000 light-years away. The pillar on the left is about one light-year in length. *Credit: Jeff Hester and Paul Scowen (Arizona State University) and NASA.*

own random velocities. Frequently small groupings of stars within a nursery will remain gravitationally bound, forming multistellar systems. The gas and dust that surrounded the proto-sun, the solar nebula, had an axis of rotation that eventually flattened into a disk with a central bulge. The majority of this material was concentrated in the central portion of the disk, where it formed the Sun. The inner regions of the cloud (just beyond the primordial Sun) were hotter than its outer edges, dissipating the lighter elements while leaving heavier ones behind to form the inner planets. The lighter gases were pushed outward to coalesce into the gaseous outer planets. Within the solar nebula, dust grains slowly aggregated into small planetesimal grains, which would aggregate to form loose collections. The process of their growth (accretion) was rapid; estimates suggest that the progression from dust grains to mountain-sized asteroids took only 100,000 years. After about a million years, the solar system contained the primordial planets as well as many moon-sized planetoids. Within a few million years the gaseous outer planets were formed. It took about 70 million years for the Earth to accumulate enough material to become its current size.

The age of the Earth was a subject of controversy until the twentieth century. Theologians made the first predictions by following biblical references to form a chronology of Adam's descendants. By the seventeenth century, the most extensive investigations estimated an age of several thousand years. In the early nineteenth century, geologists argued that geological processes required much longer periods of time to accomplish their effects than was allowed by the estimates of the age of the Earth. In fact, some geologists argued that the rock record had no clear beginning or end and was compatible with an infinite age. Eventually, two competing theories surfaced: either the Earth was as old as biblical predictions or it had always existed. Late in the nineteenth century, Lord Kelvin (William Thomson), a British physicist, proposed that the Earth was formed by the impacts of rocky material and had been cooling ever since. He calculated that the Earth was between 20 and 40 million years old. Scientifically reasonable, this calculation provided some comfort to geologists, but it still fell short of what was necessary. In 1904, Ernest Rutherford, also a British physicist, formally introduced his work in radioactivity. Its application to the aging of rocks immediately extended the Earth's age by 100 times Lord Kelvin's calculations. The age of the Earth and solar system can be accurately estimated by this technique. Meteors, remnants of the solar system's creation, almost uniformly date to 4.55 billion years.

The rate at which planetary formation occurs is based on theory. There is insufficient observational data to produce a complete description and timetable for the process. Several accretion disks have been found around young stars, which are suspect for developing planetary systems (Figure 3.2). Until recently, the theory of planetary formation originating within the solar nebula of a young star was without direct observational proof. The first definitive evidence of a planet-forming disk around another star was from direct observations of ß Pictoris, a star fifty light-years distant, found in the constellation Pictor. The extensive dust disk around ß Pictoris has a diameter of over 300 billion kilometers and has perturbations that were initially thought to represent gravitational effects from orbiting planetary bodies within the disk, but are now considered more likely related to a nearby star. The disk of ß Pictoris is seen almost edge-on from our vantage point, giving it a flattened projection. In early 1998, investigations of another star, HR4796A, have revealed a disk that is slightly more oblique from our perspective (Figure 3.3). Faintly visible to the unaided eye, HR4796A lies 220 light-years distant in the southern constellation Centaurus and is about 10 million years old, making it younger than ß Pictoris and at an age considered ripe for planetary formation. Observations of the disk around HR4796A at various wavelengths have shown a region of central clearing, which may reflect an absence of particles due to the proximity of orbiting planetary bodies that have swept up the dust in their orbital paths.

Not only does HR4796A have a possible developing planetary system; it is also part of a binary star system. Its companion, HR4796B, orbits at a distance of about 75 billion kilometers (about thirteen times the orbit distance of Pluto). Thus the presence of a second star doesn't necessarily preclude the stability of a planet-forming disk. Previously the gravitational influences of companion stars were considered incompatible with a stable planetary disk and subsequent planet formation. However, the distance between the stars of a multiple star system influences the gravitational dynamics of disk formation. Astronomical observations indicate these disks are stable for multiple star systems at least as widely separated as the HR4796 binary. It is known that most stars reside in multiple star systems, which had been considered unlikely to support planetary bodies. The discovery of a disk around HR4796A in a binary system significantly increases the possibility of planetary systems beyond previous expectations. Dust disks, which may contain planets, have been discovered around other stars, including Vega and Formalhaut, which suggests the presence of

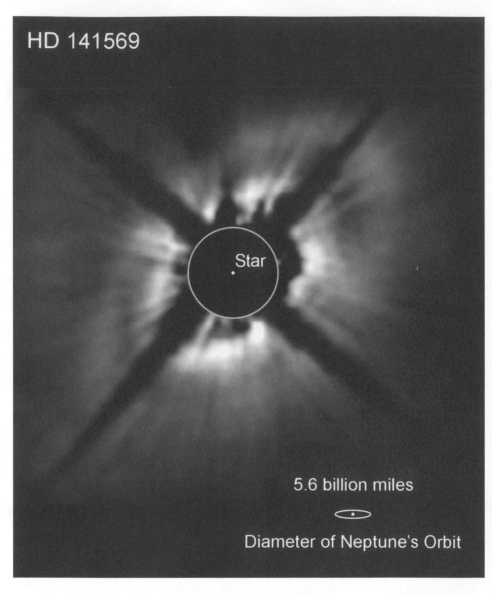

Figure 3.2 Hubble Space Telescope image in the near-infrared of the star HD 141569 reveals a 120-billion-kilometer wide disk containing a dark band separating the fainter outer portion from the brighter inner region. The dark band may represent the presence of a planet that has swept up the particles of the disk along its orbital path. *Credit: Alycia Weinberger, Eric Becklin (UCLA), Glenn Schneider (University of Arizona), and NASA.*

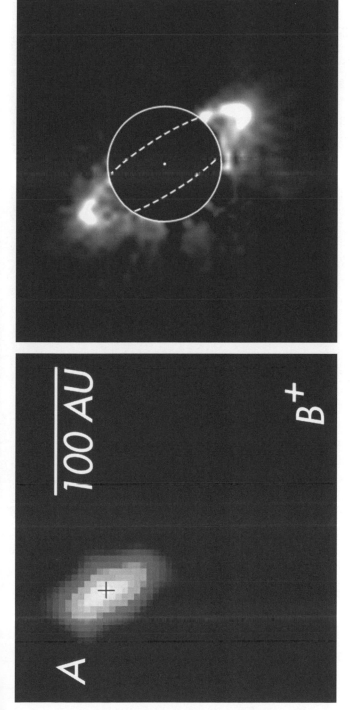 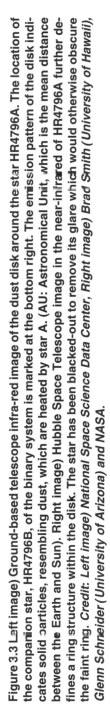

Figure 3.3 Left image) Ground-based telescope infra-red image of the dust disk around the star HR4796A. The location of the companion star, HR4796B, of the binary system is marked at the bottom right. The emission pattern of the disk indicates solid particles, resembling dust, which are heated by star A. (AU: Astronomical Unit, which is the mean distance between the Earth and Sun). Right image) Hubble Space Telescope image in the near-infrared of HR4796A further defines a ring structure within the disk. The star has been blacked-out to remove its glare which would otherwise obscure the faint ring. *Credit: Left image) National Space Science Data Center, Right image) Brad Smith (University of Hawaii), Glenn Schneider (University of Arizona) and NASA.*

77

these disks may not be as uncommon as previously thought. Recent observational studies suggest that the majority of young stars have disks, where up to 90 percent of stars under 400 million years old have dust disks. Dust disks are less common in older stars, which may reflect disk material accumulation into larger bodies (asteroids, comets, planets, etc.). The high prevalence of these presumed planet-forming disks suggests a consequential abundance of planetary systems.

The search for mature planetary systems outside of our own is an active area of astronomical investigation. Astronomers estimate that about 2 percent of the billions of sunlike stars in our galaxy may harbor planets. Direct observation of planets around other stars is not possible, as the light they reflect is very faint and easily lost in the glare of their parent star. However, several stars have been carefully scrutinized to reveal infinitesimal variations in their motion and spectra as indirect evidence of orbiting planets. Since the degree of motion is small and depends upon the mass of the orbiting body, only the effects of large planetary masses (the size of Saturn or Jupiter) with relatively short orbital periods are detectable. These recent discoveries suggest planetary formation and planetary systems are not as rare as once thought. In fact, some astronomers venture to claim that up to 50 percent of main sequence stars such as our Sun may have planetary systems. The proposed abundance of planets increases the opportunities for life to develop, but mere presence of a planet is not sufficient. Many of the recently detected planetary systems are known to contain only large planets (Jupiter-sized) with peculiar orbits creating instabilities, including extremes of temperature so vast that they are deemed prohibitive for the development of life.

In our solar system, some of the matter of the accretion disk did not end up within the planets of today. An unstable planet (or planets) may have existed between the orbits of Mars and Jupiter, which left debris along its orbital path: the asteroid belt. Beyond the orbit of Neptune is a diffuse ringlike collection of primordial material in the form of cometary bodies which forms the Kuiper belt. Short-term comets (on the order of a couple decades between trips through the inner solar system) are believed to originate from the Kuiper belt objects. Investigations of the orbits of some of the Kuiper belt objects reveal similar orbital patterns, adding credence to a theory that Pluto, which orbits within the belt (raising debate as whether it is really a planet), had a collision with another Kuiper belt object, which gouged out its moon, Charon. Of the small objects that have similar orbital trajectories to Pluto, called Plutinos, a fraction may represent fragments of Pluto from the collision.

Beyond the Kuiper belt, extending out to perhaps a light-year, lies a shell of an estimated 6 trillion cometary objects known as the Oort Cloud. The average comet of the Oort Cloud lies at about forty-four thousand AUs from the Sun and has an orbital period of 3.3 million years. These cometary bodies are weakly held in their distant orbits by the Sun but are also influenced by the gravitational force of stars passing nearby, more distant star clusters, and, to a lesser degree, the galactic spiral arms and galactic core. On average, about once a million years, a star passes close enough to nudge some of the Oort comets into a trajectory that sends them into the inner solar system. The next close approach will take place in about 1.4 million years when the red dwarf star Gliese 710 passes within 70,000 AUs of our solar system. It is estimated that comet activity will increase by about 50 percent with the passage of Gliese 710. Closer approaches cause further increases in the frequency of comet visits to the inner solar system. In October 1999, British astronomers made a controversial claim of indirect evidence of a tenth planet about one thousand times as far away as Pluto. They argue that the trajectories of some comets have been altered by the gravitational influence of a massive distant object. Their orbital analysis of these comets, which they assume originated from the Oort Cloud, reveals unusual patterns, which they attribute to the effect of an object at least as massive as Jupiter within the Oort Cloud. Critics note that such studies assume the Oort Cloud comets are homogeneously distributed, which is not necessarily the case; the Oort Cloud comets may actually form clusters, which could explain the unusual orbits. However, the presence of a planet beyond Pluto—Planet X—has been proposed for decades, but its detection has remained elusive.

Although remote, comets are of particular interest because they have a relatively high concentration of ice and organic molecules. Furthermore, perturbations of the orbits of comets may send them on a trajectory through the inner solar system with a possibility of a planetary impact (Figure 15.2). The occasional passing of a star or even periodic oscillations in the orbit of the solar system around the galactic center may lead to the repeated release of Oort comets to the inner solar system (and may ultimately be responsible for some of the major period extinctions of terrestrial life). Preparing the Earth for the development of life relied heavily on these encounters with comets. Repeated comet impacts on the primordial Earth played an important role in later events by delivering water and per-

79

haps other critical organic materials. In addition to comets, some of the Earth's water came from meteors that contained water, which was released when they melted on impact. Some of the more recent major impacts, after life had developed, played a different sort of critical role in the course of terrestrial life, as an influence on the extinction and evolution of species on a global scale.

The critical role assigned to comets in the evolution of the Earth and its life-bearing qualities has brought great attention to their study. Indeed, comets are thought to be frozen mountains of the material of the primordial disk from which our solar system arose. As comets enter the inner solar system they are heated by the Sun, vaporizing some of the material they contain. This vapor forms vast tails, which can extend for millions of kilometers. In early February 1999, NASA launched the *Stardust* spaceprobe on a 5-billion-kilometer round-trip journey that includes a rendezvous with Comet Wild-2 in early 2004. During the comet encounter, microscopic material will be collected, then returned to the Earth with a parachute landing in Utah in 2006. In the case of Wild-2, much of its original material may remain preserved, because only until recently was it diverted (by a close pass by Jupiter in 1974) from its remote orbit to more frequent trips through the inner solar system. Thus this recent change of events has made Wild-2 more accessible to spaceprobes, as well as presumably left most of its primordial material intact. The opportunity to determine the composition of comets will help investigators understand the early solar system and further define the role of comets in the development of early terrestrial life.

As the Earth formed, it was heated through a continuous shower of meteors of various sizes. As the young planet's mass increased, the impacts were of increasingly higher energies, because the meteors were being accelerated by the stronger gravitational field of a steadily more massive proto-planet. The sky was blackened with dust from the impacts of the continual bombardment. Constant high-energy impacts heated the Earth's surface into a sea of molten rock, forming a global magma ocean. Additional heat was derived from the decay of radioactive elements in the planet's core. The molten state of the young planet enabled the heavier elements to sink, while lighter elements accumulated near the planet's surface.

The Earth settled into many layers (like an onion), which have increasing density with depth. The outer layer, the crust, represents only 0.6 percent of the volume of the Earth. The majority of the Earth's volume (82 percent) resides in the middle layer, the mantle, which is composed of three zones (upper mantle, transition zone, and lower mantle). Beneath the mantle lies the core at the Earth's center. The heavier elements brought in by the meteors, in particular iron and nickel, sank in the magma ocean and permeated the underlying mantle to settle in the planet's core. The pressure of the overlying rock maintains a 2,400-kilometer-diameter solid core at the Earth's center. The solid core is "suspended" in a liquid metallic core, which is about two thousand kilometers thick. The Earth's core is about 1,000°C hotter than the surface of the Sun. Interestingly, the solid iron inner core appears to rotate slightly faster than the Earth around it. Overall, the Earth is 35 percent iron, but the crust retains only 6 percent. Lighter elements (such as silicon) make up 15 percent of the total mass but are 28 percent of the crust's composition. The segregation process that layered the Earth's material was essentially completed by 600 million years after the Earth's formation began.

The Moon can trace its origin to the early Earth. Nearly four hundred kilograms of Moon rocks brought back by the six *Apollo* moon landings and additional material collected by three automated Soviet probes provide samples from nine different lunar areas. Analysis of their composition reveals similarity to the mantle of our Earth consistent with a common history. About 20 million years after the Earth's formation, a large planetesimal (named Orpheus), perhaps three times as massive as the planet Mars, impacted with the Earth, which threw a great amount of the Earth's mantle and crust into space. Up to half of the material involved in the collision returned to the Earth over several months, leaving the rest distributed in space around the young planet. This lingering debris settled into a giant disk within a few months. During the next few thousand years, the debris gradually aggregated into larger and larger clumps, which eventually formed the Moon. Measurements of the Earth's magnetic field in the Moon's vicinity and of the Moon's gravity by NASA's Lunar Prospector spacecraft indicate that the Moon has a relatively small core, which is also consistent with the impact theory. At the time of the impact, much of the Earth's dense core would have already formed in the planet's center, so that debris released from a surface impact would consist predominantly of upper mantle material, with a relatively smaller proportion of core material. Thus once the dust settled and the Moon's layers stabilized, it would end up with a relatively smaller core than the Earth. Shortly after its formation from the inner portion of the disk of debris, the Moon was much closer to the Earth—only about twenty-two thousand kilometers away. At such close range, the Moon appeared about fifteen times larger in the sky and may have orbited the Earth every few hours! During the billions of years since it was formed, the moon has slowly spiraled outward to its current distance. The massive collision that formed the Moon would also have significantly increased the Earth's angular momentum, such that shortly after the impact a day would have been only about four hours.

The presence of the Moon may have been essential to allow the development of terrestrial life. The axis about which the Earth rotates has a slow wobble. The Moon's effect on the dynamics of the Earth's wobble is to stabilize it—without the Moon, the Earth would erratically tumble to rotate on its side or even flip upside down to rotate in reverse (as does the moonless planet Venus). The stability of Earth's axis of rotation affords stability of temperature and climate, which is necessary to maintain life. In the future, as the Moon continues to move farther away from the Earth (about four centimeters each year), its tidal effects will diminish and the Earth's rate of rotation will slow. Eventually the Earth's rotation will slow to the point where one side will permanently face the Sun while the other will be left in eternal night.

During the evolution of the Earth, its atmosphere went through drastic changes. The Earth's original atmosphere was mostly hydrogen, reflecting its origin from the lighter elements of the solar nebula disk. The extensive meteor bombardment and giant impact that formed the Moon dissipated the Earth's early atmosphere into interplanetary space. The young planet fostered intense volcanic activity, which constantly released water vapor and carbon dioxide, producing a dense atmosphere. Surface accumulation of water was prevented as the surface was a molten magma ocean for the first 200 million years. Eventually, as the surface cooled, a solid crust formed, which was drenched in torrential rains as the water vapor condensed from the dense atmosphere. The surface was eventually covered by a great ocean with scattered craters and volcanoes shrouded by dark, dusty clouds. The atmospheric carbon dioxide level decreased as some of it was dissolved in the vast ocean. A decline in the frequency of meteor bombardments allowed the lingering dust clouds to gradually clear. Three and a half billion years ago, the global ocean had begun to evaporate. As the sea level dropped, exposed portions of the crust formed the early landmasses. The content of the rocks of the continents indicate their origin is mostly volcanic. There would have been many active volcanic islands without any dominating continents for several hundred million years. It was during this time that life on our planet began in its primordial seas.

The landmasses of the Earth have undergone many changes over the aeons. As the Earth cooled and the heavier elements settled, a relatively light crust formed at the surface: the lithosphere. The lithosphere actually moves along the surface by slowly gliding over an underlying deformable layer, the asthenosphere, itself an outer layer of the upper mantle. The asthenosphere layer has a consistency of slush and lies on average about one hundred kilometers below the Earth's surface. The development of the theory of mobile continental plates marked the beginning of the study of plate tectonics.

The first speculations of continental drift were proposed in the seventeenth century by the English philosopher Francis Bacon. The first comprehensive work on the theory of plate tectonics was made by the German meteorologist Alfred Wegener in 1912. His proposal was not seriously accepted by the scientific community until the middle of the twentieth century, when the science of geology had developed the necessary techniques to verify and analyze the processes involved.

There are about twenty massive plates that cover the Earth. As the landmasses glide over the asthenosphere, they are pushed along by new crust being formed along the oceanic ridges (e.g., the mid-Atlantic ridge). Where landmasses encounter each other, one of three situations may arise: subduction, collision, or fault formation. Subduction occurs where one plate is forced under the other and is melted and recycled as it encounters the high temperatures of the underlying mantle. In some cases of oceanic subduction, melted crust may rise nearby to form an arc of islands (i.e., the islands of the western Pacific). A collision zone occurs when two plates meet head-on, as in the case of the Indian subcontinent and Asia. The force of this collision formed the Himalayan Mountains through buckling of the crust. When two plates pass side by side, a fault is formed where periodic earthquakes during brief episodes of intermittent relative movement can occur (e.g., the San Andreas Fault in California).

Reconstructing the distribution of the Earth's landmasses indicates vast changes throughout history. Currently the Atlantic Ocean is growing wider by a couple centimeters annually as the mid-Atlantic ridge continues to spread, while the Pacific Ocean basin is becoming smaller with subduction along its western boundary. Reversing the current motions of the continental plates implies they were collected into a single massive continent (supercontinent) a couple hundred million years ago. Geologists are more easily able to describe the dispersal of the supercontinent, Pangea (Greek for "all lands"), than its formation or other patterns of continental drift from times prior. In the Acasta region of northern Canada, geologists have found a rock formation dating to 4 billion years ago that may have been part of an original proto-continent that existed at least 2.9 billion years ago. Plate tectonics probably did not occur much before 3 billion years ago, as the Earth's surface would have been too hot to form rigid plates and allow subduction. In an attempt to trace the history of plate tectonics as far back as possible, geologist John Rogers has proposed that the landmasses began with a single small proto-continent (which he named Ur), which was surrounded by a vast ocean 3 billion years ago (Figure 3.4). The remnants of the proto-continent include small portions of southern Africa, Madagascar, Australia, Antarctica, India, etc. For the next 500 million years, volcanic activity and collisions with smaller landmasses increased the size of the original continent. Additional landmasses emerged during the subsequent billion years, which eventually collided to form the supercontinent Rodinia 1 billion years ago. After another 300 million years, Rodinia separated in the Northern Hemisphere to form Laurasia, while landmasses were coalescing in the Southern Hemisphere.

Six hundred million years ago, large ice sheets covered much of the land in the Southern Hemisphere and extended nearly to the equator. Under these conditions, nearly 70 percent of the life-forms went into extinction. This was followed by a warming trend as the continents began to break apart, which led to rapid radiations of new life-forms, including the first multicellular organisms. At the beginning of the Paleozoic, 570 million years ago, there were three landmasses in the Northern Hemisphere (North America, Europe, and part of Asia), with the remainder of the landmasses forming the great continent of Gondwanaland in the Southern Hemisphere. The first plants appeared on land during this period, 450 million years ago. The supercontinent Pangea was formed during the Permian Epoch, about 225 million years ago, with the combining of Laurasia and Gondwanaland, which were separated by the Tethys Sea.

The dynamics of plate tectonics had profound effects on the environment and affected the course of terrestrial life. When the landmasses combined to form a single massive continent, vast expanses of shorelines and shallow seas were eliminated in the interior. Many of the early forms of life that were in existence had flourished in the shallows, so the extensive loss of these habitats brought a massive extinction of about 90 percent of the marine life-forms. Also during the time of the formation of Pangea, the tur-

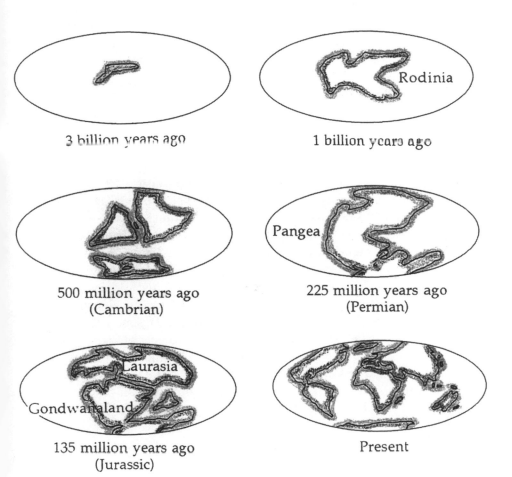

3 billion years ago

1 billion years ago

500 million years ago
(Cambrian)

225 million years ago
(Permian)

135 million years ago
(Jurassic)

Present

Figure 3.4 Plate tectonics describes the continually changing positions of the continental land plates.

moil in the shallow seas was matched on the land by massive volcanic activity in what later became Siberia. By 200 million years ago Pangea had already begun to separate again into Laurasia and Gondwanaland, which was followed by further fragmentation. The division of the continents rapidly opened new habitats and was followed by great changes in the biomass with the emergence of the true mammals from their reptilelike predecessors. With the dispersal of Pangea, the continents of today were formed.

The Earth's historic global climate has been a subject of intense investigation where studies of the geological record indicate the passage of many ice ages. Some suggest there were several ice ages as long ago as 2.5 billion years. A recent theory claims that there were at least four periods of global glaciation between 750 and 580 million years ago. These cycles would have occurred during the period of Earth's life history when the first multicellular organisms appeared. The landmass of Rodinia was breaking up by 770 million years ago, distributing its constituent pieces along the Earth's equatorial zone. The rearrangement of the landmasses led to a period of increased rainfall, which reduced the levels of atmospheric carbon dioxide. The warming effect of atmospheric carbon dioxide was therefore reduced, which led to a cooling trend that was accelerated as polar ice caps formed—ice reflects away solar heat. A massive ice age developed where the entire Earth was covered by ice for up to 10 million years—"snowball Earth." Gradually the atmospheric carbon dioxide levels were on the rise, released by volcanic activity, which set a global greenhouse warming effect in motion that was intensified as water vapor was added to the atmosphere by the melting of glaciers. The cycle of global cooling and heating may have occurred up to four times during a 200-million-year period. Furthermore, there is evidence that about 300 million years ago Pangea was covered by ice for several million years. The most recent ice age began about 35 million years ago, and in the past million years there have been about ten cycles of glaciation and retreat, without the ice ever completely disappearing. The Greenland ice sheet is a major remnant of this climactic cycle, containing layers of these previous periods of glaciation that provide an extensive record for climatologists. During the past few thousand years (the period of development of human civilization) the Earth has experienced a relatively stable warming trend. Climatologists estimate that the next ice age will not peak for another 80,000 years. Currently glaciers cover about 10 percent of the Earth's surface and contain about 75 percent of the planet's fresh water.

Various mechanisms behind the drastic global climate changes of the Earth have been proposed. An increasingly popular theory suggests the orbital dynamics of the Earth within the solar system may be responsible for historic warm and cold periods, including the ice ages. The shape (eccentricity) of the Earth's orbit around the Sun slowly varies between a relatively circular path and a more elliptical one over a 100,000-year period. The circular path would tend to minimize the temperature variations between the seasons, whereas an elliptical orbit would accentuate seasonal temperature differences. The twenty-three-degree tilt (obliquity) of the Earth's rotational axis relative to its orbit produces the annual cycle of the seasons. Variations in the angle of tilt of the Earth's axis of rotation, which has a cycle of about forty-one thousand years, will exaggerate the warm and cold seasons over long periods of time. The tilt also has an instability (precession), like a wobbling top, which brings about variations over tens of thousands of years. Astrophysical theory and computer models suggest that the perturbing forces that produce the wobble of the Earth's axis of tilt can be explained by gravitational influences from the other planets in the solar system. The presence of the Moon apparently provides stability to the Earth's tilt by overpowering much of the influence of the more distant planets.

After about the first 500 million years, the Earth had sufficiently cooled and the meteor bombardment had slowed enough to permit the emergence of life. The ingredients were available and the conditions were appropriate; all that was needed was enough time for the biochemistry to take hold.

Part II

Origin of Life

"The time has come," the Walrus said, "to talk of many things . . ."
—Lewis Carroll, *Through the Looking Glass*

4

Defining Life

The tale behind the origin of life on the Earth—biogenesis—has been a subject of speculation, religion, and science for centuries. Simply reaching a consensus of the principal qualities that define a life-form is not trivial—the task can be particularly difficult when describing primitive forms of life. The definition of life is subject to countless interpretations, including biological, spiritual, and metaphysical approaches. Following basic biological guidelines, in order for an entity to be considered a life-form it must satisfy several conditions. Fundamentally, it must be able to acquire and incorporate material from the external environment (for energy to function), release waste, and reproduce. To satisfy these primary conditions, terrestrial life incorporates a vast array of complex and diverse chemical processes, which require correspondingly complex molecules. One must realize that although there is an extraordinarily rich diversity of life on our planet, it is the result of billions of years of evolutionary change.

The distinction between one life-form and another can be blurred by biological relationships. The science of biology is teeming with examples of species that depend upon others for survival. In a broad sense terrestrial life is undeniably ubiquitous: life permeates the globe, fills the oceans and air, and blankets the land. In 1979, James Lovelock, a British atmospheric scientist, introduced the controversial concept of *Gaia* (the ancient Greek Earth goddess), where the combination of the atmosphere, soil, water, and animals living within and on the Earth's surface collectively represents a great life-form. Although each of these discrete components contributes to the stability of life, most biologists prefer to categorize life-forms as separate entities confined to a narrow shell that encompasses the globe's surface, called the biosphere.

The overwhelmingly complex and intricate interactions between terrestrial life-forms makes isolation of its individual species tedious. The multibillion-year history of the codevelopment of terrestrial life-forms has led to a vast web of interaction between different organisms. Organisms are not simply affected through their interaction with an inanimate environment but are also inundated by interaction with other organisms. These interactions come in a variety of forms, including symbiotic and antagonistic. Although biological science classically investigates life through study of the individual as a means of simplification, it is perhaps an unrealistic premise to consider an individual as existing in complete isolation (from members of its own and other species as well as the environment). Furthermore, it is believed that all terrestrial life stems from an original single form, which ultimately relates all species to a single biomass. Thus individual species lose uniqueness as they in turn represent single pieces in a great puzzle.

The relationship between the Earth and the life it harbors is clearly one of intimacy. Geochemical processes near the Earth's surface have been affected by the presence of life ever since it first appeared. The chemistry of the earliest microscopic life billions of years ago was sufficient to change the Earth's atmosphere by adding oxygen (a by-product of the biochemistry of some organisms), which in turn affected the future of life itself. In the modern era, anthropogenic alterations—consequences of man—of the environment include chemical pollution, liberation of greenhouse gases, deforestation, etc. The coevolutionary development between the Earth and terrestrial life has been taking place for billions of years, which raises concerns about the long-term effects of anthropogenic processes on the future of the planet and life.

In order for life to develop, the cosmos must provide the proper ingredients in sufficient quantity and allow enough time for the appropriate complex chemical assembly process to occur. Terrestrial life is organic—chemically based on the carbon molecule—implying the importance of the availability of carbon (and other heavy elements) in sufficient quantity. These critical elements are produced inside stars through nucleosynthesis and released into interstellar space in the death throes of supernovae. Life on Earth, including ourselves, is composed of dust formed long ago in the cores of extinct stars. Therefore, the Universe must be old enough to allow the evolution and demise of these first-generation stars to make the chemical raw materials available for life-forming second-generation star systems that follow.

Cosmology has been unable to conclusively determine whether the rate of our Universe's expansion indicates a "closed" or "open" system—the difference has significant bearing on the development of life. In an open universe, rapid spatial expansion may overcome the gravitational agglomeration of matter thereby preventing the involution of gas clouds to form stars, planets, and life. On the other extreme, if expansion is too slow and gravity predominates (closed system), the universe would collapse before enough time had elapsed for first-generation stars to produce the heavier elements required for future generations of life-bearing planetary systems. Therefore, there is a very narrow range of expansion rates permissible for the development of organic life, a balance between allowing enough time to produce life's raw materials in sufficient quantity before the universe collapses (closed) or dissipates (open).

5

Organic Life

The emergence of life has been the subject of folklore, religion, and scientific investigation for millennia. Early concepts generally assumed *spontaneous generation,* where "nonliving" materials spontaneously give rise to "living" forms. In the fourth century B.C., the Greek philosopher Democritus proposed that the interaction between the basic particles (which he called atoms) of fire and the Earth were responsible for creating life. During the Middle Ages, crude experimental investigations suggested that life-forms such as insects were the spontaneous product of decaying meat. In the seventeenth century, Francesco Redi, an Italian physician, refuted such claims by proving that maggots were larvae from the eggs of flies laid on decaying meat (rather than spontaneously produced). The results of such research stirred controversy among those in the scientific community, and it was in the middle of the nineteenth century when Louis Pasteur, a celebrated French chemist, laid to rest the theory of *la génération spontanée* through his elegant experiments. It was concluded that the production of a living form lies in the ability of self-assembly; i.e., living produce living. Ironically, we will discuss a plausible path by which a formerly lifeless Earth produced life from nonliving matter.

Following the methods of reverse engineering, where one dissects the object of investigation to determine how it works, the quintessential information to begin our study of life's origin lies in its composition. As it turns out, living matter contains nothing mystical but is a rather mundane collection of materials: water, carbon, hydrogen, nitrogen, etc. Furthermore, the elements of which organic organisms are composed are few in number and relatively abundant in the Universe. The presence of these elements in sufficient quantity is the first requirement in the development of life. The next requirement is the formation of stable simple molecules, leading to organic compounds (e.g., amino acids) and eventually more complex macromolecular agglomerations (e.g., proteins, RNA, DNA, etc.). The generally accepted hypothesis of terrestrial biogenesis begins with a period when initially simple, then later more complex, organic molecules were formed in the ancient Precambrian seas. These macromolecules are the framework used to construct the fundamental parts of living cells, enabling them to carry out their necessary functions. On occasions when the appropriate molecules were in close-enough proximity to chemically interact, the biochemistry of living organisms took place. The point at which these molecular associations became alive is subject to interpretation; however, at some point their coordinated collective biochemical processes crossed the threshold to become *life*.

It is generally agreed that terrestrial life-forms were derived from a single ancestral form. The presence of identical complex terrestrial biological processes, found in widely different organisms, suggests a common origin to all life. The similarity of amino acids, proteins, and other organic molecules among different species is certainly consistent with an underlying common developmental pathway from which the different species diverged. Even the genetic code that details the components of a human being is written in the same biological language as that used in a bacterium. Different species of an organism arise from a common origin through diversification, where modification of the original organism's structure may lead to new or better adapted function, which is inherited by individuals over the course of many generations. Species development by this mechanism allows the presence of common elements among distantly related species. Alternatively, one must assume that multiple independent life-forms arose and conspicuously developed the same (in many cases *identical*) biological constructs. Although this alternative theory assumes many fortuitous circumstances, proponents point out that the independent development of the same complex biological mechanisms cannot be excluded entirely.

The conditions on the Earth when life first appeared were far different from those found today. The dawn of terrestrial life occurred when most of the surface was covered by a global ocean, the first landmasses were just beginning to develop, and there was extensive volcanic activity. During this period, the Sun was not as bright, leaving the Earth relatively cool and enveloped by a thick, damp atmosphere composed of hydrogen, methane, ammonia, and water vapor. Oxygen was an insignificant component of the primordial atmosphere—as a relatively chemically noxious substance, oxygen would have been an impediment to the early development of life. It was later when life-forms liberated oxygen into the atmosphere, and hundreds of millions of years passed before the emergence of life-forms that could utilize oxygen's reactive properties. In addition to lacking oxygen, the primeval atmosphere also lacked ozone, which permitted ultraviolet radiation to bathe the surface. The energy of the UV light may have been a contributing factor in the formation of the early biomolecules.

Earth's organic life-forms utilize carbon as the basic atomic element that enables construction of the complicated molecules that are necessary for the reactions of biochemistry. Carbon was available for incorporation into primordial life because the carbon in the Earth's atmosphere readily dissolved into the ancient oceans, increasing its aqueous concentrations. In addition to carbon, other atoms (hydrogen, oxygen, and nitrogen) were present, and through random chemical interactions they would bond to form molecules. Occasionally, chance collisions would produce molecular combinations called amino acids. The amino acids, of which there are only twenty, are simple organic molecules that comprise the building blocks of more complicated proteins. Combining amino acids into proteins produces organic molecules, which have function (structural or enzymatic) relevant to life. In 1953, Stanley Miller exposed a collection of gases and water to electrical discharges in an experiment to simulate conditions on the primitive Earth. He was able to produce some of the proteinlike molecules found in living organisms in this elegantly simple experiment. In the years that followed, more extensive experimental models confirmed the spontaneous production of many simple organic compounds of biological relevance under the presumed conditions of the primordial Earth.

The Molecules of Life

Perhaps the foremost quality required of a life-form is the ability to extract and utilize energy from the environment—energy is essential to drive the basic functions of life. According to the second law of thermodynamics, physical processes spontaneously progress toward chaos (increasing entropy)—energy is required to overcome chaos as is required in the highly organized biochemistry of life. Terrestrial biochemistry must extract energy from the environment, process it into a chemical form, and incorporate the energy where it is required for metabolism. The broad range of biochemical processes of terrestrial life requires a ubiquitous collection of complex molecules, many of which are highly specialized for specific chemical reactions. The fundamental building block of organic life begins with the versatile carbon atom, which is able to combine in multiple ways with other carbon and hydrogen atoms to form complex molecules. Additional chemical complexity is obtained by the addition of other atoms (e.g., nitrogen and oxygen) to the carbon chain (backbone), permitting further chemical specificity.

The chemistry of life requires multiple degrees of biomolecular structural integrity. The stable combining (binding) of atoms and molecules may occur through a variety of mechanisms. Ionic bonds are formed when oppositely charged atoms or molecules are held by the Coulomb force of charge. These bonds are strong and favor forming rigidly held lattice structures (crystals) but are not readily amenable to the elaborate reactive properties necessary for life. Alternatively, when two atoms with the appropriate number and orientation of electrons are brought close together, they may share their outer (valence) electrons to hold the atoms together. These covalent bonds (sharing of valence electrons) are modestly strong and grant stability to the molecule but are weaker than ionic bonds and therefore permit the chemical reactivity required for life. Carbon atoms can form other noncovalent bonds (e.g., electrostatic, Van der Waals, and hydrogen bonds), which are weaker still and are more easily broken, providing additional levels of biochemical versatility. In an aqueous environment, water molecules affect the bond reactivity of organic molecules in solution, which is crucial in the biochemistry of living organisms—the bulk of our biomass is water. The relatively weak bonds of organic molecules may be chemically broken and re-formed to create new molecules as well as allow complex interactions, which is the basis for organic chemistry. The chemical evolution period of terrestrial life took place when the organic molecules and macromolecular precursors first appeared. This was followed by the biological evolution period, when the first cells developed, and has continued to the present.

One of the defining qualities of life is the ability to reproduce itself. In the biochemical arena, reproduction requires faithful reproduction of the biological molecules for use in progeny. The creation of copies of large complex biomolecules (macromolecules) requires structural plasticity, for which carbon has ideal characteristics. When carbon atoms form a long chain (carbon attached to carbon, etc.), a weakly held second chain may form alongside the first. Since the two chains are held together only weakly, the chains may easily come apart, which leaves two separate carbon chain molecules. Extension of this concept allows formation of in-

97

creasingly complex mirror-image molecules. The production of duplicated carbon chain molecules is an essential mechanism of carbon-based life—it is through this mechanism that the complex molecules of living organisms are duplicated for incorporation into future generations. In order for these chemical reactions to proceed in a controlled fashion, interference from ambient (potentially toxic) molecules and the harsh external environment must be minimized. In biological systems, isolation from the environment is derived from the cell wall (in plant cells) or cell membrane (in animal cells). It has been shown experimentally that certain organic molecules spontaneously aggregate into tiny spheres, reminiscent of the wall of a cell.

Terrestrial life utilizes carbon as its basic biochemical framework. Carbon has four valence (outer) electrons available to form covalent bonds with other atoms, including other carbon atoms. There are several other atoms that have four valence electrons and consequently share some chemical properties, but these atoms are heavier than carbon (more protons and neutrons in their nuclei) and are far less abundant in nature. The closest competitor is silicon, which has only one-tenth the cosmic abundance of carbon. In fact, life-forms based on silicon are a common theme for those contemplating alternate forms of life, but the relative scarcity and other undesirable properties make it a less appealing candidate. Thus it is not surprising that terrestrial life utilizes carbon as its structural unit, just as carbon is the most likely candidate for extraterrestrial forms, should they exist.

The First Cells

Bridging the gap between lifeless collections of organic molecules and living cells is the *proto-cell*. The earliest proto-cells likely began as microscopic shells with their surface composed of a layer of proteinoid molecules. Proteinoids, which are macromolecules composed of amino acids, can be artificially produced in the laboratory under conditions simulating those that existed on the ancient Earth. When placed in an aqueous solution, proteinoids spontaneously aggregate into microscopic spheres simulating the cell wall of living cells. Remarkably, the formation of proto-cells is a relatively simple chemical process and under appropriate conditions can literally occur overnight. In addition, these spheres may

98

spontaneously "grow" by incorporating other proteinoid molecules from the solution in which they are immersed (e.g., seawater) and may divide (budding) to form "offspring." Although reminiscent of living cells in outward appearance, they are inert and lack the biochemistry of a conventional living cell. They are simply organic molecular shells undergoing random chemical interactions with the environment.

In order to become categorized as a form of life, a cell must be able to incorporate and utilize energy to perform the interactive functions of life: movement, synthesis of complex molecules, growth, and reproduction. Living cells must obtain energy from the environment and convert it to a useful form to conduct these metabolic activities. Energy is readily available from the environment in the form of heat and light provided by the Sun. This source of radiant energy may be chemically "trapped" by certain molecules through modification of chemical bonds, where the energy is temporarily stored or used in biochemical reactions—absorbed photon energy is converted to chemical bond energy. The energy stored in these chemical bonds can be released to drive biochemical reactions of the cell when necessary. Those proto-cells, which acquire energy trapping molecules, achieve an independence in controlling their own biological reactions.

The energy of the Sun's photons may be captured directly or indirectly through intermediaries by different methods. Phototrophs are cells that convert the energy of light into the energy of chemical bonds, as in the process of photosynthesis by the cells of plants. Chemotrophs obtain energy through chemical reactions of ingested foodstuffs, as in the cells of animals. In both cell types, a common "energy storing molecule" is used to couple the energy acquired from the environment to the biochemical reactions of metabolism: adenosine triphosphate (ATP) (Figure 5.1). The three (relatively) high-energy phosphate bonds of this organic molecule can be broken and their energy released by chemical reactions within cells assisted by protein enzymes; an enzyme will hasten the rate of a reaction for which it is specifically catered. The energy released from the ATP molecule's bonds is used to drive the chemical reactions of cellular metabolism. Chemical reactions that require energy generally do not move forward spontaneously but can be pushed along when energy is added to the system. The energy released from ATP can be coupled to chemical reactions of the cell that *require* energy (e.g., muscle contraction) so that they may proceed. Although the ATP molecules in a cell's cytoplasm provide an immediately available source of energy, cells do not continuously drive their

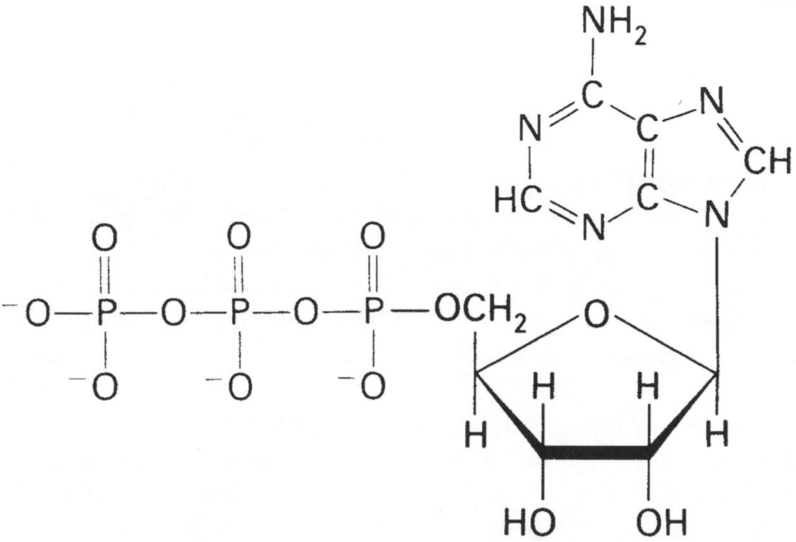

Figure 5.1 Chemical schematic of adenosine triphosphate (ATP). The symbols H, C, N, O, and P represent atoms of hydrogen, carbon, nitrogen, oxygen, and phosphorus, respectively.

metabolism by constantly recycling through their ATP; it is more efficient to have a mechanism to store the energy for variable lengths of time, since both metabolic demand and the availability of energy sources may vary. Cells have developed the ability to conveniently store chemical energy for prolonged periods in the form of sugars, fat, and some proteins.

The proteins of an ancient proto-cell's proteinoid wall would have been a (limited) heterogeneous collection, where some of these molecules served various specific functions. Eventually some proto-cells developed the ability to preferentially sequester useful proteins from the environment, giving them an advantage over contemporaries that were less selective in screening which materials enter (or escape from) their interiors. Early proto-cells that had the ability to manufacture their own proteins would be less dependent on the external environment. These privileged proto-cells would have a survival advantage over their less adept contemporaries, leading to their dominance in the population. Thus the succession to more complex cells with relatively independent function required several stages of advancement of the proto-cells from their proteinoid ancestors.

The propagation of a cell requires that it produce copies of crucial components and segregate them for release as progeny. Since proteins do not typically spontaneously create their own replicates, a mechanism to faithfully produce copies of the proteins found in a cell is necessary to ensure successful cell propagation into future generations. Terrestrial life developed a biochemical method to store and replicate the composition information of proteins as well as regulate their production. The coded instructions listing the cell's armamentarium of proteins and their composition utilize nucleic acids, which are long molecules composed of a tandem sequence of nucleotide molecules held together by a sugar-phosphate carbon chain "backbone." A biochemical dilemma arises: proteins are the functional components of a cell, yet they are formed (by proteins) from the genetic code information of the nucleic acid macromolecules—which came first, the proteins or the nucleic acids? In the prebiotic world, the first form of genetic material may have been peptide nucleic acid (PNA)—the sugar-phosphate backbone is replaced by a peptide backbone. In support of the theory, PNA molecules can be produced in lab experiments approximating conditions of the prebiotic Earth, they can spontaneously combine into chains, and they have chemical activity that may have played a metabolic role. A transition to the more versatile ribonucleic acid (RNA) ensued with the arrival of more competitive life-forms—RNA is more adept at performing the activities of a protein as well as storing genetic information with its nucleoside base sequencing. Gradually some of the original functions of the RNA were replaced by protein molecules that were more proficient.

Experiments have shown that some periodic forms (i.e., a repeated sequence of the same nucleotide) of RNA will spontaneously replicate themselves. However, modern RNA is aperiodic as the chemical and structural complexity of proteins, which is necessary for biochemistry, requires a variety of amino acids connected in specific sequences. Simple periodic sequences of amino acids do not permit formation of the intricately complex protein molecules necessary for life. Although there have not been any self-replicating forms of RNA found in nature, they can be formed in the laboratory and were likely crucial in the earliest forms of life.

In the early "RNA world," a cofactor such as an amino acid may have been encountered that enabled aperiodic RNA to replicate. Furthermore, it is likely that there was a milieu of chemistry taking place to which the RNA was exposed, and over millions of years successful combinations (of replication catalyzing proteins) were eventually secured. In fact, it is possible that the RNA molecule, in addition to catalyzing the formation of proteins, also produced the first DNA (deoxyribonucleic acid) molecules. Gradually, RNA shared the task of protein synthesis with (enzymatic) proteins, heralding the "RNP (ribonucleoprotein) world." We now live in the "DNA world," where in nearly all terrestrial life-forms the DNA molecule, similar in structure to RNA, provides the master blueprint of the molecular components within cells (including RNA) that perform the necessary chemical reactions of life (Figure 5.2).

Terrestrial biochemistry is consistent with RNA serving a role prior to the arrival of DNA. Ribose, a simple carbohydrate, which in part forms the structural backbone of RNA, is more readily formed (than deoxyribose) from the molecules that were present on the prebiotic Earth. The deoxyribose sugar of DNA is more difficult to chemically produce and in living cells is formed from ribose in a protein-catalyzed biochemical reaction. In addition, the deoxyribose-based sugar-phosphate backbone of DNA is more stable than RNA, enabling formation of longer molecular chains (providing increased genetic information storage capacity), which have long-term durability. As DNA took the primary genetic role and proteins became the catalysts, RNA was delegated an intermediary role.

The DNA molecule has several levels of biochemical binding that provide structural stability. They include the embraced twin strands of the

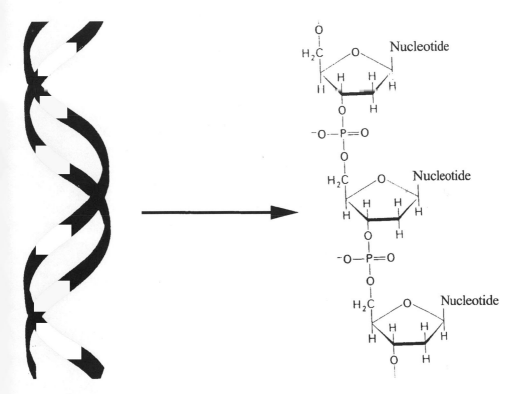

Figure 5.2 A segment of the double-helix DNA strand (left) with an expanded chemical schematic of one strand revealing the deoxyribose sugar phosphate backbone with the attached nucleotide bases (right).

double helix combined with binding proteins that secure the molecule in the form of a chromosome. The DNA molecule is sequestered in a cell's nucleus, which has a nuclear membrane separating the contents of the nucleus from the surrounding cytoplasm. These multiple levels of structural complexity and sequestration serve to protect the DNA to assure its stability and minimize biochemical harm—random damage to a cell's DNA can be fatal. In its native state, the DNA molecule forms a double helix consisting of one DNA strand intertwined with its mirror-image strand. During times of protein synthesis, when the nucleotide sequence must be "read," the strands are temporarily separated by specific proteins to allow interaction of the exposed portions of the strands with the surrounding nucleoplasm. Within the nucleoplasm of a cell's nucleus are protein molecules, which temporarily bind to the exposed sections of the DNA molecule strands and create an RNA copy of the exposed DNA nucleotides. This process produces an RNA molecule with a sequence that directly corresponds to the nucleotide sequence of the DNA from which it was fashioned. The RNA copy becomes a template upon which the specific sequence of amino acids are combined to create the corresponding protein (Figure 5.3). The collective tandem sequence of DNA nucleotides that describes a particular protein is called a gene. The earliest protein-producing cells may have had as few as a hundred genes, whereas the simplest cells of today have about five hundred genes, while most have thousands. The reliability and efficiency of this method eventually led to dominance of the DNA molecule–based cells over ancestral RNA forms. However, some RNA forms exist today, as in the case of some viruses.

Terrestrial life developed through a long process of increasing degrees of biochemical complexity to create the processes of cellular metabolism. It is generally believed that the earliest forms of life gradually acquired more complex biochemical capabilities in the process of evolving into more advanced forms. There is some support for a different mechanism, which postulates that the chemical reactions of metabolism were already present and occurring spontaneously in the prebiotic era. It was later that interdependent reactions of metabolism were sequestered from the total collection of independent reactions, through selective mechanisms of favored reaction outcomes. The extreme biochemical complexity of even the simplest life-forms lends some rationale to this theory of consolidative reduction.

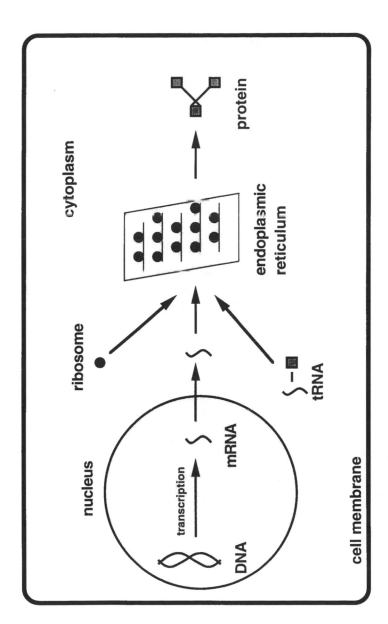

Figure 5.3 The basic steps of manufacturing proteins from DNA within the animal cell. The gene sequence of a protein is copied from the DNA in the form of a messenger RNA (mRNA) within the cell nucleus. The mRNA moves to the cytoplasm, where it combines with the ribosomes at the endoplasmic reticulum. The codons of the mRNA are sequentially read by matching them to corresponding transfer RNA (tRNA) molecules, which carry their specific amino acid. The amino acid is released by the tRNA as it is bonded to the previous amino acid of the growing chain, eventually producing the final protein product.

Extraterrestrial Life

Most scholars consider terrestrial life as purely indigenous to the Earth. However, there is speculation that life's crucial precursor molecules (i.e., various organic molecules) of the chemical evolution period may have had an extraterrestrial origin—some argue further that intact primitive life-forms directly seeded our planet. Beginning in 1962, astronomers Chandra Wickramasinghe and Fred Hoyle studied the chemical composition of interstellar dust clouds that lie in deep space and proposed that terrestrial life had its origin in such clouds. The chemist Robert Robinson supported this theory from analyses of various terrestrial organic compounds, concluding that complex biological processes were present too early in the Earth's history to have an entirely indigenous terrestrial development. Furthermore, Wickramasinghe and Hoyle propose that these findings are consistent with extraterrestrial microorganisms that exist in interstellar space, perhaps feeding on hydrocarbons, as they drift among the stars. When these organisms come under the gravitational influence of a nearby planet, they fall to its surface, populate the planet, and give rise to future life-forms following local evolutionary mechanisms. The scientists claim cellular life-forms could live in space in a condition of dormancy, covered by a thin layer of ice and dust that provides protection from the hostile environment of outer space. In addition, these layers provide protection from the high temperatures encountered during entry through a planet's atmosphere. These extraterrestrial hitchhikers and their consequences define the theory of "panspermia," in which one envisions countless biological "seeds" drifting in space seizing the opportunity to populate lifeless worlds.

Organic molecules are clearly present in some interstellar dust clouds, but the presence of organic molecules does not prove the presence of life. The formation of a planet from a dust cloud containing organic molecules would be too violent to leave any living material intact (as it would be incinerated); thus any chance of successful biological seeding must take place millions of years later when conditions become less hostile. It is known that long after a planetary system has formed material may be repeatedly deposited on its planets by meteors and comets. Recent interest in terrestrial biological seeding arose from proposed microfossils of extraterrestrial life etched in Martian meteorites that fell to the Earth long ago. Less controversial than the deposition of organisms, nonliving organic material is delivered to the Earth on a frequent basis. In fact, it is estimated that about thirty tons of extraterrestrial organic particles fall to the Earth's surface daily.

Comets are a known source of water and organic material, both of which are considered crucial to support life. Fly-bys of comet Halley by the European satellite *Giotto* and the Soviet satellite *Vega 2* uncovered some of the mysterious nature of comets. The nucleus of comet Halley (measuring eight by four kilometers) is extremely dark, with fissures through which volcanic eruptions create a cloud of gas and dust (Figure 5.4). Analysis of the dust by the *Giotto*'s particle impact analyzer indicates the presence of hydrogen, nitrogen, carbon, oxygen, and traces of other elements. Although not conclusive, there is speculation that some of these elements may be in the form of organic polymers (complex molecules found in living organisms) and may even include life itself (bacterialike forms). The spaceprobe *Stardust* is on its way for a rendezvous with comet Wild-2 and in January 2004 will collect dust from the comet's tail. The probe will bring the sample to the Earth two years later for direct study.

Just as comets have been a source of terrestrial water, other molecules could have been brought along. As a source of some terrestrial organic molecules, comets could have played a role in the development of terrestrial life. Wickramasinghe proposes that organic life was not simply brought to the Earth in a single primordial deposit, but that there have been repeated seedings. In fact, he hypothesizes that the great impact 65 million years ago that led to the extinction of the dinosaurs was by a comet that concurrently seeded the Earth and fueled the radiation of life that followed. Supporters of the theory claim that the sporadic nature of terrestrial life (periods of extensive biological radiations and of massive extinctions) follows a pattern that is consistent with occasional cometary visits. Critics cite that life on Earth is ubiquitous, flourishing at many extreme environmental niches, supported by evidence from the laboratory and archaeological records suggesting that eventual adaptation through gradual evolutionary change fostered the biological radiations. Thus extraterrestrial intervention was not necessary to repopulate the Earth following times of catastrophe.

About one hundred thousand tons of material from space falls to the Earth's surface annually. Some of this material is exchanged between the planets of our solar system. Every year, about two tons of Martian rock and dust falls to the Earth's surface after an average estimated journey of about 10 million years; a similar annual amount makes the reverse trip. The situ-

Figure 5.4 The Halley Comet nucleus as imaged by the European satellite *Giotto* on its fly-by March 13, 1986. The comet's nucleus is extremely active, producing giant plumes of gas and dust. The satellite's closest approach to the comet nucleus was 596 km. *Credit: National Space Science Data Center.*

ation raises the intriguing possibility of an exchange of microscopic life between the planets. Announced in 1997, findings suggestive of microscopic fossils of primitive wormlike cellular organisms on a meteor from the planet Mars raised controversy in the scientific community. About sixteen million years ago, a large asteroid impacted Mars, with some fragments of the explosion escaping into interplanetary space. Adrift for millions of years, a two-kilogram potato-sized fragment (ALH84001) fell to the Earth's surface to land in the ice of Antarctica's Allan Hills about thirteen thousand years ago. This fragment was found in 1984, and mineral composition analysis nine years later confirmed a Martian origin and the search for life was ignited. Scrutiny of the meteorite with electron microscopy and chemical analyses raised suspicion of complex biochemical and fossilized remains of microscopic Martian life from 3.6 billion years ago (Figure 5.5). Microscopic evaluation revealed tiny carbonate globules (200 microns across) that contained relatively high concentrations of polycyclic aromatic hydrocarbons: possible decay products of organic life. Independent analysis of the chemical components of ALH84001 and another Martian meteorite, EETA79001, both of which contain similar tiny globular features, was conducted using tedious heat of combustion experiments. Although these results suggest a terrestrial origin for much of the carbon, in the ALH84001 fragment the organic carbon (in the form of polycyclic hydrocarbons) was in too small a quantity to be analyzed, and thus an extraterrestrial origin could not be excluded entirely. Two other Martian meteorites that have fallen to the Earth and were recently examined (Makala, 1.3 billion years old, and Shergotty, 180 million years old) also contain tiny spherical and rod-shaped features that resemble fossilized life-forms. Recently formed water channels on Mars, suggested by high-resolution images made by the Mars Global Surveyor, adds circumstantial support to the possibility of Martian life. Thus the origin of life may not be unique to our planet. Admittedly, the possibility of extinct microscopic life is less spectacular than the dying technologically advanced Martian civilization envisioned by Percival Lowell at the eyepiece of his telescope early in the twentieth century.

The search for extraterrestrial life includes investigation of our planetary neighbors and beyond. It is assumed that liquid water is an essential ingredient in producing organic life-forms, making its presence in adequate quantity a critical factor. Observations and measurements made by the *Galileo* spacecraft that hint at vast liquid oceans of water beneath frozen floating sheets of ice on Jupiter's moons Europa, Ganymede, and

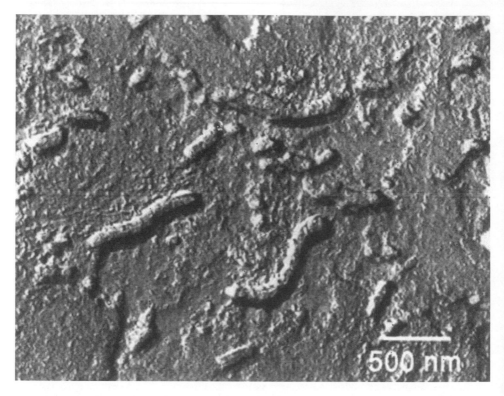

Figure 5.5 High-resolution electron microscopic image of a fragment of the ALH84001 meteorite. This image shows several of the bacterialike features that could represent ancient fossilized microscopic Martian life. *Credit: NASA.*

Callisto have prompted speculations of life on these worlds. At about 1.4 billion kilometers from the Earth lies the planet Saturn, with its largest moon, Titan, which has received attention as possibly harboring life. High-resolution telescopic imaging at infrared wavelengths—enabling penetration of Titan's dense atmosphere—hints at large continents and oceans. Furthermore, the oceans of Titan may contain methane and other organic molecules. Although some of the basic ingredients of life may be found on Titan, the weak solar radiation and extreme cold are prohibitive. A new generation of interplanetary exploration probes is under way with many missions proposed, including investigations aimed at detecting signs of extraterrestrial life. The Mars Global Surveyor project plans the dispersal of about nine probes to the Martian surface and includes delivering a sample of Martian soil to the Earth for study. One of the project's objectives is to determine if life ever existed on Mars. Several projects that will investigate our solar system are under way and include probes to Jupiter's moon Europa, the Cassini mission to Saturn, and its Huygens probe to analyze Titan; even a mission to study icy Pluto is under consideration.

In 1976, two Viking probes landed on the planet Mars where their main mission goal was to determine the presence of microscopic Martian life. The Viking probes sampled the Martian soil and conducted three separate experiments at the two different landing sites. The final results of the Viking probes investigation are interesting: one experiment was negative, one inconclusive, and the third had a positive result. However, NASA scientists concluded that the reaction product of the third experiment—the labeled release experiment—was non-biological. A few experts argue that the results should be interpreted differently: the two probes found "evidence" (but not proof) of Martian life.

In 1961 astronomer Frank Drake proposed a formula to calculate the number of detectable civilizations in our galaxy. The "Drake Equation" consists of several factors that determine the fraction of sunlike stars that might harbor a technologically advanced civilization on an Earth-like planet. The numerical values of the equation's factors rely on principles of astronomy, biology, and sociology. Although he estimates there are roughly ten thousand civilizations currently in existence in the Milky Way, the values of several of the factors are the subject of debate that substantially alters the number of extraterrestrial civilizations predicted. Using relatively conservative estimates of the factors in the Drake Equation, one finds that the number of detectable civilizations in the Milky Way galaxy is about the same as the lifetime (in years) of such a society; i.e., if an advanced civilization lasts 200 years, then there are about two hundred such civilizations in our galaxy, leaving a probability of about one chance in a billion for each star.

The predictions made by the Drake Equation must be tempered by the certainty and accuracy of the data used to make the calculation. A recently contested factor is the distribution of stars within the Milky Way galaxy which are suitable to form planets capable of supporting complex life. Planetary systems that lie too close to the galaxy center are subjected to high levels of radiation and other dangers from the increased density of nearby stars while systems that lie on the galaxy's fringes are relatively poor in the metals needed to form life-sustaining planets. Thus, only a small fraction of our galaxy is optimized to support the development of complex life. Closer to home, considering that of the billions of species on our planet only a single one achieved sufficient intelligence to create a society capable of interstellar communication, the likelihood of extraterrestrial life achieving such status may be extremely remote. Furthermore, the drive behind evolution is through survival by adaptation; it is not goal-driven toward the development of sentient intellectual creatures. Therefore, the prospect of extraterrestrial intelligent life may be exceedingly bleak.

In recent years, ground-based telescopes have collected data in support of planetary systems around dozens of nearby stars. Planets around even the closest stars are too faint to be seen directly, as their reflected light is lost in the glare of their star. Thus the current state of technology for the detection of extrasolar planets requires indirect methods to reveal their presence. Orbiting planets cause slight variations in the position of the parent star's image through mutual gravitational effects. The tug-of-war of gravity between a star and its orbiting planets causes the star to "wobble" rather than remain stationary. The wobble of a star may produce periods of relative motion toward and away from an observer, which can be detected as a small Doppler Shift of the star's spectrum. The temporal pattern and amount of relative Doppler Shift of the star's light attributed to planetary-induced wobble can be used by astronomers to estimate the mass and orbital periods of unseen planetary bodies. Ground-based observations are just sensitive enough to detect the effects on nearby stars by relatively large planets: about the size of Jupiter. The technique of interferometry can detect extremely small spectral variations by subtracting the light of a star's image formed by two separate telescopes. Slight variations caused by planets as small as the Earth in orbit around the nearest forty stars will be the target of the SIM (Space Interferometry Mission), which is scheduled to be put in orbit in 2009. It will be able to detect Jupiter-sized planets up to 200 light-years away. A few years later, in 2015, the proposed TPF (Terrestrial Planet Finder) space telescope will use an interferometric technique to detect light directly from planets orbiting the nearest several hundred stars. Mission goals include detection of planets as small as the Earth, with analyses of their light spectra to determine atmospheric chemical composition. The presence of atmospheric oxygen, carbon dioxide, or water would be a strong indicator of life. A later probe, the Life Finder telescope, will further scrutinize extrasolar planetary systems for the presence of methane and other organic materials to essentially prove (or exclude) the existence of life on candidate planets. The Life Finder telescope is not expected to be launched until about 2025.

Some of the observational time of the Hubble Space Telescope has been delegated to the "Origins" program to identify and characterize potential developing planetary systems around other stars. The program includes multiple missions of both high-altitude atmospheric and space-based telescopes to observe the Universe at various wavelengths, to provide better understanding of the creation of galaxies, stars, and planetary systems. The SETI (Search for Extraterrestrial Intelligence) program is using radio telescopes to listen to the cosmos in the hopes of detecting signals sent from extraterrestrial citizens of the Universe. Thus far, listening to the cosmos, analysis of about thirty trillion signals has revealed about 160 signals that cannot be easily explained away as natural or man-made—"extrastatistical events." These enigmatic signals come from randomly scattered locations across the sky, none have been reobserved, and none are clearly from an intelligent source. However, these locations are getting extra listening time.

6

Diversity of Life

The origin of terrestrial life was a complicated process; however, the spontaneous formation of its fundamental molecules has been repeatedly produced in laboratory experiments that simulate conditions that existed on our planet billions of years ago. Presumably, if it was a simple matter to produce the molecular building blocks of life, why did it take so long to achieve the first living microscopic organisms? In fact, life may have "begun" many times, only to be wiped out by catastrophes (dramatic climatic fluctuations, meteors, volcanic activity, etc.). There is debate as to whether the earliest fossils that have been found represent the true ancestors of terrestrial life or these fossils actually represent early life that went on to extinction as one link in a chain of "start-ups" that eventually produced the ancestors of the terrestrial life-forms that currently populate the Earth. Supporters of the second position cite the lack of resemblance of the earliest fossils to those of later periods may indicate independent origins. However, during the Archeozoic era, about 3 to 3.5 billion years ago, the ancestors of the life-forms that eventually gave rise to today's creatures were clearly in existence. The first living organisms were single-celled bacteria, which arose over 3 billion years ago. The precursors to plants and animals separated lineages about 1,500 million years ago. The fish and amphibians became separate species about 400 million years ago, while the flowering plants separated only about 200 million years ago. Throughout the geological record, terrestrial life went through repeated global life cycles of great expansive diversity separated by massive extinctions, as well as periods when certain species were clearly dominant. In order to place these phases of biological change into a fathomable categorical time line, a geological time scale was devised (Table 6.1).

The Earth was constantly bombarded by meteors and comets until about 4 billion years ago. At the close of this era, the time when life originated, conditions on the Earth were vastly different from today—we would find the atmosphere unbreathable, the Sun only about 75 percent as bright, the Moon about four times larger in the sky, and a day only about fifteen hours long. However, it was in this alien environment that the first signs of life appeared. Rock specimens from Greenland, which date back 3.9 billion years, contain specific isotopic carbon residue suggesting primitive life was present. The earliest microfossil remains of organic life, which resemble clusters of one-celled organisms, are found in specimens from Australia and South Africa dating back 3.5 billion years to the Precambrian Period. Microscopic life, which is not easily preserved or recognized by virtue of its small size, may have been present long before this period, as suggested by the ancient organic residues. The relatively brief period between the end of the bombardment and the arrival of life, about 100 million years, has raised questions about a possible role for meteors in the development of early terrestrial life. Currently, debris from comets and meteors delivers about thirty tons of organic material to the Earth daily; the amounts delivered 4 billion years ago were certainly much greater. Although life may have originated purely from terrestrial materials, organic material from extraterrestrial sources was being deposited on the Earth as it formed and continued as it cooled, providing an external source of organic matter. The point may be academic, as only a few hundred million years before the presumed arrival of life, the Earth was being formed from the very same pool of meteors and comets. Thus at this time in Earth's history the distinction between a "terrestrial" origin of organic material (implying a carbon source derived from the earlier planetary formation period) and an "extraterrestrial" origin of organic material (implying a "post-primary-bombardment" arrival of organic material) becomes a matter of semantics.

The earliest organisms were probably heterotrophs, deriving energy to support their anaerobic metabolism by feeding on organic molecules found in their environment. These organic molecules formed from the same abiotic reactions that created the biomolecules prior to the emergence of life. The metabolism of the early heterotrophs would release ingested carbon as CO_2 gas, making subsequent abiotic formation of organic molecules more difficult. The cycle of a purely heterotrophic biomass would have brought terrestrial life to a grinding halt as it used up its biomolecular energy sources. However, about 2.8 billion years ago the arrival of a new form of metabolism, photosynthesis, brought cells that were more self-sufficient: autotrophs. Autotrophic cells are able to manufacture their organic molecules, rather than extract the preformed organic mole-

Era	Period	Epoch	Millions of Years Ago	Significant Events
Cenozoic	Quaternary	Holocene		civilization
			0.01	
		Pleistocene		*Homo sapiens* arrived
			1.6	
	Tertiary	Pliocene		increase of mammalian size
			5.5	
		Miocene		earliest hominids
			23	
		Oligocene		anthropoid primates
			38	
		Eocene		early primates
			54	
		Paleocene		mammalian diversity
			65	fifth major extinction
Mesozoic	Cretaceous			radiation of insects
			144	
	Jurassic			dinosaurs
			201	
	Triassic			earliest mammals
			245	third major extinction

Table 6.1 The geologic time scale of the Earth's history from its birth 4.5 billion years ago until the present.

Geologic Time Scale

Era	Period	Epoch	Millions of Years Ago	Significant Events
			245	
Paleozoic	Permian			decline of amphibians
			286	
	Carboniferous			earliest reptiles
			345	
	Devonian			age of fishes
			405	
	Silurian			first land plants
			440	
	Ordovician			earliest jawless fishes
			515	
	Cambrian			explosive phyla evolution
			590	multicellular diversity
Precambrian Proterozoic				
	Late Proterozoic			first animals
			750	
	Middle Proterozoic			abundant cynobacteria
			1,600	
	Early Proterozoic			oxygen accumulated
Archeozoic				
			2,500	
	Late Archean			first eukaryotes
			2,700	
	Middle Archean			prokaryotic life proliferated
			3,200	
	Early Archean			origin of life
			3,800	
	Hadean			meteor bombardment
			4,500	

cules from the environment. The metabolic reactions of photosynthesis use the energy of light to produce carbohydrates (a source of chemical energy for the cell) from water and CO_2, with a reaction by-product of molecular oxygen. The proliferation of photosynthetic autotrophs reduced the atmospheric CO_2 levels and elevated the levels of oxygen. The proliferation of autotrophs affected terrestrial biochemistry on a global scale. Some of the atmospheric oxygen released by the autotrophs formed an ozone layer that reduced the ultraviolet radiation exposure at the Earth's surface. The decreased ultraviolet levels quenched the abiotic formation of biomolecules that had been instrumental in forming the first organic molecules from which life had originated.

Terrestrial life was established and flourishing by 3.5 billion years ago but may have originated as long ago as 3.9 billion years. These early life-forms were single-celled prokaryotes (i.e., the archaebacteria), of which there are a few living representatives. Prokaryotes are cells that do not contain a distinct nucleus, and therefore their genetic material (DNA) is loose within the cell cytoplasm. These microorganisms are accustomed to living under harsh environmental conditions, like those present during the early period of Earth's history. The ancient prokaryotes lived in anaerobic conditions (without oxygen), deriving carbon from methane or other organic molecules, hydrogen from hydrogen sulfide, and oxygen from minerals. Today's species of archaebacteria live in extreme conditions of high temperature (thermophiles) or high salinity (halophiles) or in highly acidic environments such as the digestive systems of animals (acidophiles).

Recent attention has focused on the thermophiles as a candidate for the first form of terrestrial prokaryotic life—the early Earth was undergoing relatively frequent meteor and cometary collisions, creating a relatively high-temperature environment. Furthermore, studies of oceanic hydrothermal vents, where heat from the inner Earth escapes along oceanic ridges, reveal a plethora of biochemical activity supporting extensive populations of primitive life (including thermophiles) under conditions thought to resemble those of the early Earth. In fact, the abundance of biochemistry around the oceanic vents raises speculation as to whether terrestrial life actually originated around vents rather than in calm, tepid shallow seas as previously thought. Probable microfossil remains of filamentous thermophile prokaryotes that lived 3.2 billion years ago were recently discovered in Australia—these ancient microbes may have inhabited hydrothermal vent environments. Genetic analyses of living bacterial species suggest that the distant common ancestor of all cells shared a common feature of living under hot, anaerobic conditions. At the time of the proposed ancestral cells, the most plausible environment for life's origin would have been found at deep-ocean hydrothermal vents. In fact, to this day it is estimated that 90 percent of all living creatures live within the oceans.

In addition to the archaebacteria, the other type of early prokaryote was the eubacteria, with many forms existing today as bacteria and blue-green algae. These ancient cells developed the process of photosynthesis for energy metabolism, which is widely used in the plant life of today. Following the establishment of single-celled life, multicellular forms followed. By approximately 3.5 billion years ago, blue-green algae (cyanobacteria) formed chains of single-celled organisms. The early prokaryotes aggregated into multicellular colonies and flourished over 3 billion years ago.

The Earth's atmosphere 3 billion years ago contained little oxygen but had developed high levels of carbon dioxide. The composition of the atmosphere was undergoing changes induced by the early organisms that would forever alter the biological future of the planet. As the early photosynthetic organisms made their own organic molecules from carbon extracted from the environment (i.e., CO_2 gas in the atmosphere), the reactions released oxygen as a by-product. Archaeological investigations in western Australia indicate that oxygen was clearly present on Earth by 2.5 billion years ago, apparently largely produced by cyanobacteria. The rise of oxygen in the atmosphere took about a billion years to manifest. The delay is thought to reflect a reaction of the liberated oxygen with ferrous iron to form iron oxide (rust), which is found in the rocks of that era. As the ferrous iron became saturated, the atmospheric oxygen levels began to rise. Oxygen is a highly reactive substance and was poisonous to many of the early forms of prokaryotic life, which led to the extinction of many of these bacterial species about 2 billion years ago. The rising levels of atmospheric oxygen allowed rapid proliferation of cells, which could utilize and benefit from its reactive nature through aerobic metabolism. Aerobic respiration became widespread about 2 billion years ago, and after another billion years atmospheric concentrations of oxygen began to level off. When the atmospheric oxygen levels stabilized, about 700 million years ago, it was a time when the seas contained both primitive plant and animal forms of cells.

Initially the prokaryotic organisms remained unchallenged for about a billion years, until the first eukaryotic cells appeared. Until recently it was thought that the prokaryotic organisms remained dominant for up to 2 billion years, leaving the arrival of eukaryotes later in Earth's history. The Roy Hill Shale of western Australia was a seafloor between 2.6 and 2.7 billion years ago, and study of its chemical composition found signature biological molecules (lipids) of eukaryotic cells indicating their presence at least 2.7 billion years ago. In addition, genetic analyses suggest a divergence of eukaryotes from the archaebacteria and eubacteria around 3 billion years ago. Eukaryotic cells, the dominant form of today, have a nucleus that contains the DNA organized into chromosomes. The nucleus protects the genetic material by sequestering it and providing a physical barrier (nuclear membrane) to prevent chemical damage from exposure to the metabolic activity of the cell cytoplasm. Dispersed throughout the cytoplasm of eukaryotic cells is a collection of organelles involved in various aspects of metabolism and function. In fact, eukaryotic cells may be the sequela of a symbiotic relationship of ancestral prokaryotic organisms. This theory claims that independently evolved prokaryotes merged into a concerted community of cells that functioned as a single organism. One line of evidence comes from the mitochondria of eukaryotic cells. A mitochondrion is an organelle involved in the energy metabolism of a cell and produces "energized" molecules (e.g., ATP) to run a cell's biochemical reactions. The mitochondria have their own DNA and undergo their own replication separately (borrowing just a few of the host cell's proteins) from the rest of the cell, consistent with an independent lineage. The permanence of the symbiotic relationship to create the eukaryotes was likely the result of a long process, requiring perhaps hundreds of millions of years to solidify.

The success of DNA-based cells, in part, arises from the elaborate complexity of the DNA replication process and cell division, which provides many avenues for genetic diversity that can lead to more successful organisms in future generations. The DNA of eukaryotic cells is separated into chromosomes, which are sequestered into the cell nucleus to provide long-term stability and minimize metabolic damage to the genetic material. During the process of cell replication, the DNA is duplicated with various opportunities for modification during the copying process, including the exchange of genes or portions of genes. As the genetic material was modified and expanded over generations, the cells progressively developed more complex form and function. In fact, there were likely many more types of ancient cells than the direct ancestors of today's cell types—the less successful forms went on to extinction. Eventually, the eukaryotes diversified into two distinct groups, thereby establishing the plant and animal kingdoms.

Animate Life Arrives

The earliest forms of terrestrial life were single-celled organisms, and until about 600 million years ago the most complicated organisms were bacteria and multicelled algae. Eventually some eukaryotic cells aggregated into groups that functioned as a single unit. The first multicellular animals were composed of two layers, with later forms developing a third layer. The three layers include an outer (ectodermal) layer, which provided a barrier from the environment, a middle (mesodermal) layer dedicated to the organism's structure (bones, musculature, etc.), and an inner (endodermal) layer mostly involved with nutrition (gastrointestinal). The early three-layer life-forms resembled flatworms. Further complexity of organisms was obtained through the segmental organization of portions of the cell layers, which led to increasingly more sophisticated creatures. Segmentation of the mesodermal layer provides the groundwork for the repeated anatomical patterns of muscles, nerves, and blood vessels, which allows formation of more complex structures at specific sites of the body plan. The genes involved with segmentation are ancient and ubiquitous and not only exist in the insects of today but also are still involved in the embryological development of humans.

Life first appeared during the Early Archean subera of the Precambrian, and by the end of the Precambrian there were soft-bodied marine and freshwater animals resembling worms, jellyfish, and sponges. Controversy remains as to exactly when the origin of the animal phyla took place, but animate life is thought to have arrived shortly before the Cambrian. The fossil record reveals an abundance of animal fossils during the Cambrian but a marked scarcity in earlier times. This scarcity may in actuality reflect a predominance of soft-bodied creatures that were not as readily fossilized and of a presumed small size, which makes them difficult for archaeologists to glean from the archaeological record. Much of the Precambrian remains a mystery, but there has been recent evidence to suggest there were a series of global ice ages between about 750 million and 580 million years ago that may have played a crucial role in the development and diversification of life at that time. The proposed global glaciations

121

took place when the dominant forms of life were prokaryotic, with some forms of eukaryotic life. The most advanced forms of life at the time were eukaryotic algae, which would have been exposed to several 10-million-year cycles of extreme climactic variation. Such conditions would have created severe pressures on survival, driving natural selection with high rates of extinction followed by adaptive radiations. In addition, at the beginning of this period the great landmass Rodinia was fragmenting and forming many new shallow seas, which provided new niches for exploitation by colonizing organisms and led to further differentiation and speciation. Thus several mechanisms may have been at work to foster rapid evolutionary change during the late Precambrian, which harbored the origin of the animal phyla.

The colossal diversification of animal life that began at the close of the Precambrian and continued during the Cambrian period must have had ancestral organisms from which they evolved. Paleontologists have tried to uncover the predecessors to these Cambrian forms but the fossil record has produced very few candidates. Some light has been shed by combining techniques borrowed from geneticists and embryologists, allowing paleontologists to extend their understanding of early animate life and enable them to propose ancestral animal creatures that predate the Cambrian explosion by tens of millions of years.

The earliest animals would have been small, with soft bodily tissues, which is unfortunate for paleontologists because such creatures do not preserve well, leaving little for the fossil record. However, in the case of the Doushantou formation in southern China, fossil preservation of ancient species occurred in rocks that contain relatively abundant phosphate. The phosphates permeate the organic material of the animal cells and provide enough rigidity to these structures to allow adequate fossilized preservation. The Doushantou formation has provided examples of animal life-forms dating back 570 million years. Among the fossils were tiny (less than a millimeter), well-preserved ancient sponges. In addition, multicellular embryos were also preserved, which indicate creatures that had bilateral symmetry: a fundamental body structure plan used in many present-day species, including humans. Animals can be divided morphologically into two forms: embryonic two-layered radially symmetric and three-layered bilaterally symmetric. Creatures with bilateral symmetry are structurally more advanced than those with radial symmetry (e.g., sponges). In order for animate life to be at this stage of complexity 570 million years ago, there must have been an original predecessor form that arose millions of years earlier.

Studying molecular evolution includes correlating the sequences and evolution of genes in contemporary species of different animal phyla to determine divergence patterns from ancestral species. By assuming that changes in a shared gene occur at a particular rate, the differences between shared genes of the different phyla may be used to determine when the species diverged from a common ancestor. The earliest split, which separated the animal phyla into two groups (the protostomes and deuterostomes), has been estimated to lie between 670 million years and possibly as far back as 1.2 billion years ago. Between 500 and 600 million years ago, from their deuterostome ancestors arose creatures with the beginnings of a centralized brain and spine representing the first members of the phyla Chordata. The body structure of chordate life-forms includes a notochord, a dorsally positioned central nervous system, musculature arranged in segments, and the presence of gill slits at some point in development. Such a body plan evolved in creatures requiring active mobility within their environment. In 1995 archaeologists working in China announced the discovery of the oldest chordate fossil, designated *Yunnanozoon lividum,* which lived 525 million years ago. About the size of a pen cap, it may represent the most ancient ancestor of the vertebrates discovered. The first chordates were invertebrate, but later vertebrate forms arrived in which a segmented bony column developed around the dorsal nervous system to serve as structural support (analogous to the bones of the spine in humans). The division between the vertebrate and invertebrate animals probably took place prior to the Cambrian, around 600 million years ago.

The earliest vertebrates lacked sufficient bone to leave evidence of their arrival in the fossil record, making it difficult to determine when they first appeared. Indeed, the earliest fossilized vertebrate specimens discovered are so structurally advanced that some scholars suggest their evolution during the past 500 million years may represent only a fraction of their entire evolutionary history. The vertebrate predecessors were likely small soft-bodied jawless fish with brains that had paired sensory organs for sight, smell, and equilibrium. The Precambrian ancestor to the bilateralians of the Cambrian probably arose from simple organisms containing a few cells. However, the direct relation to a precursor organism remains unclear. Correlating the paleontological record and the genetic molecular clock, estimates suggest that during the Late Proterozoic (the

160-million-year period preceding the Cambrian) the major animal phyla were established from a common bilateralian ancestor.

The common ancestor to the animal phyla has not been found in the fossil record. The lack of physical evidence requires paleontologists to use other means to determine the characteristics of this ancient organism. The presumed appearance of this bilaterally symmetric ancestral creature was as a microscopic tube of cells with a mouth at one end and possibly cilia along its body for movement. Analysis of genes active in embryonic development in representatives from the different animal phyla suggests this animal ancestor may have been more complex than originally thought. Evidence points toward a creature that may have had photoreceptors (primitive eyes to detect light) and antennae. However, even though such genes may have been present within the genome of this ancient creature, the expression or function of the genetic material may not necessarily have been the same as that found in contemporary species. In theory, the presence of such genetic material may have enabled the rapid period of Cambrian diversity through the expression and manipulation of the genes in different ways. Furthermore, it is well known that minor alterations in genetic material can produce major alterations in the morphology of an organism.

The establishment of the major animal phyla in the Late Neoproterozoic period set the stage for the Paleozoic era. The Cambrian (named by English geologist Adam Sedgewick) is the first period of the Paleozoic, in which the oxygen content of the atmosphere and oceans rose, animal life was dominated by the invertebrates, and the only plants were marine algae. By 550 million years ago, hard-shelled sea creatures had developed, which left an excellent fossil record for archaeologists. During this period, the trilobites appeared and gave rise to a variety of forms that dominated the ancient oceans for 100 million years. Around 544 million years ago a rapid evolutionary proliferation (Cambrian Explosion) was set in motion, such that within a short period many representatives of all of the major animal phyla appeared. The most rapid species diversification occurred during the Tommotian and Atdabanian stages (530 to 525 million years ago) of the Cambrian period. Thus over about a 50-million-year period all of the basic animal groups alive today appeared, while during the 3 billion years that preceded nothing more complex than jellyfish had evolved.

The force behind the rapid expansion and distribution of life during the Cambrian is unclear. Most scholars favor the idea that drastic global climatic change altered the environment in such a way that certain species were vulnerable while others were better able to adapt, leading to extensive changes in the balance of species populations. The Cambrian Explosion occurred at a time when the Rodinia landmass was breaking up at an exceptionally rapid rate of about 20 cm/year. The separating components of Rodinia would have produced an overall increase in shorelines and shallow seas as the water of the great ocean flowed into the gaps that were forming between the dividing landmasses. These tectonic changes occurred over about a 15-million-year period coinciding with the Cambrian Explosion. In addition, sea levels apparently rose and fell during the same period, which would further alter habitats and heighten species selection pressures. Furthermore, these circumstances would also incite climatic alterations affecting the external environment, inflicting additional adaptive pressures on life-forms. Alternatively, it has been proposed that the hard-shelled creatures enjoyed an evolutionary advantage over the softer creatures by virtue of their structural rigidity and basically consumed their softer contemporaries!

The seas of the Cambrian were full of invertebrate life. It was during the Late Cambrian, about 515 million years ago, when a massive underwater landslide created the Burgess Shale formation of the Canadian Rockies. During the landslide, Cambrian sea creatures were carried to deep waters and buried, providing astonishingly exquisite preservation for archaeological study. Among these creatures was the *Anomalocaris,* nicknamed the Monster of the Cambrian Seas, which was an intimidating predator. *Anomalocaris* had an appearance reminiscent of a lobster, nearly a meter in length, with a large frontal feeding appendage that preyed on the small animals living on the seafloor and midocean. The anomalocaridids may have been closely related to today's arthropods (insects, arachnids, and crustaceans). Also swimming in the Cambrian seas was the controversial primitive vertebrate *Pikaia gracilens,* similar to the sand eels of today, which some biologists argue may be the ancestor of nearly all vertebrate animal life of today, including man.

The Cambrian Period ended by 505 million years ago and was followed by the Ordovician, when the first primitive vertebrate fishes appeared. Animal life remained restricted to the oceans during this period, with only a few brief unsuccessful attempts at occupying the land—until at least 360 million years ago. It was near the end of the Ordovician period, about 438 million years ago, when the first known mass extinction of terrestrial life took place. The period of extinction was accompanied by a long ice age, with an unclear relationship between the two events. There were five major extinctions and at least two minor extinctions in Earth's past (Figure 16.1). Although the extinctions are marked by an abrupt great loss in species diversity, they were followed by periods of extensive species diversification and repopulating to fill the vacant niches.

Plant life made its first appearance on land about 450 million years ago, during the Ordovician period. A few simple plants, mosses and lichens, sparsely covered the landscape at the beginning of the next period, the Silurian, but by 400 million years ago much of the Earth's land was covered with great forests. Within the sedimentary rocks that formed during this period and in the Carboniferous period later, the organic material from these ancient forests was compressed and aged, creating the reservoirs of fossil fuel that give energy to the machines of today. Between 50 and 100 million years after plant life moved onto land, the (invertebrate) animals followed.

The first conclusive evidence of true vertebrates has been uncovered from the early Silurian, over 400 million years ago. The immediate vertebrate ancestors were small fish, only a few centimeters long, with complex brains and sensory organs. They were probably physically active creatures in the ancient seas. During the late Precambrian or early Cambrian, the vertebrate radiations presumably took place, although the fossil record is lacking in clear evidence, because these early forms produced poor fossils from lack of sufficient bone. It was during this period that *Pikaia* swam the seas, but not all archaeologists are convinced it was truly a vertebrate ancestor. About 430 million years ago, one of the earliest true vertebrates, *Ostracoderms* (an armor-plated fish), appeared. It had a brain encased within a skull and also had developed an effective cardiovascular system, including a heart that bears resemblance to that seen in the early embryonic growth of humans. These fish had developed jaws and were active swimmers and predators.

The Devonian period, from 405 to 345 million years ago, was characterized by the rapid evolution of the fish—with armored, bony, and scaly forms found in abundance. Sharks appeared during this period and rose to dominate the oceans by the Carboniferous period, which followed. The sharks of today have a form similar to that of their ancestors hundreds of millions of years ago. However, it was only 25 million years ago that the notorious seventeen-meter-long Megalodon shark appeared and terrorized the oceans as the largest predatory fish that ever existed from then until its disappearance just 2 million years ago. Some of the early bony fish had

126

plates derived from scales to provide support and protection to the brain. In later forms, these scale-derived structures gave rise to the flat bones of the skull. Some fish developed skeletons; among them was the lobe-finned fish, which is believed ancestral to the first land vertebrates. The second massive extinction took place near the end of the Devonian, about 354 million years ago, with a major loss of species followed by rapid diversification and repopulation.

A little over 400 million years ago the first invertebrate land animals appeared, comprised mainly of arthropods such as small insects and spiders. It was many more millions of years before the first vertebrates successfully occupied dry land. The tetrapods, "four-legged" vertebrates, were among the first vertebrates to spend time on land, about 360 million years ago. Their ancestral forms, aquatic tetrapods, arose from the bony fish about 30 million years before. The tetrapod lineage can be traced back to the lobe-finned fish, where some species had developed muscular limbs, which apparently enabled these fishlike amphibians to increase their mobility in shallow water by pushing themselves along the bottom. Eventually these creatures led to the tetrapod amphibian vertebrates, which ventured onto the land. Their land excursions were brief, because the moisture within their porous skin would quickly evaporate in the dry air. Furthermore, their time on dry land was also limited because their eggs could not maintain moisture and had to be laid in an aqueous environment. Eventually, by 310 million years ago, reptiles emerged from amphibian ancestors and had developed a water-tight skin covered with scales, enabling them to contain their body's moisture. In addition, reptiles (amniotes) were successful on dry land because they were able to produce young through the protected environment of an egg. The egg suspends the embryo in a fluid environment surrounded by a shell with an inner lining of membranes (amnion and chorion) that provide a porous barrier through which life-sustaining oxygen from the atmosphere can pass to nourish the developing embryo. This offered a tremendous advantage in that the reptiles were no longer tied to the seashores and waterholes and were thus able to venture forth to conquer the land.

A recent discovery in Scotland of *Casineria,* a vertebrate that existed 340 million years ago, may represent a transitional form between the amphibians and reptiles or, alternatively, a separate predecessor species of reptiles, birds, and mammals. *Casineria* was about fifteen centimeters long, walked on four limbs, and had a tail. It is believed that *Casineria* may have been an amniote or, at least, ancestral to the amniotes. The small size of *Casineria* may have been suited to life on land in the early Devonian, as the only food available would comprise of small invertebrates—the ability to eat plant material had not yet developed. The other Devonian tetrapods, averaging one meter in length, would have required a substantial dietary intake of invertebrates to survive on land. Thus the smaller creatures, such as *Casineria,* may have been the dominant land-dwelling vertebrates that eventually gave rise to the reptiles.

The third massive extinction of terrestrial life took place about 245 million years ago, during the Permian period—about 96 percent of all species vanished from the fossil record. The mechanism behind this event remains elusive, but it occurred over a very brief period, with estimates ranging from several thousand years to a few million years. Some recent investigations suggest the root cause was an increase in the carbon dioxide content of the atmosphere, leading to a greenhouse effect and global warming. At the time of the extinctions, the continents were coalescing to form Pangea, which would have created global climatic disturbances. Severe storms may have produced extensive erosion of surface layers, and in some regions this may have uncovered large coal deposits. Chemical reactivity between the exposed coal and atmospheric oxygen releases carbon dioxide. As atmospheric levels of carbon dioxide increase, the stage is set for global warming. There is also evidence of extensive volcanic activity in the region of present-day Siberia, where massive lava flows were formed with release of sulfur dioxide and carbon dioxide into the atmosphere. In another line of evidence, analysis of sediments laid down at the time of the Permian extinctions reveals high concentrations of fullerenes—spherically shaped carbon molecules—which contain gas atoms (helium and argon) in isotope concentrations suggesting an extraterrestrial origin. Furthermore, there is evidence of a large circular gravitational anomaly of the south Atlantic sea floor west of the Falkland Islands which may be the residual of an impact site from the end of the Permian Era. These findings may represent the consequence of a large (about nine-kilometer) meteor or comet impact that brought the global devastation of the Permian extinction. The sea creatures, amphibians, and reptiles would have been vulnerable to any of these environmental stresses.

Whatever the cause of the Permian extinctions, in the aftermath, the surviving species rapidly flourished to fill the vacant niches. The extensive species differentiation and proliferation particularly facilitated reptilian advancement, including the predecessors of the dinosaurs. The term *dino-*

128

saur was first used in print in 1842 by the English anatomist Sir Richard Owen to describe recently found fossils that could not be classified into existing animal groups—from Greek *deinos,* which translates as *terrible,* and *sauros,* as *reptile.* The Permian extinction marks the beginning of the next era, the Mesozoic, also known as the "Age of Reptiles."

Shortly after the beginning of the Mesozoic, in the early Triassic period, the fossil remains of the reptile *Eoraptor* are found. *Eoraptor* lived at least 230 million years ago and has features consistent with what archaeologists predict the common ancestor to the dinosaurs must have had. An adult *Eoraptor* was about a meter in length, walked on two hind legs, and had paired clawed hands. It was an agile meat-eating predator that fed on small lizards. It is generally believed that the ancestral lineage to the dinosaurs contained reptilian species that were small, agile carnivores of the early Triassic, and it was later in the Jurassic period when the herbivores and larger forms developed and dominated the land. The Mesozoic also fostered many new forms of flying creatures. Perhaps one of the most famous was the *Pterodactyl,* which were neither birds nor dinosaurs but actually flying reptiles. *Pterodactyls* primarily ate meat and fish, with most species having a range in wing span from about thirty centimeters to over three meters. The largest *Pterodactyl* species was *Criorhynchus,* which may have had a wing span of up to twelve meters. Some archaeologists argue that the larger species would have been able to fly with little effort by taking advantage of their light weight and large wing spans and deriving lift from rising drafts of warm air (as do modern gliders). It has been estimated that they could easily fly hundreds of miles—some speculate that certain species were capable of flying completely around the globe. They flourished for nearly the entire Mesozoic era, becoming extinct at the end of the Cretaceous after dominating the skies for 160 million years. The late Mesozoic harbored an extensive radiation of another class of flying creatures that we are all too familiar with today: insects.

About 225 million years ago, Laurasia (from the Northern Hemisphere) and Gondwanaland (from the Southern Hemisphere) were combining to form the supercontinent of Pangea. During much of the first half of the Mesozoic era the Earth contained only one great landmass in the Southern Hemisphere. Thus dinosaur artifacts are found on all continents. By the end of the Triassic period, the breakup of Pangea was in progress, which isolated populations of plant and animal species as they drifted apart on their respective continents. In the later Mesozoic many new trees, grasses, and flowering plants species developed, which affected the diversification of dinosaur species. During the Cretaceous, the last period of the Mesozoic era, some dinosaur species became extremely large. The largest species thus far discovered measured up to fifty meters long, stood as tall as an eight-story building, and weighed up to 100 tons. The dinosaurs roamed the Earth for about 160 million years, almost a thousand times longer than *Homo sapiens*.

It was approximately 201 million years ago when the fourth massive extinction (the Triassic-Jurassic boundary) occurred. The extinctions affected both terrestrial and marine life. Although more than half of the reptilian species perished, most of the dinosaur species survived. As the continents were colliding and moving past one another in the gradual breakup of Pangea, there were massive volcanic eruptions and lava flows across the central part of the great continent. The magmatism process (massive lava flows) occurred over a few-million-year period, which peaked at the close of the Triassic period, about 200 million years ago. The margin of uncertainty of the dates of the magmatism and the Triassic-Jurassic boundary mass extinctions overlap, which raises the possibility of a causal relationship. The cause of the massive lava flows is unclear but may have been a large upwelling of magma from the mantle, which had formed a giant sealed-off pocket beneath the lithosphere. The vast underground pool of magma erupted to the surface during a period of a few million years to cover an area of over 7 million square kilometers. The lava eventually became a black basalt rock formation the size of the continental United States. The interior of the supercontinent began to break apart during the Jurassic period, which divided the volcanic rock among the separating landmasses. One of the deposits of this basaltic rock faces New York City from the west bank of the Hudson River as the Palisade cliffs. The massive lava flows of the Triassic-Jurassic boundary may be directly related to the force behind plate tectonics, because Pangea split through the site of the magmatism to form the Atlantic Ocean. In addition to the havoc wreaked on the land, the atmospheric carbon dioxide levels may have had a rapid rise through the increased volcanic activity, as evidenced by studies in paleobotany. The fossils of plants living at the time of the Triassic extinctions have a significantly decreased density of leaf pores—a response to increased carbon dioxide levels. Interestingly, the 100-kilometer-wide crater of Quebec's Lake Manicouagan was formed by a celestial impact dating to nearly the same period as the Triassic-Jurassic boundary, about 190 million years ago, raising question of an association with the Triassic-Jurassic boundary extinctions.

During the Cretaceous period, a time dominated by large dinosaurs, with small mammals on a slow rise, there was a minor global extinction about 91 million years ago. The minor extinction may have been the result of a brief period of exaggerated volcanic activity. However, the most infamous of the period extinctions was to come at the close of the Cretaceous period.

It is generally accepted that the end of the dinosaurs, and many of their contemporaries, came abruptly. All over the Earth, high concentrations of iridium have been found in the sediments of the geological record that were laid at the end of the Cretaceous period and marked the beginning of the Tertiary period (the K-T boundary; *K* from *Kreide*, German for chalk, referring to Cretaceous, and *T* from *Tertiary*). The time at which the iridium layer was laid down in Earth's history, 65 million years ago, coincides precisely with the fifth major period extinction. Meteorites are a source of iridium, and if large enough, such a source would explode on impact with the Earth's surface and send debris into the atmosphere that would be spread all over the globe. Thus iridium dispersed in such a scenario would settle in the sediments of the K-T boundary layer found around the globe. Although the impact of a comet may produce a similar globally devastating event, investigators recently claimed to have found fragments of rock at the K-T sediment layer that are consistent with a meteorite origin.

Measurements of the amount of iridium worldwide at the K-T boundary suggest the meteor was between twelve and sixteen kilometers in diameter, released an energy of about 100 million hydrogen bombs upon impact, and formed a crater nearly two hundred kilometers across. Although no such crater can be seen today (since such a feature would be eroded by geological activity over the intervening millions of years), indirect evidence points to an impact site nearly two kilometers beneath the Earth's surface along the northern coast of the Yucatán Peninsula of eastern Mexico—the impact site has been named Chicxulub. Analysis of the crater's shape suggests the meteor had a northwesterly trajectory such that the impact scorched much of the North American plate with a wave of incinerating gas and debris. The extreme heat at the impact melted the nearby rock into tektites, which are found in high concentrations at the Chicxulub site—supporting it as the location of the meteor impact. The explosion of the impact sent surface rock debris into space, some of which promptly returned to Earth in the form of a massive global shower of meteorites. The lighter material would have eventually coalesced into an orbiting ring complex, similar to Saturn's, over a period of a hundred thousand years. The shadow cast on the Earth by the rings would have had significant climatic effects. This orbiting material gradually fell to Earth, incinerating in the atmosphere over the following 2 or 3 million years.

The downpour of blazing meteors immediately following the Chicxulub explosion ignited extensive global fires, producing vast amounts of soot, which also settled into a layer found within the sediments of the K-T boundary. The smoke and dust from the collision would have distributed throughout the atmosphere, blocking sunlight and leaving the Earth's surface in a "nuclear winter" for months. In fact, it would require incinerating all of the forests of today on the entire Earth to equal the estimated quantity of soot laid down at the K-T boundary. In addition to the soot, the impact would have led to acid rain and giant tsunami waves. It was under these extreme environmental conditions that the extinctions occurred: about 76 percent of all terrestrial species perished. Among those that perished were 90 percent of all land vertebrates, including the dinosaurs. Frank Kyte, a geochemist at UCLA, was examining marine sediments dating to the K-T boundary when he found a significant lack of microfossils (consistent with a massive extinction on a planetary scale), and in one sample he found an unusual rock within a core of ocean floor sediment. Analysis of the rock indicated a high concentration of iridium, chromium, and iron, suggesting an asteroid origin: perhaps a piece of the meteor that produced the massive Cretaceous extinction! Interestingly, several mass extinctions of varying degrees have occurred in Earth's history and some, but not all, correspond with the dates of known large impact craters.

The evolutionary changes of ancestral reptiles that led to the highly successful dinosaurs also left them susceptible to the effects of the meteor impact and its aftermath; an abrupt twist of fate enabled mammalian proliferation in a world that had been dominated by the reptiles for many millions of years. There were mammal-like reptiles in existence perhaps as early as 250 million years ago, but it was the K-T period extinction, with its vast changes in the Earth's ecosystem, that set in motion the mammalian rise to dominance. The first true mammalian species may have come into existence as recent as only 100 to 165 million years ago. The earliest mammals were likely ground-dwelling, with some later forms developing arboreal lifestyles. At the time of the K-T extinction, the largest mammals were about the size of small rodents.

During the late Carboniferous period, about 300 million years ago, the mammalian ancestors were reptiles in the subclass Synapsida. The first order of the synapsids was the Pelycosauria, from which a second order developed, the Therapsidia. The pelycosaurs resembled the large lizards of today. They were the first true carnivores and probably favored larger prey than their contemporaries, which led to their increased body and cranial sizes. During the early Permian, the therapsids arose from pelycosaur ancestors. By the late Permian, about 250 million years ago, the cynodont group of therapsids arrived. Unlike their predecessors, the cynodonts likely developed a warm-blooded metabolism. Many varieties of cynodonts existed during the Triassic period, and although they are considered ancestral to mammals, the direct lineage remains uncertain. This group of mammal-like reptiles arose early in the era of reptilian domination but initially lacked specialization. They resembled small rodents and probably adopted a nocturnal lifestyle. Gradual evolutionary rearrangement of the extremities brought their limbs beneath the torso from their original reptilian lateralized position. The scales of their reptilian ancestry were modified into body hair. Along with these morphological modifications, they maintained a warm-blooded metabolism. Evolutionary change also affected the skull with alteration and enlargement of the cranial vault and brain. The jaw mechanism was modified with incorporation of part of its structure into the vestibuloauditory system (middle and inner ear) for hearing, balance, and spatial orientation. In the spring of 2001, excavations in southwest China, a region known for Mesozoic fossils, uncovered a tiny mouse-like animal (not much larger than a paper clip) with skeletal evidence of already significant development of the middle ear—some archaeologists argue that this creature, *Hadrocodium wui,* may be a direct or closely related ancient ancestor to mammals and ourselves. *Hadrocodium* lived up to 195 million years ago.

Although there is suggestion of the presence of mammal-like creatures by the late Triassic, it is fairly certain (by the fossil record) that by 165 million years ago a small squirrel-sized creature, probably arising in the region of Eurasia, marked the divergence of mammals from reptilian ancestors. However, some genetic analyses of living mammalian and reptilian species point to a more recent divergence of about 100 to 120 million years ago.

It is interesting to consider an intriguing evolutionary path of some mammals during this period. Present-day dolphins and whales had land-dwelling mammalian ancestors, the mesonychians, which resembled hoof-footed wolves. About 50 million years ago this dolphin ancestor was foraging in swamps and estuaries, but over millions of years of evolutionary change it eventually returned to an aquatic lifestyle. Thus after leaving the ocean during an amphibious stage of evolutionary development, it evolved into a mammalian form, then returned to its ancient aquatic existence as a mammal. Human dependence on the ocean has not entirely vanished since our ancestors moved onto land 360 million years ago—we spend our first nine months of life suspended in an oceanlike environment of amniotic fluid.

7

Origin of Man

In retracing the steps of primate evolution, we begin with small rodentlike animals scurrying underfoot of the reigning dinosaurs, tens of millions of years ago. From these inconspicuous creatures evolved the dominant land-dwelling species of today, including *Homo sapiens.* After the early hominids left the trees to stand erect, with subsequent remarkable development of the brain, they ventured from their African home to populate the rest of the world. Coincident with these advances was the creation of language, culture, society, and the ability to manipulate the environment creating shelter, agriculture, and domestication.

A review of the path of our evolutionary heritage is recapitulated during embryonic development. Not every stage is expressed, but various vestigial structures develop, then give way to those structures expressed in the final form. Furthermore, our distant past in the ancient seas is temporarily revisited by gestational suspension in (amniotic) fluid during our first nine months of life. Even blood, which nourishes the cells of our bodies, maintains chemical similarities to the salt water of the ocean from which it was fashioned. In the developing embryo, around the fourth week after conception gills appear along the neck, which are reorganized into the jaw, larynx, and middle ear in the weeks that follow. An embryonic tail vestige is present by six weeks but regresses a week or two later. Such structures would have persisted in our earlier evolutionary forms, as they served relevant functions in our predecessors. In addition, the embryos of humans in their early phases are nearly indistinguishable from those of other animals (not closely related), attesting to a common evolutionary origin long ago.

The primates gradually arose from the small mammals that survived the K-T period mass extinction. By 50 million years ago, primate ancestors were arboreal rodent-sized animals. Study of similar primates that live today suggests these early primates likely had few offspring and provided extensive parental care during the early phases of their offspring's development. Survival depended strongly on the ability to evade predators (or other dangers) and obtain food, both of which benefit from improved dexterity of the extremities. Structurally, these early primates had developed an opposable thumb, which significantly improved the ability to grasp. Another 15 million years brought forms similar to the nonhuman primates of today. Excavations in northern Egypt have uncovered early prehominids from 35 to 31 million years ago, including *Propliopithecus* and *Parapithecus.* The largest of these weighed no more than ten kilograms and lived in the dense canopy of a tropical rain forest. Unfortunately, the fossil record of the hominid ancestors for the next 10 million years is scant. By 25 million years ago, our predecessors no longer had a tail but did have a vestige (coccyx), which is still expressed in modern humans. Approximately 20 million years ago, a new genus of hominoid, *Proconsul,* was present in Africa. *Proconsul africanus* lived in a rain forest environment and had a fully opposable thumb with essentially modern human finger proportions. They lived on a fruit-based diet and were clearly a hominid ancestor. It was around this period in hominid evolution that the environment of ancient Africa was changing from a tropical rain forest to a drier climate. Changes in the hominid diet to harder foodstuffs required stronger teeth and jaw musculature, which the emerging Afropithecines had developed by about 17 million years ago.

Defining the evolutionary path to modern man from predecessor anthropoid primates and hominids is largely based on the morphological appearance of fossilized remains. Another scientific tool recently added to the anthropology arsenal is the discipline of genetics, which historically has been used to identify similarities among living species to determine the (genetically) closest living relatives. In the case of primates, the three major groups are the African apes, Asian apes, and humans. Previously, the Asian and African apes were considered more similar to each other than either is to humans. However, genetic evidence indicates African apes are more closely related to humans than they are to the Asian apes. Within the African apes, humans are most similar to chimpanzees, with only a 1.2 percent genetic difference—implying the DNA sequences are 98.8 percent identical between the two species. Among the other African apes, human DNA has a difference of 1.4 percent from gorillas but has a genetic difference of 2.4 percent from orangutans of Asia. The greater genetic difference between humans and Asian apes suggests the Asian ape lineages separated prior to the split of the human and African ape lineages. The gradual change in DNA sequences over time occurs at a relatively constant rate, permitting an estimation of the length of elapsed time from a species divergence to produce the measured genetic difference. Such interpretations of the human and ape DNA imply that about 12 million years ago the Asian ape lineage separated from the common ancestor, followed by a division in the ancestral path of humans and African apes between 5 and 10 million years ago.

The earliest hominid fossils (found in East Africa), which clearly suggest an ancestral divergence from the African apes, date over 5 million years ago. This divergence took place during a brief period of global cooling that was sufficient to dry up the Mediterranean basin. Although the period of cooling may have been a coincidental event with hominid species divergence, environmental fluctuations are often implicated as a mechanism to alter evolutionary selection pressures that lead to speciation, as in the extreme examples of major period extinctions and subsequent species radiations. The earliest hominid lineage ancestor may be *Ardipithecus ramidus kadabba* which lived between 5.2 and 5.8 million years ago. Thus far, the fossil record has provided little remains of this species, but there is suggestion that it walked upright. The Ardipithecines were replaced by the Australopithecines beginning about 4 million years ago. These early hominids, Australopithecines, were small by today's standards, measuring under one and a half meters tall—their brains were only half the size found in modern man. There was a strong sexual dimorphism in the Australopithecines—the males were up to twice as large as the females. Anthropologists consider this difference a reflection of a lifestyle of roaming bands rather than permanent family groups.

According to the fossil record, by five and a half million years ago, the early hominids may have become bipedal, i.e., walked erect on two feet. At the time of the emergence of bipedalism, the hominid brain was not much more advanced than that of today's great apes. The upper body and head of these early hominids was undoubtedly apelike, but the legs and pelvis were different in that they had acquired structural adaptations suited for bipedalism not found in previous species. The pelvis became more flattened and wider; the femurs (thigh bones) became angled inward as they are in modern humans (in monkeys the femurs are directed vertically). However, even 3.2 million years ago *Australopithecus afarensis* (including the famous "Lucy") maintained some features suitable for tree climbing (Figure 7.1). Hominid lifestyle was influenced by the gradual environmental changes taking place, particularly the recession of the extensive forests in Africa, which were giving way to the grasslands of a savanna. It has long been held that with this change of habitat the arboreal lifestyle was replaced by an increasingly ground-dwelling existence.

In a savanna, without trees to climb, survival can be improved by assuming an upright posture to maximize the elevation of the eyes, enabling more effective surveillance of the surroundings. Erect posture not only improves surveillance but also frees the hands, which led to the development of crude stone tools by 2.6 million years ago. The ecosystem of the ancient African savanna was crucial to the evolution of humans. The biomass of a savanna is concentrated at or near ground level, including the durable plants, grasses, and grazing animals. Most hominid predators would have occupied the same environmental stratum. The level terrain enables visual detection of both sources of food and potential dangers. The occasional cluster of trees would afford sanctuary for hominids from their predators. It was under these environmental conditions that the majority of human evolution took place.

In the spring of 1999 archaeologists announced the discovery of the fossilized remains of a previously unknown hominid ancestor's skull in East African sediments that date to 2.5 million years ago. The hominid, named *Australopithecus garhi,* was found among flesh-carving and marrow-extracting stone tools, suggesting the australopithecine was probably a scavenger with a diet that included a significant proportion of meat (Figure 7.1). *A. garhi* has larger teeth than the more ancient *A. afarensis* but does share a forward projecting face, small cranium, and skull crest. Less than three hundred meters away from the *A. garhi* remains, archaeologists uncovered the partial remains of an upper and lower extremity of another hominid within sediments dating to the same period. If we assume they belong to the same species, analyses of both specimens suggest that *Australopithecus garhi* was a transitional form between *Australopithecus afarensis* and the first species of *Homo.*

There has been a recent—and still controversial—addition to the hominid evolutionary tree. In November 2000, a few fossilized bones of what would later be called *Orrorin tugenesis* (nicknamed "millennium man") were discovered in Ethiopia. *Orrorin* is touted by its supporters as a hominid that lived six million years ago and, furthermore, was bipedal. In fact, they claim it was more similar to recent hominids than *Australopithecus afarensis*; suggesting *Australopithecus* was a dead end of the human evolutionary pathway. The date of the find is of importance as the divergence of hominid ancestors from the African apes falls near the time of *Orrorin* making it an original hominid candidate.

There is controversy as to whether the environmental changes from forests to grasslands in Africa were directly related to humans' becoming bipedal. A variety of alternative theories have been proposed for the development of bipedalism, including behavioral and food-gathering advantages. The fossil record indicates that humans became bipedal, freeing the hands, before the brain began to increase in size. Coincident with these modifications, the human diet gradually became more carnivorous. Perhaps through advancement in intelligence, tools advanced as well as hunters' cunning, which led to a diet richer in meat. The relative structural frailty of hominids (relative to other carnivores) suggests they were more adept at scavenging than hunting. Their tools would enable them to effectively carve morsels from a carcass for rapid consumption.

The evolutionary progress of hominid intelligence and dexterity with complimentary increase in the size of the brain required anatomic and behavioral modifications. The fully developed human brain, encased in the skull, is too large to pass through the bony birth canal of the human female. Although some laxity is provided by softening of the ligaments (near the time of infant delivery) that hold the pelvis together, the opening is still insufficient. Nature's compromise was to deliver the infant with a premature brain, which is smaller than the adult brain, reducing constraints on the female skeleton. However, birth in a relatively premature stage requires continued growth and development of the brain after the child has been born. As a result, the human infant is functionally immature (relative to other primates) and requires extended parental nurturing long after birth.

The exact lineage to modern humans, *Homo sapiens,* remains unclear from the incomplete fossil record. It may seem difficult to understand why there were so many different human ancestors. Early work in human paleoanthropology attempted to explain human phylogeny as a continuous progression from the earlier forms that gradually led to modern man. However, the evolutionary path to *Homo sapiens,* as in other species, was the result of sporadic changes, with some forms having survival advantage while others faded into obscurity and extinction. Thus a fossil record containing a variety of ancestors, some coexisting, is not unexpected. In fact, there may be many more ancestors in the human phylogeny tree that are yet undiscovered. This variety of early hominids creates a formidable challenge for anthropologists to derive a clear and direct lineage to modern humans. In the case of human evolution, the disappearance of some lineages may have been through competition or conflict. The coexistence of modern humans and Neanderthals is well documented in the fossil record, but

139

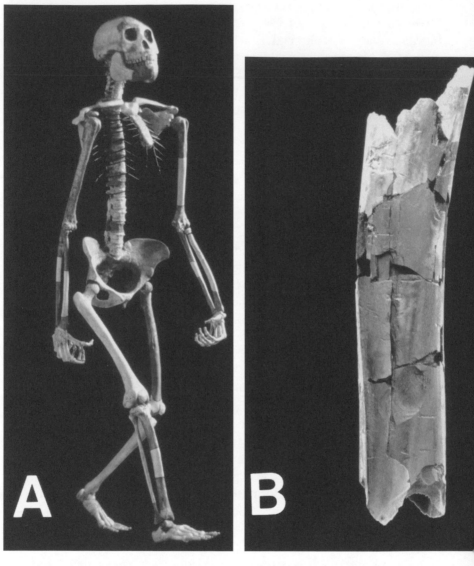

Figure 7.1 A) A reconstruction of *Australopithecus afarensis* illustrates the bipedalism proven by the 3.2-million-year-old original fossil skeleton. However, features suited for an arboreal lifestyle were also present, particularly in the upper extremities. B) A 2.5-million-year-old bovid tibia shaft (antelope leg bone) which has been intentionally shattered and carved, presumably to extract marrow, provides the earliest fossil evidence of hominid tool use.

140

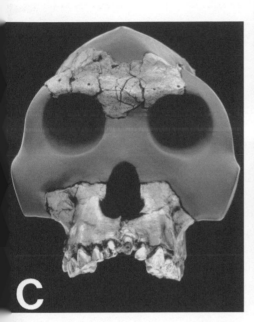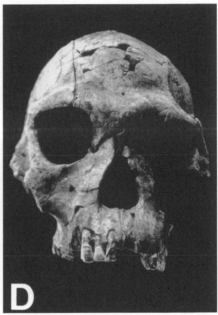

C) *Australopithecus garhi* was a contemporary with the tool use illustrated in (B). D) *Homo habilis*, the first *Homo* species, appeared almost 2 million years ago. The fossil record indicates bipedalism preceded brain expansion by well over one million years and the manufacture and use of a small variety of stone tools arose 500,000 years before a *Homo*-size brain appeared. *Credit: David L. Brill.*

the abrupt disappearance of the Neanderthals is also evident. Although it is difficult to prove such a theory, history has repeatedly shown that human behavior is capable of subjugation and genocide. Indeed, the game of survival is one with very high stakes, and therein may lie the source of such behavior.

The pattern of human phylogeny that has developed contains multiple evolutionary branches where some went on to extinction while others continued and new forms arose (Figure 7.2). Most, if not all, speciation of hominids took place in eastern Africa. Some of the co-existent hominid forms may be explained by migrations of subpopulations to other parts of Africa where they eventually vanished. A plausible interpretation of the hominid fossil record suggests that around three million years ago, the primitive African ancestors *Australopithecus afarensis* were followed by either *Australopithecus africanus* or *Australopithecus garhi*. Although *A. africanus* and *A. garhi* are consistent with hominid evolution, they do not appear to be directly related to each other and may have been co-existent species. Historically, *A. africanus* had been considered a link in the hominid evolutionary chain, despite its remains being located on the other side of the African continent from the other hominids. The recent discovery of *A. garhi* remains in Ethiopian sites, where both ancestral and later species are also found, support it as a candidate in our ancestral line.

Nearly two million years ago, the australopithecines were replaced by the first *Homo* species, *Homo habilis* (a primitive tool maker). There was a coincidental global decline in temperature of up to 15° C during this period of species transition, which may have been influential through alteration of the hominid ecosystem. *Homo habilis* arose in East Africa, stood a little over a meter tall, and had a cranial capacity of 680 ml (modern man's is about 1450 ml). The relatively larger cranial volume that *Homo habilis* had over prior ancestral hominids is one guideline used by anthropologists to categorize *Homo habilis* into the genus *Homo* by assuming an increased brain volume implies improved intellectual capabilities. Analysis of the tools made by *Homo habilis* suggests a dexterity of the hands had developed that was not seen in previous ancestral hominids. In addition, some anthropologists argue that subtle impressions made by the brain on the inner surface of the skull bones within the cranial vault of some *Homo habilis* specimens may represent the development of the brain's speech center (Broca's Area, usually found in the left frontal lobe). A developing communication center of the early hominid brain may reflect increased intellectual and behavioral requirements to function within an increasingly

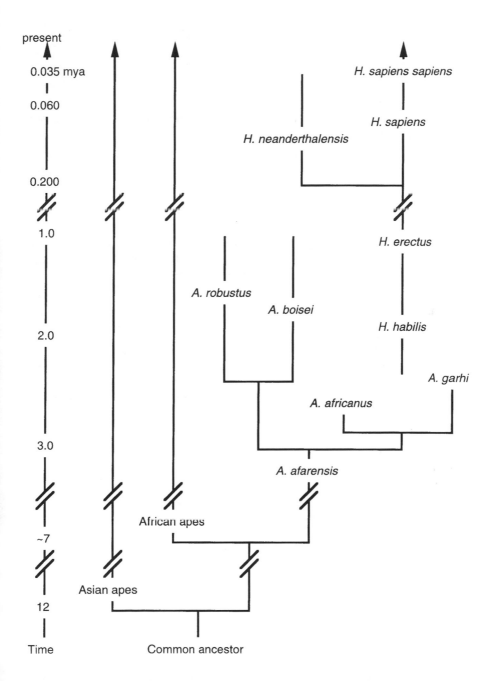

Figure 7.2 Human phylogeny, a proposed ancestral lineage of modern humans, *Homo sapiens sapiens*. (mya = millions of years ago)

more complex social environment. Presumably, *Homo habilis* lived in semipermanent encampments and maintained a food gatherer and scavenger lifestyle but was largely an herbivore. Contemporary hominids of *Homo habilis* include *Australopithecus robustus* and *Australopithecus boisei*, both of which are probably not ancestral to humans.

Between 1.9 and 1.7 million years ago, *Homo erectus* appears in the African fossil record. The simultaneous waning presence of *Homo habilis* remains in the archaeological record suggests that the arrival of *Homo erectus* represents an evolutionary continuum in human phylogeny. In any event, with the appearance of *Homo erectus* the earlier *Homo* types vanished within 300,000 years. *Homo erectus* clearly reveals a divergence from the apelike body habitus of its predecessors, with skeletal structure similar to modern humans, including relatively shorter arms and longer legs. The skull of *Homo erectus* retained rugged features, including an underdeveloped chin, prominent brow ridges, and evidence of a thick, muscular neck. They had a larger brain than prior hominids, were excellent stone tool crafters, and made use of fire. Their diet consisted mostly of plants; however, evidently they increasingly supplemented it with meat, as fragments of animal bones are common at their excavation sites. Estimates of the entire population of these early hominids vary between about 125,000 and 500,000 individuals. They were nomadic, and some of the *Homo erectus* population left their native African lands to venture into Europe and Asia. Many European and Asian sites confirm hominid presence dating back 1 million years, with several possible occupations as far back as 1.8 million years. The Boxgrove site, a quarry in southern England, contains clear evidence of hominid activity at least 500,000 years ago. A variety of cutting tools and axes were used there with a high degree of precision, as evidenced by cut marks on carcasses. The skill of these ancient hominids suggests an intelligence rivaling that of modern man.

By 500,000 years ago, the remains of *Homo erectus* indicate continued enlargement of the brain, with some retention of the coarse skull and neck features. Some paleontologists refer to this form as the "archaic" *Homo sapiens*, which were intermediate between *Homo erectus* and modern *Homo sapiens*. Others categorize them as "advanced" *Homo erectus* or even *Homo heidelbergensis*. In Europe, these "ancients" may have been the predecessors to the Neanderthals.

In 1856, quarry workers in the Neander Valley near Düsseldorf, Germany, found an incomplete skeleton including portions of a skull and other bones. It was originally thought to be the remains of a bear but shortly afterward was recognized as humanoid. The name given to the new species was Neanderthal, designating its location, where *Thal* was German for valley (now spelled *Tal*). Neanderthals, classified as a subspecies of *Homo sapiens* (i.e., *Homo sapiens neanderthalensis*), appeared in Europe and western Asia about two hundred thousand years ago and probably originated from the *Homo erectus* lineage that preceded them in these territories. The Neanderthals were spread throughout Europe by 70,000 years ago, and much of their existence was under the harsh conditions of repeated cycles of ice ages. Their population was relatively sparse, as they probably never numbered over 100,000. Their remains are often in caves, which may have been living space or places of burial. It has been suggested that the Neanderthal populations were displaced to the margins of Europe as their numbers diminished before they vanished. Interestingly, at the time of their disappearance (about thirty-five thousand years ago), the ancestors to all modern humans, *Homo sapiens sapiens* (also a subspecies of *Homo sapiens*), arrived.

Study of the fossil remains of the early *Homo* species has shown a gradual increase in overall body mass, which began about 1.8 million years ago and continued until 36,000 years ago. The brain size remained relatively stable prior to 600,000 years ago. However, between 600,000 and 150,000 years ago the *Homo* species underwent periods of rapid increase in brain size relative to body mass. In the era of modern man, *Homo sapiens sapiens,* a gradual parallel decrease in body and brain size ensued.

Some anthropologists have suggested that the evolutionary path to modern humans was strongly influenced by the invention of tools. The first tools, basically rocks that fit in the palm fashioned with sharp edges to cut, predate modern humans, *Homo sapiens,* by more than 2 million years. Among the earliest recognized tool use was by *Homo habilis,* a small hominid that was largely an herbivore. Despite a brain less than half the size of modern man's, it was advanced enough to recognize the importance of tools. Although their tools were not sufficient to effectively hunt large game, they had the ability to rapidly carve up a fallen carcass for transport and consumption elsewhere—permitting a quick departure before any hungry carnivorous competitors arrive to claim the flesh. Thus *Homo habilis* was probably an herbivore who was also an opportunistic scavenger. It was not long after this period in hominid evolution that the teeth and jaws began to regress, as plants were becoming a smaller portion of the diet. This was followed by an increase in the size of the hominid brain. Stone age tools changed a little over the millennia as they became larger, with broad, sharp cutting edges (axes) by 1.5 million years ago. About five hundred thousand years ago stone tools became smaller and more refined. Compound tools, such as a wooden handle combined with a stone blade, arrived with modern humans and Neanderthals. The stone age came to a close about six thousand years ago as the ability to utilize copper and bronze developed. The ability to forge iron, from heating hematite, was developed by 1100 B.C., with iron enjoying widespread use from the Roman Empire onward.

The early forms of *Homo sapiens,* known as Cro-Magnon (named after a site in France), spread throughout Europe about forty thousand years ago. Although less advanced by physical stature than Neanderthals, the Cro-Magnon utilized superior tools and weapons. In fact, the two groups apparently coexisted for thousands of years, but there is no archaeological evidence of interbreeding to any significant degree. The exact nature of their interaction is unclear: avoidance, assimilation, or conflict? However, some recent investigations provide insight into the mystery. In 1997, Svante Pääbo and colleagues published results of successful extraction and examination of a segment of mitochondrial DNA from the right arm bone of a Neanderthal skeleton. The variations in the sequence as compared to contemporary DNA suggests a genetic isolation between 550,000 and 690,000 years, which precludes any significant cross-breeding with modern human ancestors. These findings, as well as those of other recent genetic studies, suggest the Neanderthals were an evolutionary dead end; their contribution to the gene pool of modern humans was negligible at most.

Since the archaeological and genetic evidence suggest that the Neanderthals and Cro-Magnon were not directly related, then from where did the Cro-Magnon originate? The most widely accepted theory claims that in Africa modern humans first appeared about one hundred thousand years ago and then later ventured (as *Homo erectus* had done a million years before) into Europe and Asia. This theory is supported by the genetic evidence of the populations of Africa and the other continents of today. Genetic variations of modern populations date no further back than about two hundred thousand years, with some data suggesting a divergence closer to one hundred thousand years (which is consistent with an exodus of modern man from Africa at the time of the genetic divergence).

A popular competing theory postulates that *Homo sapiens* are direct descendants of their ancestral *Homo erectus* lineage, who had already left native Africa nearly a million years before. Thus the morphological differences between the two species reflect local evolutionary and adaptive changes, rather than an abrupt repopulation by a second nomadic wave from Africa. Although this is a simpler model in concept (which has weak archaeological support in that the remains of the various hominid ancestral species are similar), the genetic evidence does not support a million-year gradual change from *Homo erectus* to *Homo sapiens.*

As man endured the forces of nature to evolve into his modern anatomical form, the next challenge was to expand his territory as populations prospered and grew. Modern humans originated in Africa at least 100,000 years ago, and their nomadic lifestyle led them to populate regions rich with prehistoric wildlife and vegetation. Theoretically, one might expect the *Homo sapiens* species to be the result of one or two individuals, with the entire population being related through generations of descent. However, theories of population biology suggest there is a minimum-size breeding group that can lead to a species and propose that modern humans originated from a community of up to 200 individuals. By 10,000 B.C. (when global migrations had brought man as far as South America) the global population had risen to between 2 and 20 million.

Genetic analysis of contemporary human populations has produced a family tree of the world populations. Comparisons between different subpopulations (e.g., Pacific Islanders, Europeans, etc.) provide a genetic "distance," through differences in the genetic code at target loci (e.g., blood groups). There is a primary split between African and non-African populations, followed by a secondary split between Australians, Pacific Islanders, and Southeast Asians and the rest of the world's population. These differences are consistent with genetic isolations as a consequence of population migration. The primary split of African and non-African populations indicates genetic isolation consistent with an exodus from Africa. Subsequent branching of genetic distance of the various subpopulations is indicative of further migrations.

Following the major global migrations, the more recent variations in modern human physical characteristics have produced the different ethnic groups. The isolation of ancient peoples in their respective environments led to reproductive (and genetic) isolation. Breeding isolation may also arise within a population through nonenvironmental mechanisms, as with cultural restrictions. Eventually, through genetic drift, the various local populations acquired traits that were beneficial for survival (or culturally favored) in their respective environments. These minor genetic adaptations reflect phenotypic characteristics that are the morphological features

147

used to categorize the various races of humans. The physical characteristics that distinguish the races are superficial—skin and hair color, facial features—and are limited to a small collection of genes. In fact, the genetic material is about 99.9 percent identical in everyone. Despite the individual physical characteristics of races, the genetic differences between members of different races fall within the range of variations of individuals within a race. The traits that characterize the races largely reflect selection pressures induced by local climates to which the populations were exposed over tens of thousands of years. Concurrently with the isolation of various populations, local practices and superstitions laid the foundations for the development of indigenous cultures—creating an association between race and ethnicity.

Once the ancients had sufficiently developed the capacity to control their environment such that not all of their time was devoted to survival, the beginnings of culture arose. There is evidence from the archaeological record that as long as 1 million years ago human ancestors had the ability to plan for future events and even convey cultural knowledge to subsequent generations. Such practices would facilitate progress in tool-making and hunting techniques. Nonetheless, the early primitive human ancestors of the African savannas left little record of their culture, because for them survival was the primary effort of their existence. Ancient sites reveal that these primitives were predator almost as often as prey to their contemporary carnivores (including the saber-toothed tiger).

Man's gradual rise to dominance over other species was manifest through his ability to adapt and thrive in a wide variety of environmental situations. This environmental versatility enabled a successful global occupation through vast migrations over thousands of years. During these nomadic times, villages and communities were established and early primitive cultures arose. Through the dispersal of populations, division and segregation eventually led to distinct cultural and ethnic subpopulations that developed over many generations.

Civilization

Since the beginning of recorded history and into the present era, humanity has been a complex interplay of countless cultures. While many cultures developed independently, influences and assimilations occurred

148

as populations interacted more frequently. The roles of politics, agriculture, and even the natural forces of climate are some of the crucial elements that shaped early societies. Influence from neighboring or nomadic societies can alter, incorporate, or decimate developing societies. The interactions of these forces is complex and not completely understood but seem to play a role in the formation, stability, and even demise of civilizations.

The evolution of civilization is a complicated long-term process involving innumerable factors that create pressures on a society, requiring adaptive measures to ensure its continued existence. Cultural anthropology and sociology have been able to construct a general scheme describing the evolution of culture through a series of distinct stages. Although these stages can be distinguished in several of the classic ancient civilizations, not all clearly trace these steps as they develop. Early human history was a period of nomadic hunter gatherer lifestyles. Food collecting gradually gave way to an increased level of cultivation and domestication. In communities which are on the way to civilization, typically a formative period arises with village farming, followed by the creation of a bureaucracy. Continued growth of the community and cultural progress lead to a classic period marked by florescence as a mature civilization.

8
History of Modern Man

As the loose associations and small groups of ancient peoples were able to coalesce into communities, opportunities for great change arose. In an ancient community where the daily struggle with the environment became less of a primary concern, individuals were able to actively pursue more intellectual interests. Increasing population density brought more frequent social interaction among members, which encouraged development of effective and more elaborate communication between individuals—in the absence of communication, it is difficult to form a cohesive social structure. The stability afforded by a community with effective communication among its members enabled primitive man to develop more complex societies and lay the foundations of trades and cultures. As the populations of communities grew, division of labor through allocation of certain tasks enabled certain members to dedicate significant time to master a trade that could be taught to future generations.

Accurate description of the society of ancient man is made difficult by the scarcity of material left in the archaeological record. When mere survival was the primary concern, little time was dedicated toward cultural development. The earliest artifacts include simple tools, consisting of bone and chipped stone, which mark the beginning of stone age culture. Cultural anthropologists have developed a cultural index of the stone ages, defining three stages. The first level is the Old Stone Age (i.e., Paleolithic), which is defined by the use of crude stone tools. It began with the first stone tools, which date back over 2 million years, at least to the time of *Homo habilis*. The Middle Stone Age (i.e., Mesolithic) is an intermediate stage that marks the transition to the final stage, the New Stone Age (i.e., Neolithic). Tools of the Neolithic are characterized by their polished quality. In addition, Neolithic communities practiced plant and animal domestication. Since these definitions are cultural, they occurred at variable times in history depending on the technological level of the population under investigation. In fact, several surviving stone age cultures were discovered worldwide well into the twentieth century. Their study by cultural anthropologists has been crucial in the understanding of the development and practice of primitive cultures.

Study of the *Homo habilis* sites suggests a scavenger rather than hunter lifestyle. *Homo habilis* was not a very large hominid and would have encountered difficulty with both competition and avoidance of contemporary predators. The transition to a hunter gatherer developed with *Homo erectus,* who were larger (similar to present-day *Homo sapiens*) and had developed more advanced tools. Recent archaeological findings associated with *Homo erectus* include advanced wooden spears and other weapons, which suggest they lived in more advanced hunter societies than had been previously suspected.

Humans apparently developed into their modern anatomic form before they began to acquire the intellectual capacity to create culture (benevolent community living, effective communication, development of trades, etc.). These cultural advancements began between 60,000 and 40,000 years ago, as modern humans began to spread across the continents. Scholars argue that a certain minimum intellectual capacity must be surpassed in order to comprehend certain abstract concepts prior to the development of a culture. Beyond instinct and cognitive abilities required for survival, intellectual concepts necessary for appropriate social behavior and culture must be present. Akin to intelligence, cognitive abilities of self-awareness, communication, cause and effect, and authority must be understood to allow cultural development and stability through social interaction. Anatomical characteristics related to intellectual development may present as changes in the morphology of the brain, which could leave evidence behind for archaeologists. However, the true (anatomic) nature of such changes remains elusive, because the soft tissues of the brain are not readily preserved in the fossil record. In general terms, the majority of the difference between the human brain and that of other animals is the substantial expansion of the cerebrum (the dominant structural feature of the human brain) and its cortex (surface layer of nerve cells). It is predominantly the cerebrum that is responsible for the highly refined cognitive capabilities that separate man from other species. Interestingly, the division is not as clear as once thought, as primate research has provided evidence of other species having capabilities that were once considered limited solely to human intelligence.

One of the primary cognitive functions commonly used in defining man's higher intellectual abilities is the development of language and use of verbal communication—a characteristic that is not necessarily restricted to humans. The capability to utilize language is derived from specialized areas within the brain. The formation of speech also requires the presence and appropriate configuration of certain anatomic structures outside the brain, such as the larynx (voice box). There is much controversy in placing a date in human evolutionary history when the brain had obtained the centers required for language—determining these anatomic features from the archaeological record is a formidable challenge. The identification of some of the structures required in the articulation of speech may be less difficult to ascertain from the fossil record, but the presence of such structures does not prove they were actually used for verbal communication. Studies of the anatomic remains of *Homo erectus* suggest the larynx was developed enough to support speech, although such conjectures remain controversial. Among the remains of a Neanderthal found in northern Israel was a hyoid bone (a small bone located in the throat) that is nearly identical to that of modern man and is used by some archaeologists as support for the theory that Neanderthals had a simple spoken language at least 60,000 years ago. Human ancestors probably had the capability to create language and verbally communicate by the time *Homo sapiens* arrived (100,000 years ago), but speech in human ancestors may have been possible as long as 500,000 years ago, according to some fossil anatomical studies.

The discipline of historical linguistics can offer some insight to the first languages. Although most linguists believe that languages rapidly change, limiting the ability to trace origins beyond about five thousand years, some argue that certain core elements of language may be traced much further back. Such analysis has led to a collection of fourteen ancient language superfamilies, which were altered with time and human migrations. Although the language used by ancestral humans will likely never be known, the most ancient languages are probably the "click" languages of southern Africa. Furthermore, mitochondrial DNA studies indicate the Khoisan people, the primary speakers of click languages, have among the most ancient human lineages. As crude oral and pictorial forms of communication progressed, pictorial written languages followed. Phonetic styles of written language were developed several thousand years ago as verbal communication continued to progress to more sophisticated levels.

The people of the early Paleolithic era hunted game, collected plants, and lived in small groups of perhaps five to twenty individuals. Later in the Paleolithic, as humans spread into Europe and Asia, the abundant game they encountered supported larger communities. The growth in population brought the individuals of communities into closer proximity, occupying open encampments or caves. The Neanderthal community maintained a hunter-gatherer lifestyle, including seasonal migrations to follow local food sources. Their existence was harsh, with the typical Neanderthal living until the late thirties or midforties. They lived through a period of ice ages, and for that reason it seems likely they were the first humans to wear clothing. Little remains of their cultural beliefs; however, they buried their dead and sometimes left them with artifacts, which suggests belief in an afterlife. There are many examples of Neanderthals who lived with debilitating injuries or handicaps for many years. The long-term survival of such individuals suggests they were cared for by others, implying a compassionate community. Musical instruments (e.g., a flute carved from the femur of a cave bear) dating as far back as 80,000 years ago have been found at some Neanderthal sites. Despite the culturally advanced Neanderthal society, they fell into decline during the Middle Paleolithic and are no longer found in the fossil record by 35,000 years ago. By the late Paleolithic (20,000 years ago), individuals were wearing animal furs or leather clothes sewn together to make shirts, pants, and moccasins. The Paleolithic period comes to its close with the glacial recession at the end of the Pleistocene epoch, 10,000 years ago. With the close of the last glacial period, there was a period of stable global warming during which the birth of the first ancient civilizations took place.

Survival in prehistoric times was undoubtedly a challenge, and foremost it depended on the acquisition of food. Human ancestors met this challenge with their intelligence through invention of tools. The oldest evidence of tool use by hominids was stone tools found in the fossil record dating back over 2 million years. About a million years ago wooden spears were used for hunting. It wasn't until about twenty thousand years ago that the bow and arrow was invented; however, the oldest conclusive evidence dates back only twelve thousand years. Until about ten thousand years ago, early humans largely depended on hunting and gathering of wild sources of food for survival. Such activities demanded significant physical exertion by the hunters, as well as cooperation among the individuals within a hunting party. Eventually a much more easily accessible and dependable supply of food was obtained through the domestication of animals and plants about ten thousand years ago—the first Neolithic cultures. The more advanced communities of this period were living in shelters composed of wood and animal hides, wore clothes sewn from animal skins, managed herds of wild animals, traded with neighboring communities, painted and sculpted, performed primitive religious rituals, and buried their dead.

One of the particularly intriguing aspects of ancient stone age cultures was the practice of trephination. This procedure is considered the oldest surgical practice and involves the removal of part of the skull bone (craniotomy) without damaging the underlying brain and its protective tissue covering (meninges). Several specimens have been found dating as far back as 5,000 years and a recent find in France dates back 7,000 years. Experts argue that the practice may have been performed thousands of years earlier, as evidenced by the advanced level of technical skill in the specimens studied. Trephination was widely practiced during the Neolithic in Europe and continued into the Middle Ages but was perhaps most widely performed in ancient Peru, where it continued for a thousand years beginning in the fifth century B.C. Motivations varied from the treatment of crushing skull injuries to the relief of headaches and seizures or the performance of rituals including the release of evil spirits. Despite its injurious potential and the presumed unsanitary conditions of ancient times, skeletal remains suggest over 50 percent survived the practice by evidence of healing of the skull. Forms of trephination are practiced in contemporary industrialized nations in the treatment of increased intracranial pressures, e.g., cases of severe head trauma with internal bleeding. Outside the medical establishment are special interest groups that promote trephination in the treatment of various ailments, such as depression and lethargy.

During the Neolithic period, humans used polished tools, made pottery, practiced horticulture, and domesticated animals for both work and food. These activities allowed a community to support much larger populations. By 5000 B.C. there were village farming communities throughout the Near East with populations of up to 500 individuals. It is difficult to cite a specific time period for the Neolithic period in world history, as early humans were at various stages of cultural development at different locations around the world.

The archaeological record (with support from genetic studies) suggests modern humans originated in Africa by 100,000 years ago, and soon afterward migrations to western Asia began (Figure 8.1). Continued easterly migrations may have populated eastern Asia and Australia by 60,000 years ago. DNA analyses of contemporary populations suggest a

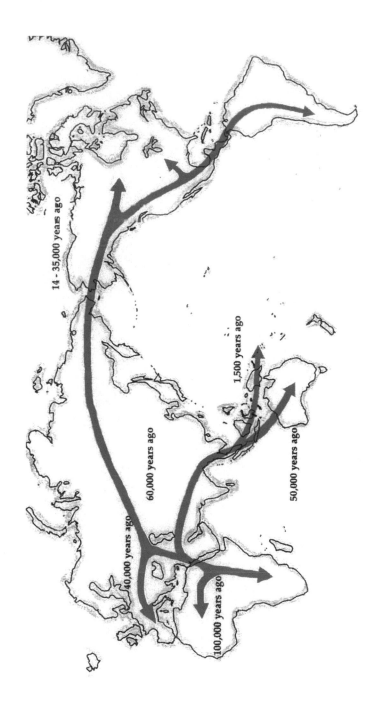

Figure 8.1 The major human migrations out of Africa to populate the Earth, which began approximately 100,000 years ago.

155

back-migration of individuals from Asia into Africa between 50,000 and 10,000 years ago. The populating of Europe by modern man was more recent (about forty thousand years ago), predating the most recent glacial maximum of the Pleistocene epoch, which occurred approximately eighteen thousand years ago. The Americas were populated later, presumably by migrations across the Bering Strait land bridge from eastern Asia. The islands of the Pacific were the destiny of the last great migrations, which began only a few thousand years ago.

Western Asia

Modern man arrived in western Asia shortly after migrations from the African continent, 100,000 years ago. Gradually small Neolithic farming communities averaging only a few dozen people developed throughout the region. Some of the scattered communities fostered larger populations forming villages. By 8000 B.C., at Jericho (of ancient Palestine) the first proto-city of record developed, which had a population of 2,000 to 3,000 people. The ruins of Jericho reveal a series of external fortifications including a six-meter-high stone wall, suggesting warfare was a constant threat. Other early proto-cities developed, including Jarmo (in Iraq) and Çatal Hüyük (in Turkey) around 7000 B.C. These early cities supported their populations through farming of grains and developed textile industries, art, and systems of trade with neighboring communities.

The earliest known system of writing was developed by the people of Uruk, a city-state of Sumer, in southern Mesopotamia (Figure 8.2). During the late Neolithic, a method of record keeping was devised through a system of variously shaped tokens that were usually stored in clay vessels. Apparently such tokens were used to keep count of various items such as the number of animals in one's flock. By 4000 B.C., the Mesopotamians replaced this cumbersome record-keeping technique with a system of written images called ideographs. In the millennium that followed, ideographs were modified to a series of wedge-shaped marks (cuneiform), which could be easily imprinted into clay tablets that would then dry to provide long-term storage of the information for later reference. This written language contained about 350 symbols, including both pictorial and phonetic characters. It was also during this period that new methods of widespread efficient transportation were developed (animal-drawn carts and sailing

Figure 8.2 Mesopotamia and surrounding areas where the first civilizations arose.

vessels), which were combined with improving agricultural techniques. These advances in technology contributed to the arrival of the earliest civilizations of Mesopotamia, Egypt, and India.

Mesopotamia (translates as "between rivers") was located between the Tigris and Euphrates Rivers, which originate in the mountains of eastern Asia Minor and drain into the Persian Gulf. It is roughly in the same location as modern-day Iraq. This strip of fertile land is referred to as the Cradle of Civilization by historians—where the first civilizations developed.

The rivers of Mesopotamia were important in the development of their culture in that they provided irrigation, commerce, and the ability to communicate between the neighboring communities along their banks. As early as 7000 B.C., there was a Neolithic farming community, which continued for a thousand years until it developed a mixed economy of farming and herding. Irrigation techniques were developed by 5000 B.C., and the population grew to 15,000 over the next millennium. By 3500 B.C., the Sumer plain experienced a rapid development of culture, including its cuneiform written language, which was adopted by neighboring communities. About fourteen hundred years later, a phonetic alphabet was developed by Phoenician merchants of the eastern Mediterranean. The phonetic alphabet became the foundation for all European and Middle Eastern and some Asian alphabets.

The Mesopotamians lived in complex societies occupying several large cities. A theocracy developed in the form of kingly dynasties, where the governing power was channeled through religious bureaucracies. The theocracy continued for almost eighteen hundred years. By 3500 B.C., large temples (ziggurats) were being built to provide multiple functions, including a place of worship, a sanctuary, and a food storage and redistribution center. The open plains of Sumer invited invasions, and in response a hereditary military kingdom emerged, which evolved into an era of empires. By about 2340 B.C. (after a period of conflicts between the independent city-states), Sargon the Great of Akkad united the cities of Mesopotamia to form the first empire, with its capital in Sumer. A little over a hundred years later the Akkadian empire collapsed and was followed by the rise of the Ur III dynasty to the south. The seat of the empire would change every hundred years or so, depending on which urban center had acquired control.

Around 2200 B.C. there was a curious change in the population distribution of the Mesopotamian civilization. Some of the city-states in the northern areas were abandoned, with indications of population migrations to the southern districts. Generally this has been attributed to consequences of invasions, disrupted routes of trade, or even political instability. A more recent theory invokes climactic change; a period of drought led to a collapse of agriculture, forcing the inhabitants of the northern districts to migrate to more fertile lands in the south. The seasons of drought continued for about three hundred years, according to studies of soil samples from the region. By 1700 B.C. control of the region was established from Babylon, with the next major civilization united and controlled by Hammurabi. He has received historical notoriety for the famous code of rules, including the often-paraphrased passage:

> If a man has destroyed the eye of a patrician, his own eye shall be destroyed.
> If a man has knocked out the teeth of a man of the same rank, his own teeth shall be knocked out.

As with the previous empires, Hammurabi's succumbed to invasions and ended within a century of its establishment. In the seventh century B.C., Babylon came under the rule of a succession of monarchs—the most famous of the rulers was Nebuchadnezzar. During this seventy-year period, there was a cultural revival, the Neo-Babylonian period, with a return to the splendor enjoyed a millennium before. The city was revitalized with the construction of the grand palace of Nebuchadnezzar, a nearly-one-hundred-meter-high ziggurat, and the elaborate adornment of the Ishtar Gate, which marked the entrance to the city. The hanging gardens of Babylon, one of the seven wonders of the ancient world, were allegedly built by Nebuchadnezzar to appease one of his wives. In 539 B.C., Babylon fell to the Persians.

To the east, in the region of present-day Pakistan and northwest India, around 2600 B.C. an urban theocratic civilization developed in the Indus Valley extending to the Arabian Sea. Originally this area supported farming villages, from which the first cities began to form by about 3300 B.C. The civilization that eventually arose was well organized and benevolent. The Indus Valley civilization independently developed a pictorial written language, perhaps by 3300 B.C., which still remains undeciphered.

The walled cities of the Indus Valley (Harappan) civilization contained large brick buildings with sanitation systems laid out in a grid pattern. There was evidence of trade with Sumer, which also brought cultural influence. Despite some linkage to Mesopotamia, the Indus Valley civilization was largely isolated geographically and developed a large and uniform society indicative of political stability, without any clear record of military conflict. Perhaps the lack of neighboring competitor cultures enabled a long-term stability of the civilization. Sociological models suggest that civilizations flanked by competitive cultures tend to undergo more rapid cultural change.

After about seven hundred years of a prosperous existence, the Indus Valley civilization fell into decline. It was originally thought that the Indus Valley civilization was dominated by invasions from the northwest (Aryan); however, this scenario is unlikely, as none of the major centers have archaeological evidence of a military takeover. Recent investigations indicate that the farming industry and trade routes had declined. The culture began to disintegrate as the written records were lost over the years. The smaller townships were gradually abandoned. Eventually neighboring cultures (Aryans) began to prosper as the original Indus Valley civilization disappeared. The mechanism behind the decline of the Indus Valley civilization may have been a consequence of climate and alterations in the flow pattern of the Indus River, which supported the trade and farming industries along its course. The civilization's fertility cult religion, which was replaced by Brahman (Hindu) rituals with establishment of the caste system, remains a vital part of the culture of the region to this day. However, not all was lost from the Indus Valley culture, as some of the ancient techniques in the crafting of jewelry are still practiced in India.

Africa

As the birthplace of man, Africa is the site of the most ancient human remains. The oldest stone tool artifacts were fashioned over 2 million years ago, with the richest collections of ancient stone tools found in eastern Africa. The vastness of the African continent, its many internal natural barriers (deserts, rain forests, and mountains), and unpredictable growing seasons are among the reasons the early inhabitants were not able to establish successful urban communities. The exception is in northeastern Africa, where the Nile River is a reliable water and nutrient resource, permitting stable longitudinal civilized habitation and development.

Until the nineteenth century, the copious Egyptian hieroglyphic writings found at historical sites paid silent tribute to their ancient culture.

Most of the original descriptions of ancient Egypt came from observations by contemporary Greek and Roman writers. A new era of understanding was opened beginning in July 1799, during Napoléon's failed attempt to conquer Egypt. One of the French soldiers noticed curious inscriptions on a rock that formed part of the foundation of an ancient fort they were in the process of dismantling. Napoléon, a student of history, was particularly intrigued with Egyptian history. Along with his invading forces to Africa he brought many scholars to conduct archaeological investigations. The importance of the stone discovered by the soldier was quickly realized, and it was taken to France for further study. Years later it was found that the 700-kilogram Rosetta Stone, as it became known, displays a tribute to Ptolemy V (a ruler of Egypt circa 196 B.C.) inscribed in linguistic triplicate, including hieroglyphics. The other two languages, Greek and another Egyptian script, were used to translate the ancient hieroglyphics. Egyptian hieroglyphics came into existence shortly after the cuneiform was invented in Mesopotamia. It is unclear whether the development of a pictorial language by the Egyptians was a result of contact with the Sumerians, but the ancient Egyptians never produced a simple written language.

Early civilization developed in Egypt under similar conditions to those in Mesopotamia (Figure 8.3). The arid conditions of Egypt required a reliable source of water, which the river provided. Just as the Mesopotamian predecessors settled along the banks of the Tigris and Euphrates Rivers, the predynastic (Badarian) people of Egypt settled in sedentary villages along the flood plain of the Nile River. These villages date back at least as far as 4000 B.C. Genetic analyses of contemporary populations are consistent with migrations from the northern regions southward along the Nile River. These trends also follow the historical records that indicate a southward progression of colonization during the various pharaonic kingdoms. However, unlike in Mesopotamia, discrete urban centers did not develop.

In time, the ancient agricultural villages of Upper and Lower Egypt were unified by the warrior chief Narmer in 3115 B.C. He became the first pharaoh, marking the beginning of the Old Kingdom, with its six dynasties, and the start of the Egyptian civilization. It was a peaceful time in Egypt's history, as foreign influence was minimal due to geographical isolation. The limited exposure to neighboring cultures in part enabled political stability. The isolation and extended peaceful period fostered the development of a ritualized and traditionalized civilization. There is archaeological evidence suggesting some trade interaction with the

Figure 8.3 The ancient civilizations of the African continent.

Mesopotamians; however, cities did not develop in Egypt until late in the second millennium B.C. Construction of the Great Pyramids was begun by the third dynasty, around 2700 B.C. The political system deteriorated during the sixth dynasty, which brought a period of ruling chaos (the first intermediate period), which lasted from 2200 to 2050 B.C.

Eventually political control of Egypt was established by those in power at the ancient city of Thebes. The arrival of the Thebian political system began the eleventh dynasty, marking the beginning of the Middle Kingdom (2050–1800 B.C.). Afterward, another period of political disintegration (second intermediate period) occurred, followed by restoration of a unified Egypt in 1567 B.C., heralding the New Empire, marked by the beginning of the eighteenth dynasty. Among the pharaohs of the New Empire were King Tutankhamen and the Ramses dynasty of twelve kings. Tutankhamen was a minor ruler, but in 1922 his was the first intact pharaoh tomb uncovered. An autopsy of the body revealed a fracture at the skull base, raising accusations of an assassination across the centuries—however, some archaeologists argue the injuries were more likely the result of shoddy embalmers. Autopsies of other Egyptian mummies indicate that arthritis and atherosclerosis (hardening of the arteries) were common, and DNA analyses hint at a high prevalence of incest in ancient Egypt. Following the eighteenth dynasty there was a gradual decline of political control into the twentieth dynasty, which ended the New Empire. After periods of Persian domination and a brief allegiance to Alexander the Great, the Roman Empire seized Egypt in its territorial conquests.

The remainder of Africa had a much different history. The earliest groups consisted of hunter-gatherer communities, and by 5000 B.C. farming and domestication were developing. It was later that agriculture spread to southern Africa. The transition from hunting to farming was gradual and probably delayed by the scarcity of native crops and lack of metals for farming tools. Iron was brought by the Phoenicians, who established Carthage in the ninth century B.C., and the Assyrians, who conquered Egypt in the sixth century B.C.

The Bantu, of West Africa, had acquired the skills of farming and metallurgy. Their expanding population led to migrations throughout sub-Saharan Africa in search of new lands to cultivate. During the two millennia of migration, the Bantu arrived in eastern Africa during the early first century A.D. and reached southern Africa about three centuries later. The small kinship communities were governed by chieftains, which eventually led to true chiefs and kings as the groups became more domesticated and formed larger villages by the first century A.D. In the eastern Sudan, the Kush kingdom developed. It was dominated by the Egyptians initially, but with Egypt's decline the Kush gained power. By the eighth century B.C. Assyrian invasions ended the Kush dominance of Egypt, and eventually they moved their capital south to Meroë. The culture of Meroë gradually lost its Egyptian influence, becoming a more diverse society, which peaked around the first century A.D. Decline came in subsequent centuries, with disruption of trade routes and the decreasing demand for goods by the crumbling Roman Empire. The end of the Kushites came in the fourth century, when they were conquered by the Ethiopians.

By the first millennium B.C., the farming communities of the ancient Ethiopian region began to assimilate with migrating Semites from Yemen to form the culture and language that define Ethiopia. The capital city was Axum, a major trading center. The Axum kingdom flourished to dominate East Africa by the fourth century A.D. Contact with the Byzantine Empire brought Christianity to Axum, and it was made the state religion in A.D. 350. The Muslim aggressions of the seventh and eighth centuries isolated Axum, strangled its trade, and destroyed the integrity of their society. Preserved by splinter groups that scattered throughout the mountainous regions, their culture survived to the twentieth century.

Western Africa, the Sudan, is historically noteworthy for a succession of great kingdoms. By the fifth century A.D., farming communities had coalesced to form the Ghana kingdom, which became a powerful trade center. The kingdom fell to Muslim invasions and was replaced by the Mali Empire in the thirteenth century. The Malians greatly prospered by supplying gold to Europeans.

Many of the ancient African communities suffered hardships from their harsh climates and delayed acquisition of metals. In addition, it was common to exploit the environment through overutilization of the fragile topsoil, grazing land, and forests. Some communities became dependent on slave trade rather than developing products for commerce. The influences of Islam may have produced an early decline in slave trade, but the trade never ceased and became more prevalent by the fifteenth century as the demand for slave labor increased.

Europe

In Europe, at the close of the Pleistocene (15,000 to 11,000 years ago), the environment provided plentiful food and other natural resources to allow development of densely populated communities. Gradually enlarging local populations led to formation of more complex societies, where individuals were delegated responsibilities (food gathering, ceremonial tasks, etc.) by community leaders. Despite the arrival of more prosperous times, the life expectancy of men was only about fifty years and of women just forty years (the higher female mortality was related to childbirth).

These stone age people had developed language, wore jewelry, and made elaborate cave wall paintings. The oldest known paintings of animals, found in the Chauvet cave in France, are approximately thirty thousand years old. Archaeological finds at the European sites suggest religious practices were already present in these societies by analysis of some of the ancient art forms. The meaning of much of the art remains obscure, but it may have been used to enhance the skills of hunters, reflect local tribal rituals, or illustrate the visions of the community shaman. It was also during these times that the production of stone and ivory sculpture flourished.

Around 10,000 B.C., in northwestern Europe, the dawn of the Mesolithic period occurred, the transition between the Paleolithic and Neolithic. During this period, the big-game animals of the Paleolithic period disappeared and human communities became more sedentary. Advances in tools and weapons included the development of bows, arrows, harpoons, and canoes. By 3000 B.C., the Bronze Age had begun to spread to Europe. The earliest dominant civilization of Europe, the Minoan, was prospering on the Mediterranean island of Crete, with a commerce based primarily on seaborne trade. Minoan influence spread through Europe to the cities of Greece, including Mycenae, and Troy of Asia Minor.

After years of struggle, by 1000 B.C. the Greek civilization began its rise to become the center of culture of the European continent. It combined elements from contemporary great civilizations, including the Mesopotamian, Mycenaean, and Egyptian cultures. At about this time the great epic poet Homer was creating his famous works (c. 750 B.C.). Ancient Greece was divided into many independent city-states, including Athens, which eventually dominated the region.

In 508 B.C., a civil revolt against centuries of oppression by the Athenian aristocracy led to the first democracy, championed by Cleisthenes, himself an Athenian nobleman. Representatives were chosen from among the population and met every nine days to discuss policy and public works and put them to a vote by casting stones into a vase: a black stone to vote against and a white one to vote in favor. For the first time in human history, the common people were able to choose their leaders and could even have them removed by a vote. Not only was this the birth of a new political system that would serve as a model for millennia, but it also marked the beginning of a period of great cultural and economic prosperity. Furthermore, Athens was undisputed as the center of Greek philosophical thought. Socrates (c. 470–399 B.C.), disheveled and barefoot, roamed the streets of Athens debating with locals, pointing out the importance of criticism and the use of intelligence and reason. Unfortunately, his candid criticisms of Athenian politics led to his arrest and execution. Plato (c. 428–347 B.C.), a follower of Socrates, tried to formulate the ideal society. Aristotle (384–322 B.C.), himself a pupil of Plato, studied nature, history, and politics.

Although the city of Athens was destroyed by a Persian invasion in 480 B.C., the Athenians defeated the Persians in a decisive naval battle at the straits of Salamis shortly afterward. Among the ruins in Athens was the temple to Athena on top of the Acropolis (a mountain of rock around which the city stands). The rubble, which had been left as a memorial to those who had died in battle, was removed and the Parthenon built in its place. It was a costly project (estimated at about a billion dollars in today's currency) and took fifteen years to complete. However, it helped establish Athens as the cultural center of Greece. Although much more highly adorned in its heyday, the Parthenon remains as a tribute to the excellence in architecture the ancient Greeks had achieved. Athens' role as a dominant force in Greece fell into decline as it engaged in a long series of conflicts with the Greek city-state of Sparta. Athens eventually surrendered to Sparta in 404 B.C. Both territories were weakened by the long confrontation, from which neither ever fully recovered.

During the mid–fourth century B.C., King Philip of Macedon (a territory north of Greece) defeated the combined Greek armies and became the ruler of Greece. In 336 B.C., he was assassinated and his son Alexander succeeded him. Alexander, an educated man (tutored by Aristotle), was interested in expanding the realm of Macedon beyond Greece. In 334 B.C., Alexander the Great, as he became known, left Macedon with his army to conquer the vast Persian Empire, which merely took just over a decade.

During his campaign, he entered Egypt, where he established Alexandria, which in later years became one of the major cities of the ancient world. He spread Greek culture throughout the territories he conquered as he encouraged the local populations to adopt Greek ideals. The practice of adopting Greek cultural attributes was followed years later by the Roman Empire.

Among the diverse population groups of ancient Italy were the Latins who inhabited a small village on the Tiber River about 1400 B.C. From these humble beginnings of a community largely composed of farmers and shepherds with a tribal chief political body, Rome became the center of dominance of the entire Western world. The culture of early Rome was influenced by Greeks from southern Italy and by the Etruscans from the northern regions. The Etruscans probably originated in western Asia before migrating to northern Italy. In the centuries prior to the rise of Rome, Greek adventurers and explorers who encountered Etruscan communities found a strange liberal culture where women held status similar to men (unheard of in ancient Greece), abundant art and sculpture depicting the beauty of women (the male body was favored by the artists of ancient Greece), and shocking erotic celebrations (adopted by the Roman culture that would follow).

According to legend, the city of Rome was founded April 21, 753 B.C., by the twin brothers Romulus and Remus, who were themselves raised by a she-wolf. Romulus became the first ruler of Rome after killing his jealous brother. Legend claims that Rome was ruled by a series of seven kings for the next two centuries. The early Romans lived in city-states throughout Italy that were ruled by the wealthy, who were generally of Etruscan lineage. The Etruscans occupied and ruled Rome by 616 B.C. However, the Etruscan rulers were harsh toward the Roman people, and in 509 B.C. a group of Roman aristocrats overthrew the Etruscan dynasty and replaced the monarchy with a republic. The early Roman population was divided into a minority (10 percent) of patricians, the governing class, and a majority class of plebeians. Gradually the plebeians were able to gain more political control.

Through military engagement and annexation, the Roman Empire accumulated land and population with a vast military force. The quest by the Roman people to rule the known world spurred a series of conquests, and by 267 B.C. the entire Italian peninsula was conquered. Rome continued expansion to the east by conquering Greece and Asia Minor by the second century B.C. Following the three Punic Wars with Carthage and the Illyrian Wars with western Asia, the entire Mediterranean basin came under the rule of the Roman Empire. For the next 500 years, the majority of the Mediterranean area remained under the control of the Roman Empire (Figure 8.4).

The empire was initially run by a series of kings. By the fifth century B.C. this system was replaced by an aristocratic government, which eventually deteriorated, and after a series of civil wars a monarchy arose. Through aggressive political means and military force, Julius Caesar was named dictator in 46 B.C., but he was assassinated two years later. Turmoil followed, with order restored when Augustus Caesar, the nephew of Julius Caesar, was appointed as the first emperor. When he acquired Egypt (during which his former coruler Marc Anthony committed suicide and Anthony's wife Cleopatra VII took her own life by the legendary poisonous bite of an asp), its wealth was used to finance the notorious Roman games, which openly displayed violence and torture to the public. The empire developed enormous resources through its conquests, taxation, and its vast population of slaves. Order and stability were assured by military support and Roman engineering, which is best known for its production of a network of roads (85,000 kilometers, more than twice the distance around the Earth), architecture, and aqueducts. Many of these structures remain intact to this day.

The Roman Empire was a pagan society that in some respects was tolerant and would assimilate the religious beliefs of conquered territories into its culture. Romans worshiped a variety of gods borrowed from other cultures, with the most popular including the Persian god of light Mithras, the Egyptian goddess of nature Isis, the Greek goddess of agriculture Demeter, and the Phyrgian goddess of nature Cybele. The intolerance of Christianity was initially difficult for Roman citizens, as it required they devote themselves to one God and relinquish the others, to which they had grown accustomed. However, many Romans found the ability to receive forgiveness for mortal sins an appealing aspect of the Christian religion.

During the mid–third century A.D., the Roman Empire was under growing tension from constant threat of barbarian invasions in the provinces. The aristocracy became fearful of the increasing numbers of Christians among the Roman citizens, and eventually a law was passed making the practice of Christianity a capital crime. This measure proved ineffective, and by the early fourth century, with about 10 percent of the Roman population practicing Christianity, it became the official religion of the empire under Emperor Constantine with the Edict of Milan. In an attempt to unify the deteriorating empire, Constantine moved the center of rule

Figure 8.4 The Mediterranean Basin at the time of the Roman Empire. The bold line indicates the empire's boundaries at the time of the rule of Augustus Caesar, during the first century A.D.

from Rome to Byzantium, which he renamed Constantinople in the year 330. Although the Roman Empire would not last through the next century, Constantinople remained as the capital of the succeeding Byzantine Empire for a millennium. It fell to the Islamic Ottoman Turks in 1453 and was renamed Istanbul.

In the twilight years of the Roman Empire, the provinces became increasingly independent. Thus the empire became less politically centralized, which weakened it the most at its outlying border provinces. The Roman Empire was eventually overtaken by the barbarian invasions of Germanic troops from the north and the Huns of Asia from the east. The last Roman emperor, Romulus Augustulus, was removed in A.D. 476 by the Germanic chief Odoacer.

After the fall of the Roman Empire, the successor states of Europe developed. This transition marks the beginning of the Middle (or Dark) Ages of Europe. In the early eighth century, the call of Islam to "fight those who resist the will of Allah" brought the subjugation of northern Africa and southern Europe, creating an empire that was as large as that of Alexander the Great (over a thousand years before). In 1095, Pope Urban II called on Christians to regain control of the Holy Land from the Muslims of Turkey, leading to ten major crusades, which lasted nearly three hundred years. Ultimately, the crusades were a failure, but eventually Europe was dominated by Christianity, which in turn became a cohesive political force. Darkness once again fell on Europe with the arrival of the bubonic plague in the mid–fourteenth century, during which about 40 million of Europe's 90 million population perished in just two years. The plague continued to recur in Europe over the next 300 years, but never again with the degree of devastation of the original "Black Death" of 1348.

The sixteenth century brought theological reformation. The sciences and arts were allowed to blossom with the awakening of the Renaissance, which laid the foundations of rational thought for the modern era. Coincident with these advances, the manufacture of weapons also flourished, which facilitated the conflicts that shaped the political boundaries of the world.

Eastern Asia

In Asia, independent Neolithic cultures spanned the northern and

170

southeastern territories of the continent. Chinese tradition claims there was a succession of kings forming the Hsia dynasty of northern China beginning in 2205 B.C. It was during this period, around 2000 B.C., that the first forms of writing were developed. The ancient script consisted of symbols carved on bones and shells. The Hsia rule was overthrown in 1766 B.C. when China was organized into a unified political body under the Shang dynasty. By then a written language was already in use, similar to the written language of China used today. The Shang dynasty was overthrown by a massive slave revolt with establishment of the Chou dynasty around 1050 B.C. The ruling dynasties exercised a more spiritual than authoritative control under which dozens of states and small kingdoms developed. Feuding among the various territories became more intense by the eighth century B.C., marking the beginning of the Age of Warring States.

In attempts to understand the course of humanity during this period, Chinese philosophy underwent a period of expansion. In the late sixth and early fifth century B.C., K'ung Fu-tze (the founder of Confucianism) traveled throughout China with his disciples, pursuing a simple life. His teachings promoted the idea that harmonious order and political stability could be obtained through appropriate social conduct (regardless of social class). He taught during the deterioration of the Chou dynasty, which was finally overthrown after a period of political turbulence in the eighth century B.C. Naturalist and Taoist philosophy sought solutions through the balance of opposites—yin and yang—opposing pursuit of active aggression. However, the dominant state, Ch'in, adopted a harsh totalitarian system of rules, punishments, and rewards. By the third century B.C., there were seven Chinese states left as Ch'in began an aggressive takeover of its neighbors.

At the age of thirteen, Ch'in Shi Huang Ti became the ruler of the kingdom of Ch'in and promptly escalated the subduction of neighboring states to mark the beginning of the Ch'in dynasty. The successful conquest unified China, with its name ultimately derived from the Ch'in dynasty. Ch'in Shi Huang Ti, a ruthless leader, declared himself the first emperor of China in 221 B.C. Many massive civic projects were implemented under his rule, including construction of the Great Wall (although it was in reality a piecing together of smaller walls made by the feudal states in previous times), construction of a road system that was more extensive than that of the Roman Empire, and standardization of the Chinese language, currency, and systems of measurement (based on the number 6). He was notorious for using forced labor, burning historical and philosophical texts, and the live burial of 400 Confucian philosophers over a disagreement. Construction on the Great Wall exemplifies the feudal oppression of the era, as forced conscription of millions provided the labor force and countless thousands died under the poor working conditions; many were buried as landfill within the structure. Although brutal, Ch'in Shi Huang Ti was not without insecurity, as he was plagued by paranoia concerning his own death. He became obsessed with the search for an agent of eternal life and would consume "magical" potions thought to extend his mortality. He did manage to survive three assassination attempts but died suddenly at the age of forty-nine in 210 B.C.; some historians suggest his death was hastened by the noxious potions he consumed in his desire for immortality. His mausoleum (about a kilometer from his burial mound on Mount Li), Ch'in Shi Huang Ling, contains thousands of clay statue warriors, the terra-cotta army, enclosed in a room large enough to hold three large airliners. It was about eight years after the first emperor's death that peasant revolts overthrew the Ch'in dynasty and replaced it with the Han dynasty.

The Mandarin bureaucracy, which lasted 2,000 years, had its roots in the Han dynasty (202 B.C.–A.D. 220). Government officials were trained in Confucianism and educated at an imperial university. During the Han dynasty, the expanding population covered nearly all of present-day China. Following the Han dynasty, there were centuries of turbulence, which saw the expansion of Taoist and Buddhist tradition. By the end of the first millennium, improvements in the cultivation of rice had doubled the population of China in only two hundred years.

The largest land empire in history was created by Genghis Khan and his immediate successors in the thirteenth century. Genghis Khan began a bloody sixty-year conquest to take control of China. His grandson, Kublai Khan, completed the conquest and was declared the emperor of China in 1271. China became more cosmopolitan as trade routes were opened, exposing China to the world it had tried to avoid through its policy of foreign isolation. Although the Mongol rulers impressed Marco Polo during his extended visit in the late thirteenth century, the Chinese people were disillusioned and mounted frequent uprisings. In 1368, a revolt led by a peasant successfully overthrew the Mongols, marking the beginning of the Ming dynasty. The Ming dynasty rebuilt the empire and was a period of economic and cultural prosperity. The Ming dynasty, which was the last native Chinese dynasty, lasted nearly three centuries until it was overthrown in 1644 by Manchurian invasions from the north. The Manchus were actually "invited" into China by the Ming military to assist in subduing an up-

172

rising in Peking. The last emperor of the Ming dynasty watched the revolts and fall of his regime from a hill in the Forbidden City, where he promptly hung himself. The Manchus maintained control of China for 267 years until overthrown by a democratic rebellion in 1912. After World War II, the ruling party was driven from the mainland to Taiwan. The mainland became the communist People's Republic of China on October 1, 1949.

The Americas

During the Pleistocene, immense glaciers covered much of northern Europe and North America and contained so much of the Earth's water that the sea level was lowered about 140 meters. The falling sea levels exposed a shallow continental shelf in the northern Pacific that created a land bridge (Beringia) between Alaska and Siberia along the Bering Strait. The land bridge provided a path for human migration from Asia to North America. However, glaciers on the North American continent would have obstructed migrations beyond the Alaskan region. The amount of territory covered by the Alaskan and Canadian glaciers of the Pleistocene fluctuated over the millennia. Between 11,000 and 13,000 years ago there was an ice-free corridor through western Canada that would have allowed human migrations from eastern Asia. The first excavations of ancient human sites in North America were consistent with an arrival of humans from Asia during the period of the open corridor. However, further discoveries of older sites of habitation in the Americas clearly predate migrations via the Canadian ice-free corridor. Although the presence of glacial ice sheets would have created a formidable barrier to human passage through Alaska before 13,000 years ago, further studies indicate that between 14,000 and 16,000 years ago there may have been a relatively ice-free path along the western Atlantic and Canadian coastline. Furthermore, prior to 22,000 years ago there was a narrow corridor along western Canada, free of glacial ice, which may have provided a path for even earlier nomads.

It has been generally accepted that the migrations to the Americas originated from Asia and utilized the Bering Strait land bridge. Native Americans physically resemble the people of China, Mongolia, and Siberia, as do their fossilized remains. However, the fossil record of individuals that may have originated from the earlier migrations, those that predate 12,000 years ago, are less clearly Mongoloid, and some actually have more

Caucasoid features. It is believed that ancestors of the modern populations of eastern Asia originated from Siberia near the end of the last ice age. This population spread throughout eastern Asia, displacing the previous inhabitants and possibly contributing to the early migrations across the Bering Strait land bridge to the Americas. Thus the first migrations to the Americas may have been made by populations that were not directly related to the ancestors of modern Asians. These people would have had more Caucasoid features, which is consistent with the early American fossils. Later migrations from Asia would involve people with features similar to those of modern eastern Asians.

Until recently, the oldest New World artifacts were spearpoints discovered near Clovis, New Mexico, which were used by big-game hunters about 11,500 years ago. The spearpoints of the Clovis culture are found at ancient mammoth kill sites all across North America. Recent excavations in Chile at Monte Verde have revealed human habitation possibly as long ago as 14,700 years. The Monte Verde site was a camp for a small band of about twenty to thirty foragers. They lived in tents, hunted mastodons for meat with wooden spears, and ate berries, nuts, and tubers. The occupation of the Monte Verde site in southern South America would have required a migration of almost fifteen thousand kilometers from Alaska. Studies of ancient human population migrations suggest it would have taken at least a thousand years. Thus the pre-Clovis culture of Monte Verde contradicts the previous popular theory of a migration through the Canadian ice-free corridor 12,500 years ago. Alternatively, a migration along the west coast of the Americas primarily by boats may explain a lack of pre-Clovis sites found inland. Furthermore, ancient coastal settlements may currently lie offshore in the ocean depths, since sea levels were lower at the time of the ancient migrations. In addition to the Monte Verde site, there are other archaeological investigations that support the contention of pre-Clovis cultures in the Americas.

The migrations to the Americas cannot be completely explained by the original theory of a massive Bering Strait transit 12,500 years ago by a Clovis culture people. Excavations at multiple sites in the Americas raise questions concerning the validity of the theory. Excavations at the Meadowcroft Rock Shelter of western Pennsylvania have uncovered man-made artifacts that date back at least 12,900 years, and some may be up to 14,250 years old. However, critics of the dating of the Meadowcroft artifacts claim that the site is contaminated by ancient carbon deposits, which invalidates radioactive dating techniques. Campsite artifacts from

Cactus Hill in Virginia date back to between 16,000 and 18,000 years ago. There are cave paintings in Pedra Furada, Brazil, that may be 17,000 years old, and stone tools within the caves that date as far back as 33,000 years ago. In addition, artifacts found in Pendejo Cave in New Mexico may be up to 30,000 years old. Interpretation of the archaeological record may be further complicated as the early migrations to the New World may have been by small groups of individuals over many thousands of years (both before and after the last glacial maximum). Such a pattern may leave only a few sporadic sites from the earlier migrations.

Support of a pre-Clovis presence of humans in the Americas can be found in genetic and linguistic studies. Comparison of the differences between certain DNA markers found in Native Americans and Asians suggests a divergence occurred approximately forty thousand years ago. Furthermore, studies of the languages used by the various native American cultures imply a separation from Asia about thirty thousand years ago.

Intriguing theories of a transatlantic migration perhaps 18,000 years ago have also been proposed. Some artifacts found in Virginia that are thought to predate the Clovis culture by a few thousand years may represent a predecessor culture that has similarity to the Solutreans of western Europe: a Paleolithic culture found in Europe between 16,500 and 24,000 years ago. Although the possibility of an Atlantic crossing may seem unlikely 18,000 years ago, the occupation of Australia by southeast Asians required boats over fifty thousand years ago. Alternatively, ice flows bridging Canada to Greenland and northern Europe may have provided a migration pathway.

In the Americas, there was plenty of game, abundant plant life, and vast expanses to support the prehistoric human immigrants. Shortly after the human migration to the New World, there was a drastic change in the wildlife populations. Many species, including the saber-toothed tiger and mastodon, vanished. Theories attempting to explain these rapid extinctions include overhunting by humans as well as the climatic changes brought by the recession of the last ice age.

By 7200 B.C. there were small nomadic hunter-gatherer tribes on the central plateau of Mexico. Agriculture of various plants, including corn, is evident by 5000 B.C. in Mesoamerica. The invention of agriculture led to a general increase in population. The moderate climate and abundant fresh water from lakes supported the growing populations. The agricultural lifestyle gave communities more time for labor in the construction of civic structures, including elaborate temples. Pyramid building was first devel-

175

oped along the Peruvian coast around 3500 B.C.—800 years before the first Egyptian pyramids. The early South American pyramids were made of dirt, rather than stone as in Egypt, and had flat tops which were probably reserved for use as temples. The oldest city in the Americas may have been Caral, located in the Supe Valley of Peru, which was inhabited as early as 2627 B.C.

The earliest dominant civilization in the Americas was that of the Olmecs, along the Gulf coast of Mexico, by 1500 B.C. (Figure 8.5). They developed a theocracy with cities centered around temples, and had a primitive system of writing. Initially their civilization was based in the ceremonial center at San Lorenzo, which was destroyed about six hundred years later. A new center was established at La Venta but was abandoned about three hundred years later. The last center was at Tres Zapotes, where the Olmec culture remained until the first century A.D. The Olmec culture is well known for its production of elaborate ceremonial centers, sculptures, pottery, and jewelry. The Olmecs worshiped a jaguar god with human features, a common element of other Mesoamerican religions. There is archaeological evidence supporting the practice of human sacrifice and cannibalism by the Olmecs.

In the Valley of Mexico, about forty kilometers north of present-day Mexico City, the great city of Teotihuacán ("City of the Gods") was built by a sedentary agricultural society. The rise of the Teotihuacán civilization nearly coincided with that of the Roman Empire on the other side of the Atlantic. The city began during the first century A.D. and grew to an estimated population of two hundred thousand at its peak 500 years later, when it covered twenty square kilometers. Along the major north–south road, the Avenue of the Dead, were the dwellings of the aristocracy. Craftsmen and laborers lived in more primitive conditions on the outskirts of the city. Mysteriously, sometime between the sixth and eighth century A.D. the Teotihuacán civilization was destroyed and burned—probably the aftermath of nomadic invaders. Study of the remains of the populace suggests a general decline in health of the inhabitants, perhaps related to climatic changes, which may have contributed to the end of their civilization. The Teotihuacán culture survived, as it had spread to neighboring communities, and is found in successive dominant civilizations, including the Mayans, Aztecs, and Toltecs.

The Mayan civilization originated in what is now Guatemala by 500 B.C. Perhaps as early as the fourth century A.D., the Maya developed a complex written language as a form of hieroglyphics. Between A.D. 700 and

Figure 8.5 Mesoamerica, which hosted the development of the earliest civilizations of the New World.

900, there were many wars and migrations among the native Mesoamerican cultures, which disrupted their continuity and development. Debate continues as to the cause for the rapid decline of the Mayan civilization, which began in the ninth century. Some postulate that ineffective agricultural techniques led to diminishing food resources, while others argue hostile interactions with rival communities as a consequence of the collapse of the Teotihuacán civilization a hundred years earlier. Study of human skeletal remains from the period indicate that malnutrition was prevalent. It has been proposed that coincident with the decline of the Mayan civilization was an extended period of drought in Mesoamerica from A.D. 800 to 1000, which contributed to declining agricultural yields. According to the Mayan culture, the cosmos goes through a 5,000-year cycle of destruction and re-creation. The next destruction phase will occur during the winter solstice in 2012, when the Universe will come to an end, but will be followed by a profound new order.

From the northern region of Mesoamerica, the Toltec Empire arose. The Toltec Empire, which had cultural similarities to the Teotihuacán, oversaw multiple smaller kingdoms, which were required to pay tribute. The most famous leader of the Toltecs, Topiltzin, died around the year 1000. At one time he had promised he would return in 1519, which became a point of consternation with the arrival of the Spanish that pivotal year. The Toltec Empire fell in the twelfth century due to a variety of influences, including overpopulation and conflict with neighboring states. The Toltecs were followed by the Aztecs, who were the last of the early major civilizations of Mesoamerica.

The Aztecs were originally an agricultural community from western Mexico. Following migration to the Valley of Mexico, they built the city of Tenochtitlán on an island at the location of present-day Mexico City by the early fourteenth century. About a century later, the Aztec civilization began to dominate neighboring states, creating the Aztec Empire, which lasted a century. At its peak, the city of Tenochtitlán harbored a population of up to two hundred thousand, with another four hundred thousand in the surrounding areas. The Aztec Empire conquered much of Mesoamerica, amassing up to 15 million subjects. The Aztec religion warned its followers that the Universe would end unless sufficient human sacrifices were made. Thus many of those captured during the Aztec conquests served as human sacrifices. The Aztec civilization remained the dominant force in Mesoamerica until the early sixteenth century.

The vast expanses of North America supported the development of hundreds of different tribal communities. Several tribal populations grew significantly in size and dominance. Two of the most prominent tribes, the Apache and Navaho, originated from the subarctic forest regions of North America and migrated south shortly after the close of the first millennium A.D. In the region of present-day Canada, the Eskimo (Inuit) culture developed and spread from Alaska to Greenland. The early population centers of North America developed later and were less concentrated than those that arose in Mesoamerica and South America.

One of the earliest major population centers of the North American mainland was forming around A.D. 1000 as the city of Cahokia. It was located near present-day Saint Louis and rapidly developed over a matter of several years into a community of thousands. The population was an integrated mix of many tribal cultures from the surrounding areas. The city was laid out with large mounds, plazas, and post henges (circular arrangements of wooden posts that align with various celestial events) that correlated with the cardinal directions. At its peak, circa A.D. 1050, Cahokia covered about fifteen square kilometers. Much of the culture of Cahokia is unknown, but human sacrifice was practiced. Many similar smaller communities developed along the Mississippi Valley extending southward toward the Gulf of Mexico. A period of decline in Cahokia set in by A.D. 1200, which may have been related to consequences of local deforestation. The gradual loss of surrounding wooded territory brought a decline in the animals that lived in those habitats, reducing food resources. Agriculture also fell into decline. The population of Cahokia gradually dispersed into its tribal groups, leaving the city deserted by A.D. 1400. It was a century after the fall of the Cahokian civilization before the Europeans arrived and began exploring the mainland.

The first known contact between the Western Old World and New World was made by the Vikings. As a young man while living in Greenland, Leif Eriksson encountered a sailor who just returned from an intriguing voyage. Bjarni Herjolfsson was traveling to Greenland from Iceland, but bad weather interfered with navigation, allowing them to drift south of the Greenland coast. After days of sailing, they caught sight of unfamiliar tree-covered shorelines. Realizing they had sailed too far west and anxious to return home, they decided not to go ashore and promptly turned around. About fifteen years later, in A.D. 1000, Leif Eriksson made a voyage from his settlement on southern Greenland to the east coast of Canada. He traveled south along the coast as far as Maine, erecting temporary camps without long-term settlements. They made camp for the winter at L'Anse aux

Meadows in Newfoundland. The camp was used by later groups of Vikings numbering up to five hundred individuals over a ten-year period until it was finally abandoned. Although there is archaeological evidence of trade, the relations between the Vikings and the native Inuit were less than friendly. The Vikings were known as fierce warriors, but their swords were outmatched by the deadly bows of the Native Americans. A later Viking voyage to North America that brought a small group of settlers ended in disaster, as many were killed by the Native Americans. Leif Eriksson's brother, Thorvald, fell victim to a mortal chest wound from the arrow of an Inuit. It would be nearly five centuries before the next contact would occur, but it would lead to a far more lasting European presence.

After nearly ten weeks at sea, Christopher Columbus crossed the Atlantic and made landfall in the Caribbean on San Salvador Island September 12, 1492. His expeditions explored the islands of the Caribbean in search of material riches, but insufficient quantity was secured. Eventually an established European presence was made with continued exploration of the surrounding topography. The interaction between the Europeans and Native Americans was exploitive and destructive, with the loss of many lives from conflict and European diseases (against which the Native Americans had no natural defense).

In South America, the development of major civilizations took place later than in Mesoamerica. Peruvian coastal fishing villages developed a stable agricultural existence by 2500 B.C. Over the next 500 years, these villages aggregated into chiefdoms, with a gradual inland expansion. During the ninth century B.C., the Chavín culture developed in the northern Andes Mountains, where a great ceremonial center was built. The Chavín had a religion similar to that practiced by the Olmecs, which included a central jaguar deity. By 300 B.C., the Chavín mysteriously disappeared, but their influence on neighboring cultures persisted long afterward. A long civilizational phase developed along the northern Peruvian coast, followed by military empires including the Huari, Tiahuanaco, and Chimu.

The Inca were originally a nomadic tribe in the Andes region and eventually settled in the Cuzco Valley. They began their expansion by the mid–fourteenth century and eventually established an empire estimated to encompass 16 million subjects by the time of the arrival of the Europeans. Their civilization, unlike the civilizations of Mesoamerica, had no large central urban centers. It was ruled by a divine emperor, whose authority was dispersed through a series of officials to the individual communities. At the time of the arrival of the Spanish, in the sixteenth century, the Incan

180

Empire was struggling with internal civil wars, which weakened the integrity of their civilization. In addition, some of the conquered tribes were not sympathetic to Inca authority and assisted the Spanish in the subjugation of the Inca. The last Inca rulers fled, and their state continued to decline until it ended in 1572.

The final chapter of the great native civilizations of the Americas came swiftly. The Aztec culture and their capitol city of Tenochtitlán were obliterated by the Spanish out of greed in August 1521, by the conquistador Hernando Cortés. In South America, the Incan Empire was overthrown by another conquistador, Francisco Pizarro, a decade later. As Europe enjoyed the dawn of the Renaissance, the occupation of the Americas led to a plantation economy. It was supported through the acquisition of approximately 10 million Africans as slaves between the early sixteenth and late eighteenth centuries. In the centuries that followed, the predominantly European migration to the Americas shaped the populations of today, with simultaneous subjugation of the indigenous cultures.

Australia and Oceana

Australia and Oceana were the last regions occupied by the major ancient human migrations. The isolation of the Australian continent led to the evolution of unique animal and plant species. During the last glacial age (approximately fifty thousand years ago), a transient natural land bridge developed, which allowed migration from southeast Asia to New Guinea and Australia. During the Pleistocene, the islands of Indonesia were accessible by simple rafts, as the sea level was much lower, making the voyages relatively short. Inhabitants of Australia became isolated as the sea levels rose about ten thousand years ago. The earliest human artifacts in Australia, including rocks with carved and painted designs, may date as long ago as 60,000 years. The early Australians were foragers, making seasonal migrations to secure edible plants and animals. The Aboriginals of Australia represent descendants of these early migrations, and much of their primal culture has been preserved in various forms. They developed a close spiritual relationship with the land and its resources, through their strong dependence upon it. The few isolated primitive cultures that have survived to the present provide valuable information in the understanding of ancient

181

customs. Currently about two hundred thousand Aborigines reside in Australia. However, most are of mixed ethnicity and largely detribalized.

After settlement of Australia by the Aborigines, there were many visits by other cultures. The first foreign visitors may have been Chinese explorers, who were followed by the Portuguese and French in the fifteenth century. The Dutch arrived in the sixteenth century but never established a permanent settlement. The celebrated British Capt. James Cook crossed the treacherous Great Barrier Reef to be the first to explore the east coast of Australia in 1770. Records of Captain Cook's voyage describe encounters with a variety of indigenous tribes in Australia. One of the prime objectives of Cook's first voyage of discovery was to encounter the legendary continent of Terra Australis, which was thought to be a great southern landmass that balanced the landmasses of the Northern Hemisphere. The existence of the great continent had been proposed centuries earlier and had never fallen completely into disfavor. Even though the mythical continent was not found on the first voyage, a second voyage was undertaken to specifically disprove the existence of the undiscovered continent. Captain Cook, during these famous voyages, discovered and charted many of the islands of Oceana and provided invaluable records of the inhabitants and their cultures.

Shortly after Captain Cook's brief exploration of Australia, a growing European presence was established. Settlement of Australia by Europeans began in 1788, with a community of soldiers, government officials, and transplanted convicts. England did not have an effective system to maintain prison populations and had been sending its prisoners to North America, but after the American Revolution it turned to Australia. The European population in Australia began to expand more rapidly by 1830, when Britain claimed the entire continent as a British territory. At the beginning of the twentieth century, the Commonwealth of Australia was established.

The forefathers of today's Polynesians sailed from the shores of southeast Asia to populate the islands of the Pacific. Northeast of New Guinea, the Bismarck Archipelago was inhabited by 1500 B.C. and may have been the site from which the subsequent expansion across the Pacific was launched. Later, the Tongan Islands were populated by 1300 B.C., and settlers had arrived in Samoa by 1000 B.C. This was followed by a period of relative stagnation that lasted until about A.D. 300, when there was another push for expansion eastward across Oceana. At that time, there was a migration (from Samoa) to inhabit the Society Islands of Tahiti and the Marquesas. By A.D. 500, the Hawaiian Islands and Easter Island were settled by people originating from the Marquesas. Later migrations from Tahiti populated New Zealand and the Cook Islands by about A.D. 900.

The process by which the widely dispersed islands of the Pacific were populated has been a subject of controversy. Previously it had been assumed that the Polynesians simply "stumbled upon" the various remote islands of the vast Pacific Ocean as they aimlessly roamed around the open sea in their boats. However, it seems more likely that this expansion was a deliberate undertaking, as the routes of approach to many of the islands are against the prevailing winds and currents. Under such circumstances it is unlikely that either adventurers sailing randomly across the ocean or shipwrecked travelers would drift to the shores of these islands. Furthermore, these early settlers brought crops and livestock, which would have been essential cargo in establishing a permanent settlement. Thus it is plausible that the Polynesians utilized their refined skills of sailing and navigation to explore and populate Oceana.

Eighteenth-century rationalists created the popularized visage of the "noble savage," portraying the Polynesians living in idyllic harmony with nature. Although some groups were benevolent, such characterizations overlooked the cannibalistic and warring nature of many of the island cultures. Equally disturbing were the results of contact with early European explorers. Such contact exposed the islanders to diseases for which they had no natural immunity, and devastation of entire populations resulted. In addition, the European explorers brought alcohol and weapons, which fueled further disintegration of Polynesian culture. The twentieth century brought the nuclear age, with its own form of destruction unleashed on the island environments and cultures.

Contemporary Global Community

Once the establishment of a global population had been accomplished, the expansion into unclaimed territories was no longer possible. Most cultures have an underlying territorial premise, which has led to boundaries between ethnic populations based on geographical and political differences. Such containment of societies provides an element of stability both within and between groups. In the modern era there have been many instances of small-scale migrations for a variety of political, social, and environmental reasons. In recent times, largely due to the advances in communications technology, the historic divisions between many of the numerous global populations seem to have faded. However, the differences between people and populations can lead to mutual distrust—a "fault" of human nature. Much of the suspicion between diverse peoples can be lifted through improved mutual understanding, which often is simply a matter of communication. Besides, as the archaeological record and population genetic data suggest, all humans originated from a common population a mere 100,000 years or so ago. Unfortunately, not all sociopolitical conflicts have been resolved and new ones will surely arise. Indeed, human nature was forged from hundreds of millions of years of competition (for survival).

The population of the Earth is truly one global community, and the activities of one nation affect not only itself but all nations to some degree, necessitating cooperation. Individual national political, economic, and social instabilities can rapidly escalate to global concerns. Political crises have repeatedly required interventions of global political bodies to implement worldwide opinion. Major political instabilities and global environmental crises may require a concerted effort of many nations to avoid large-scale disaster. Mutual cooperation, perhaps assisted by technology, may bring the notion of a consonant global community to reality someday. Despite inevitable (temporary) setbacks, this is an exciting time in human history, one which holds much promise, but it will require a great deal of effort to ensure a harmonious future for the global community.

Part III

Sentient Ascension

What a piece of work is a man! How noble in reason! How infinite in faculty! In form and moving, how express and admirable! In action how like an angel! In apprehension, how like a God!
—William Shakespeare, *Hamlet*

9

Humanity

Man has risen to a position of dominance over the other creatures of the Earth within a few million years. It may be arrogance that man considers himself the most highly evolved species—a claim based on basic attributes such as tool use, language, complex behavior, and society. It has come to light in recent years that other primates in fact share these same qualities. Although the level of accomplishment in these facets by man is perhaps more advanced than that of other species, the distinction between humans and other members of the animal kingdom by these measures is becoming less clear.

The "nature" of being human has fueled volumes of philosophical debate. Some authorities suggest human behavior is rooted in "evil" drives where aggression dominates—arising from a long history of the struggle to survive. Perhaps competition with prehistoric carnivorous beasts and even fighting battles among themselves, ancient humans and their ancestors found it necessary to rely on aggressive behavior to enhance survival. Others argue that man's central pedestal is intellect, which has led to the great accomplishments in which we revel. Human motivation may, in part, include elements of these two extremes (instinctive versus intellectual) but is tempered by innumerable other influences to create a complex pattern of behavior. Clearly our ancestors began their ascent in a harsh world where survival was a constant struggle. Perhaps when some measure of control over the natural environment was obtained, the benefits of "civilized" behavior allowed such conduct to become more highly developed.

When a layperson beholds a large stone, he thinks, *What a large piece of granite.* The same stone examined by a sculptor prompts visions of infinite forms. The human body is by far mostly water, which in part reflects our birth from ancient seas; however, it also contains a handful of other elements and minerals. Like the inerts of a stone, the variety of the atomic building blocks of life is relatively few, but the vast complexity with which the building blocks can be combined creates the reality of life.

Mechanisms of Evolution

Until recently, it was the general consensus that humans have been in the same anatomical form since they first came into existence. Such notions seem natural, as man has not perceptibly changed in physical form throughout recorded history. The original concepts of human origins, largely through tribal or formal religious doctrines, assumed humans arrived in the same form as modern man through a mysterious event called creation. Until relatively recently in the history of civilized man, the social and political climate prevented further intellectual investigations because such endeavors were considered blasphemous and threatened the stability of society. Those who openly questioned traditional dogma were considered criminal and subject to persecution. In addition, the systematic techniques of the scientific method of investigation were not widely practiced and accepted until at least the time of the European Renaissance. It wasn't until the nineteenth century that the first credible claims that countered creation themes were made public. It was eventually realized that nature derives its diversity through species modification over a period of time and that man was not immune to this process. Furthermore, this evolutionary process is far from congenial. In order for a species to exist, it must be able to survive the hardships of nature. Despite the beauty and quiescence one may conjure while contemplating the "natural world," in the game of life (and evolution) the stakes are high and compassion has little role.

The survival of a species begins with the survival of the individual. In order for a species to continue, there must be progeny; some individuals must survive to reproduce. It is a direct consequence of terrestrial biology that those who survive to reproduce will contribute inheritable characteristics to their progeny through the transmission of copies of their genes. The survivability of a gene requires its incorporation into the genome of future generations—clearly requiring the survival of its host to reproduction. Those who tend to survive to reproduce are generally those favored by the rules of natural selection, "survival of the fittest."

In 1859, Charles Darwin published his landmark work *On the Origin of the Species by Means of Natural Selection.* Natural selection is a process that fosters long-term change in a species defined by survival experience over generations. The process of natural selection can be broken down into four principal elements. First, organisms will produce more offspring than are expected to survive to reproductive competence. Second, the characteristics of the offspring resemble those of the parent through genetic inheritance. Third, offspring that are better able to adapt to changes that arise in the environment tend to survive to produce their own descendants. Fourth, over many generations this process propagates favored characteristics into future generations, which collectively define the changing (evolving) species. Darwin put forth his ideas without knowing how the traits he was describing were transmitted from one generation to the next. Ironically, Gregor Mendel, an Austrian monk trained in botany and a contemporary of Darwin, was simultaneously laying down the foundations of genetic inheritance through cross-breeding of pea plants in his garden—the work would go unnoticed for decades.

Richard Dawkins, in his book *The Selfish Gene,* has provided an updated version of Darwin's theories through the introduction of replicators. A replicator is basically something that can make a copy of itself. In a competitive environment, the more successful replicator will become the dominant form. The earliest form of terrestrial life may be thought of as a successful replicator molecule. As terrestrial life became more complex, the modus operandi of the replicator became a little more abstract. In essence, a particular replicator is analogous to a gene complex in animal life-forms, with the replicator's survival linked to the animal's survival. Thus the replicator that survives is the one that is successful in displacing its competing replicator variants, rather than a "species" displacing another. In fact, a successful replicator is not necessarily beneficial to the species or individual. Hence it has become popular among biologists to consider such genes "selfish."

The complete set of genes available for inheritance among the members of a breeding population is the gene pool. Individuals within the population will produce offspring that inherit their genetic material through combinations of the genes that are part of the gene pool. If a breeding population encounters a barrier that isolates a subpopulation (e.g., a mountain range, a large river, etc.), the original gene pool becomes divided as well. Over generations, the two gene pools will begin to differ in content through slight incremental alterations of their constituent genes: "genetic drift." The isolated subpopulations may experience different conditions of natural selection, which influences their respective gene pools. Such genetic stressors include environmental (e.g., climatic), social (e.g., breeding practices), and genetic (e.g., mutation) contributions. These genetic variations may produce phenotypic alterations yet preserve the majority of the genetic constitution, allowing the subpopulations to remain the same species (capable of producing fertile offspring from pairing between members of the different groups). The various human races are a consequence of breeding isolation over tens of thousands of years, but with all races remaining of the same species. Prolonged genetic isolation leads to speciation, where crossing of members from different subpopulations no longer produces completely viable offspring.

Through a variety of mechanisms, genetic information may be permanently changed by direct biochemical alteration of its constituent nucleotides. The nucleotide molecules that make up DNA are held together by molecular bonds. The transfer of energy from radiation (e.g., cosmic rays) or chemical exposure may cause these structural bonds to break or different ones to form. Such changes to the DNA may result in an exchange, addition, or deletion of nucleotides. Should the DNA strand be broken, there are cellular mechanisms to affect repair, but these are not flawless and can result in the replacement of a damaged nucleotide by an entirely different one. In some situations, a different nucleotide may be formed through a radiation exposure event, which may go undetected by the DNA's repair system. In addition, DNA may be incorrectly copied or damaged during cell division and replication. These genetic alterations constitute mutations.

A genetic mutation occurs when the product of a gene is somehow modified from the original form in its structure and can affect expression of the original gene. When a nucleotide within a codon is altered, the corresponding amino acid may be substituted by another amino acid according to the altered (mutated) codon. Therefore, the protein product of a gene may be altered by virtue of a change of just a single nucleotide in the coding sequence of its DNA segment. There are many ways in which other nucleotide changes, including alteration of noncoding sequences, may affect protein function and expression. At the beginning of a gene, there is a stretch of nucleotides that control the production of the gene product. Altered regulation may yield too much, or too little, of the protein, thereby affecting its functionality.

What are the consequences of mutational changes? Proteins are generally very complex molecules composed of many amino acids. They are crucial for the normal biochemical function of the cells and organism. Statistically, random changes to such complicated molecules lead to a less functional protein. Therefore, most mutations are not beneficial to the individual (or species). In fact, their effect may be detrimental, whereby the individual is significantly disadvantaged or may even fail to survive. Clearly lethal mutations will not be carried on to subsequent generations through inheritance mechanisms—unless they express lethality after reproduction. While some mutations may have no discernible effect, in the rare case a mutation is beneficial. Individuals receiving a beneficial mutation may have an improved rate of survival, with the opportunity to pass on their ("improved") genes to future generations Thus, mutations are a source of the "raw material" for evolution.

It is assumed that most speciation occurs from allopatric divergence, where population subgroups are somehow separated. The classic example is the situation of geographic isolation, where an ancestral population is divided into subgroups by intervening rivers, mountains, or extensive distance. Over time, the various reproductively isolated populations gradually (genetically) diverge to form new species. Speciation may also occur without geographic isolation. In sympatric speciation, creation of new species within a population can occur through biased mate selection for different traits. In particular, through assortive mating, where pairing between similar individuals is favored, their common traits tend to become incorporated into the genome of future generations. Thus if this divergent trait selection continues over generations, speciation may occur within a single species population—the original population contained members expressing different traits, of which a subset of those traits were selected for expression in the new species.

Although the individuals of a species are nearly identical genetically, individuals and their experiences are unique. Furthermore, the individual members of a species vary in their physical characteristics (phenotype), a consequence of their unique genetic inheritance. In general, genetic inheritance is an "averaging" of the parents' genetic material, whereby offspring are genetically (and phenotypically) similar to their parents: longitudinal inheritance. Each gene has two copies (alleles), one originating from each parent (with the exception of some of those located on the two sex chromosomes). One-half of the genetic material from each parent is sequestered into the respective mature germ cell, a gamete (a sperm in males and oocyte in females), and with the fusion of the gametes (fertilization) the normal complement of genetic material is restored. In a simplistic approach, essentially half of the genes expressed by an individual come from one parent and the remainder from the other. However, genetic material is exposed to several processes that can modify it (e.g., produce mutation) such that the phenotype may have characteristics not found in either parent.

Individuals resemble their parents because their inherited genetic material (genotype) is derived from their parents and therefore reflects a fusion of the parental genetic complements. However, the inherited genes are generally not identical to those of the parents from which the genetic material originated. Imperfections in the process of copying genetic material (through the repeated cycles of cell growth and division) may lead to incorporation or deletion of variable amounts of DNA. Portions of a gene may be "swapped" between the maternal and paternal alleles, creating a hybrid gene. Furthermore, genetic material may be directly altered through mutation. In fact, it is estimated that every person has on average four or five (recessive) lethal genes: the presence of a dominant (nonlethal) allele of the gene overshadows the lethality of the recessive allele. However, should one inherit a recessive lethal allele of a gene from each parent, the outcome is fatal.

Through the evolutionary process, over many generations, a species' characteristics are modified through genetic transmission of favored inherited traits. The genetic modifications that underlie the expressed characteristics occur from a variety of mutation mechanisms. Often, in lieu of creating new genes, it is genes already present within the genome that are subjected to modifications—however, there are ways to gain new stretches of genetic material, such as errors in the replication of DNA or the insertion of new genes by certain viruses. Thus, through modification of existing structures, species undergo change in response to environmental pressures governed by natural selection. A classic example in population biology is that of evolutionary modification of the neck in a community of giraffes consuming the leaves of tall trees. The taller giraffes have a selective advantage over the shorter ones in that they graze on the plentiful leaves of the treetops and therefore are more likely to obtain sufficient nutrition. Their height advantage improves their chances of survival to reproductive age, thereby producing more offspring than their shorter cohorts. The offspring of the taller giraffes will tend to express the characteristics of their parents (i.e., grow tall).

There is clear evidence of organs in the human body derived from predecessor structures that served a completely different purpose in the distant past. Embryological development exemplifies this concept through

192

transient appearance of structures that are subsequently altered into new forms that would otherwise serve a vastly different function in the adult creature. These atavistic characteristics represent transient expression of part of the body plan of distant ancestors from our evolutionary history. The transient appearance of these structures is thought to represent the re-activation of a gene or gene cluster that is still present in the genome dating to an earlier evolutionary stage. Humans continue to express certain vestigial organs that no longer serve a relevant function but were useful in previous evolutionary stages. There are over a hundred vestiges described in humans, including the appendix, body hair, coccyx, wisdom teeth, ear muscles, etc.

Early in the nineteenth century, French naturalist Jean Baptiste de Lamarck proposed a theory of gradual change in a species over time, which became known as Lamarckism. Continuing with the example of grazing giraffes, Lamarck proposed that ancient giraffes had shorter necks and gradually acquired longer necks by their own intentions. As the giraffes spent their lives stretching to reach the abundant leaves of treetops, their necks would increase in length. Somehow, this characteristic would be expressed in the offspring. Thus, by the theory of Lamarckism, structures of value can be improved, while structures that are "unnecessary" may disappear. It has since become clear that acquired traits are not inheritable; the genome is not altered by the individual during life but is predominantly determined at the moment of conception with the fusion of the parental genetic material.

One might anticipate that evolutionary changes to a species occur as small incremental modifications over many generations. Returning to the giraffe analogy, it is not necessary that each successive generation of giraffes become a fraction of an inch taller than the previous until we have the modern long-necked species. It is more likely that evolutionary modifications result from more abrupt alterations of structure. Although, in theory, the process of natural selection may provide incremental benefit through gradual change (as increasing height with giraffes or brain size with hominids) as the intermediate forms may also offer an advantage, not all gradual modifications of an organism's structure have intermediate forms with clear survival advantage.

The resources of organic substrate available for structural adaptations of an evolving organism are limited. This places a constraint on nature's ability to confront the pressures of natural selection by this method—reassembly of tissues to serve a new function may alter or prevent its original function, which could lead to dire consequences. Furthermore, the process of species modification is not directed toward a predetermined goal but rather is continually responding to environmental pressures, which can vary with successive generations. A beneficial trait tends to be incorporated (the genetic material is added to the gene pool) into future generations: the result of chance modification rather than a predetermined event. In the case of humans, many anatomical structures have been identified that had different functions in the past but have been reengineered through the generations. In fact, the original function of a structure may have been vastly different through subsequent adaptation to a new function; e.g., the feather originally provided insulation rather than accommodating flight. Another example of modified function is gills, still expressed in human embryological development, of which some were modified to form the larynx (used in forming speech rather than purely respiration). This modification of gill function, which spanned hundreds of millions of years, had intermediate forms that were still involved with respiration, but the entire process was not a smooth, continuous transition of gill tissue into a larynx. Considering a situation where evolutionary modifications of a characteristic are a smooth, gradual process, there would arise intermediate steps without benefit to secure the "partially modified" structure's expression in future generations—one can argue, on a fortuitous premise, that as long as an intermediate form is not detrimental it may continue to be expressed and inherited until further modifications yield a useful function. However, most evolutionary biologists believe the acquisition of new traits typically involves discrete modifications.

It may seem that the ongoing process of evolutionary change implies that all species are aspiring to an ultimate superior form. The pressures of evolution favor a species form that maximizes survivability guided by natural selection in their environment. The optimum form may change with time according to the predominant contemporary selective pressures of the environment. One of the dangers in the attainment of high specialization is vulnerability. The dinosaurs became specialized with their massive size, which led to their dominance, but a change in the environment (to which they were vulnerable) quickly extinguished them. Furthermore, one should not consider ancestral forms of a species as necessarily more "primitive." Prior forms may have been exceptionally suited for their environment, but with changes in evolutionary pressures new morphological characteristics were favored.

Genetics and Biochemistry

In 1953, American biochemist and geneticist James Watson and British biophysicist Francis Crick revealed the fundamental molecular structure of the DNA molecule. In the years that followed, the language of nature's genetic code—written within the structure of the DNA—was deciphered. The field of genetics continues to provide significant advances, particularly in its roles of understanding disease and applications in various treatment modalities. The human genome, which codes for about thirty-five thousand functioning genes, is among the largest known. Although the human DNA is tightly compacted into twenty three pairs of chromosomes in each cell nucleus, if unfolded its 3 billion nucleotides measure about two meters in length.

The genetic code, recorded by the tandem sequence of nucleotide bases within the DNA molecule, dictates the sequence in which amino acids are placed in the production of proteins. In order to uniquely identify each of the twenty amino acids, the genetic code utilizes combinations of the four nucleotide bases of DNA, which requires a minimum of three sequential bases (Figure 9.1). The genetic code is therefore triplet where the three bases of each coding group constitute a codon. Since more combinations than needed are available ($4^3 = 64$), the genetic code is "degenerate." Thus most amino acids are coded for by several different codons. There are also "nonsense" codons, which are used to signal the beginning and ending points of a protein's genetic code sequence.

The origin of the genetic code is a mystery. Utilizing four bases requires at least a triplet code to permit unique identification for the twenty amino acids. Why not four or five bases for a codon? The original genetic code was likely simpler than the modern one and may have first appeared in the prebiotic era, predating the RNA world. It was derived from the biochemical interactions between the amino acids and bases. It has been suggested that the relative sizes of amino acids and the chemical cofactors involved in protein production biochemically favor an interaction with three nucleotides.

The Genetic Code

Codon	Amino Acid	Codon	Amino Acid
UUU	Phenylalanine	AUU	Isoleucine
UUC	Phenylalanine	AUC	Isoleucine
UUA	Leucine	AUA	Isoleucine
UUG	Leucine	AUG	Methionine
UCU	Serine	ACU	Threonine
UCC	Serine	ACC	Threonine
UCA	Serine	ACA	Threonine
UCG	Serine	ACG	Threonine
UAU	Tyrosine	AAU	Asparagine
UAC	Tyrosine	AAC	Asparagine
UAA	STOP	AAA	Lysine
UAG	STOP	AAG	Lysine
UGU	Cysteine	AGU	Serine
UGC	Cysteine	AGC	Serine
UGA	STOP	AGA	Arginine
UGG	Tryptophan	AGG	Arginine
CUU	Leucine	GUU	Valine
CUC	Leucine	GUC	Valine
CUA	Leucine	GUA	Valine
CUG	Leucine	GUG	Valine
CCU	Proline	GCU	Alanine
CCC	Proline	GCC	Alanine
CCA	Proline	GCA	Alanine
CCG	Proline	GCG	Alanine
CAU	Histidine	GAU	Aspartic Acid
CAC	Histidine	GAC	Aspartic Acid
CAA	Glutamine	GAA	Glutamic Acid
CAG	Glutamine	GAG	Glutamic Acid
CGU	Arginine	GGU	Glycine
CGC	Arginine	GGC	Glycine
CGA	Arginine	GGA	Glycine
CGG	Arginine	GGG	Glycine

Figure 9.1 The genetic code (of the messenger RNA) uses three nucleotide bases (A = adenine, C = cytosine, G = guanine, and U = uracil) to code for an amino acid. The gentic code is redundant, as most amino acids are specified by several codons. AUG, which codes for methionine, is the START codon to indicate the beginning of a protein. UAA, UAG, and UGA are STOP codons, which indicate the end of a protein. The genetic code as stored in DNA differs from messenger RNA by the replacement of thymine with uracil.

In the process of creating a protein, the corresponding segment of the DNA is copied through the manufacture of a messenger RNA (mRNA) molecule intermediary within the cell's nucleus (transcription). By keeping the DNA sequestered in the cell's nucleus, it is protected from the biochemically active cytoplasm, where it would be much more susceptible to chemical damage. The DNA to mRNA transcription process is rapid; about thirty nucleotides of the mRNA are placed each second. After copying a gene sequence from the DNA, the mRNA leaves the nucleus to enter the cell's cytoplasm. In the cytoplasm, the mRNA secures itself to a ribosome on the endoplasmic reticulum where the copied sequence is read, three bases at a time, by transfer RNA (tRNA) molecules. The tRNA molecules are amino acid–specific and must recognize their particular codon before releasing the amino acid they carry. Once the appropriate tRNA is in position, according to its specific codon, it will release its amino acid to add to the growing chain of amino acids of the nascent protein (translation). As the amino acid chain is manufactured, it may undergo further biochemical modification with the addition of other chemical groups or changes in shape as it folds upon itself. The protein may proceed to excretion by the cell, incorporation into an organelle, or undergo further processing.

Surprisingly, almost 99 percent of the human DNA doesn't code for anything. The noncoding segments of DNA are termed junk and are located between, and even within, regions that code for proteins. The noncoding segments are removed biochemically through "splicing," a process that utilizes specific proteins to snip out the noncoding pieces of the mRNA template prior to translation. Various theories have attempted to explain the significance of junk DNA: e.g., ancient gene fragments that are no longer transcribed, a reservoir of substrate for insertion into existing genes to create new ones, etc. Approximately half of the junk DNA is in the form of transposons (transposable genetic elements), which are gene sequences that have the ability to either biochemically "relocate" in the genome or induce copies of themselves that are inserted in the genome. Their properties are similar to some viruses and may even have an origin in common with some viruses (including the AIDS virus). Depending upon where a transposon gets inserted, it can produce a variety of effects on expression of nearby functional genes. In fact, it has been suggested that the presence of transposons may explain the apparent discrepancy between the striking genetic similarity between humans and chimpanzees and the distinctly different physical characteristics of the two species. A fraction of the remainder of the noncoding DNA is involved with gene expression.

Some of the noncoding sequences are known to have regulatory functions of certain genes. Promoter and enhancer sequences are generally located upstream of a gene, where they are able to functionally assist RNA transcription by forming complexes with specific proteins to regulate their transcribing function.

The complexity of life is staggering. The number of genes in a given species is a crude measure of complexity. A genetic census of terrestrial life can identify those life-forms that contain relatively small genomes. Among the simplest life-forms are certain species of bacteria known as mycoplasma, with some species having as few as 500 genes. It is estimated that the minimum number of genes required for a life-form is about 300. Somewhat humbling, the human genome is not much larger than the 25,000 genes found in the small flowering plant *Arabidopsis thaliana*, less than twice the size of the genome of the tiny worm *C. elegans*, but only about half that of *triticum aestivum* (wheat). Until recent genetic research indicated the human genome consists of between 26,000 to 40,000 genes, most estimates had placed it closer to 80,000 to 100,000 genes.

Through growth, multiplication, and specialization, a single cell differentiates into the many tissues that form a human being. In fact, the human body contains approximately 100,000 billion (or 100 trillion) cells; about a billion are formed each hour, while a similar number are lost. The complexity of life has a high premium, where minor structural and biochemical flaws can be lethal. In fact, it is estimated that five of every six human embryos die by eight weeks, with the majority succumbing to genetic incompatibilities.

The mapping of the entire human DNA sequence is the primary objective of the human genome project. On June 23, 2000, completion of the first draft of the human genome sequence was announced jointly by the human genome project and a private biotech firm, Celera Genomics—many considered the announcement premature, as the sequence was not entirely organized and adequately tested for accuracy. Although the human genome sequence will be completely deciphered relatively soon, the data will keep geneticists busy for many more decades, as they must tease out and identify tens of thousands of genes (recalling that most of the DNA is "junk"). Approximately eight thousand of the genes in the human genome have been identified, leaving the majority unknown. Once the human genome has been fully sequenced, it is conceivable that genetic screening of individuals for variances from the "normal" sequence will uncover genetic abnormalities—the diagnosis of genetic ailments can be made in the absence of symptoms. However, there is concern that an individual's DNA information may provide fodder for genetic discrimination used by health insurance companies and employers. Outside the genetic role of DNA, it has applications in personal identification and forensics, as it is almost entirely unique between individuals when considering the complete sequence: coding and noncoding segments.

The field of genetics has been making vast contributions to biology and medical science. It has been clearly established that many diseases that afflict humans are genetic (inherited). It has been estimated that up to 60 percent of people will develop a genetically related disease during their lifetime. Some recent discoveries include genetic links to obesity, breast cancer, mental disorders (epilepsy, retardation, addictive behavior), and progressive debilitating disorders, including Alzheimer's disease. Alzheimer's disease, which afflicts about 4 million Americans, most over sixty-five years of age, is characterized by abnormal buildup of amyloid protein in the brain tissue (senile plaques). In October 1999, scientists at the biotechnology company Amgen announced they had identified a protease (ß-secretase) that is involved in Alzheimer's disease. Amgen researchers suggest the amyloid protein represents fragments released from the surface of brain cells by the action of the ß-secretase enzyme. The gene for the ß-secretase enzyme has been isolated, enabling scientists to produce large amounts of the enzyme using genetically altered bacterial colonies. Armed with sufficient quantity of the enzyme, scientists can use it in research designed to identify medications that can safely inhibit the production or activity of the enzyme in hopes of treatment. Alzheimer's disease is also marked by accumulations of abnormal material within the brain's neurons, called neurofibrillary tangles. The tangles, composed of tau proteins, are an active area of research, with hopes of finding ways to prevent their accumulation as another means of treatment.

The ability to deliberately modify genetic material invites the opportunity to alter an individual's genome. If the genetic material of the first cells from which a human is formed is altered (i.e. during embryogenesis), the modified genes will be present in all the cells of the body—including the cells responsible for reproduction, thereby enabling transmission of the new genes to future generations. Such genetic manipulation has the potential to eradicate genetic abnormalities that may be otherwise transmitted to subsequent generations. The DNA sequence may be manipulated in many ways, leading to countless outcomes: different sex, altered phenotypical characteristics, perhaps even increased intelligence. The production of genetic clones of animals has been accomplished, leaving such work in humans clearly feasible. The potential of such work seems unbounded and its consequences profound. Ethical issues concerning the applications and future course of gene technology task society to define its position.

Are we simply the product of our genes? If such were the case, we would expect individuals with identical genes—identical twins—to be exactly the same; they are generally distinguishable. Humans have an element of developmental plasticity influenced by environmental factors. These factors include natural elements (e.g., food), biological elements (including the nine months of gestation prior to birth), and social exposures. The long-term effects of nutrition are well known, as malnourished populations tend to be smaller and more susceptible to diseases than adequately nourished populations. Exposure to certain environmental factors elicits changes in the human body; e.g., exposure to pathogens, which induce activity of the immune system, afford protection from future exposure (immunization), and exposure to sunlight, which induces changes in the skin. Social and cultural habits can affect an individual's exposure to elements of the environment. There are a variety of adaptations the human body can make to environmental situations, which add another venue of complexity to the individual. In fact, such plasticity proved crucial in man's spread across the planet to improve chances of survival in the vastly different environments encountered.

Sentience

During the fourth century B.C., the celebrated Greek philosopher Plato embraced the belief that the human mind was contained within the head. He arrived at this conclusion based on the observation that the head was roughly in the shape of a sphere, which he considered the most advanced geometrical form. However, Plato, like his contemporaries, believed diseases of the human mind were the consequences of the wishes of pagan deities. Although some continued to question the location of human consciousness, by the Middle Ages it was generally agreed that the mind was derived from the brain.

Unlocking the secrets of the human brain requires both anatomical and physiological investigations. The discipline of neuroanatomy was marked by slow progress in its early stages. The earliest descriptions of the brain's anatomy were written between 400 and 500 B.C. and simply described the presence of optic nerves and the division of the brain into the two cerebral hemispheres. In the early fourth century B.C., Polybus popularized the theory of the "four humors," which dominated the disciplines of anatomy and physiology for nearly two millennia thereafter.

The early philosophers proposed that nature had a goal of achieving a balance between its components. It was believed that pairs of qualities would balance other pairs, giving special significance to the number 4. In addition, the proposed four basic elements of nature (water, air, fire, and earth) had specific qualities (wet, dry, hot, and cold) and represented the substance of which the Universe is composed. The balance between these elements maintained the harmony of nature. The theory of the four humors was an extension of the four-element concept applied to human health. As with the elements of the natural world, a balance must be maintained among the humors to assure a harmonious and disease-free existence of the body. Any imbalance between the humors (blood, phlegm, yellow, and black bile), which may manifest as an excess or deficiency, would lead to illness. It was noticed that when individuals became ill, they may react (e.g., fever) or discharge substances (e.g., vomit, sweat, etc.), which was felt to represent the release of excess humors. It was believed that if the victim's purge was insufficient, the recovery would be incomplete and the consequences could prove fatal. Thus the physicians' role was to intervene by assisting the purge process through bloodletting or other means. Despite such misleading theories of human biology of the time, study of the brain through careful dissection was introduced in Alexandria by 300 B.C. Gradual progress was made, primarily by Galen in the second century B.C., but like his predecessors, his conclusions of anatomy and physiology were strongly influenced by humoral concepts.

There was a long period of stagnation and even decay in the understanding of neuroanatomy and function following the work of Galen. The domination of the social establishment by religious authority (which delegated little importance to corporeal concerns) persisted until the European Renaissance. During centuries of scientific paralysis, the ancient works of Galen and his predecessors remained unchallenged. In the late fifteenth century, the limited works of Leonardo da Vinci in neuroanatomy raised new interest in the discipline. It was in the mid–sixteenth century that the Belgian anatomist Vesalius brought anatomic appreciation of the brain into the modern era. Technological achievements borrowed from other fields of scientific investigation accelerated advances in neuroanatomy

201

and neurophysiology with the invention of the microscope and discovery of electromagnetism.

The nervous system, which includes the brain, spine, and nerves, enables coordinated activity of the mind and body. The nervous system coordinates a variety of processes including internal stability of the body, such as coherent function and activity of organ systems: cardiovascular, respiratory, digestive, etc. In the more neuroanatomically complex species, the nervous system dedicates relatively more activity to interaction with the external environment, manifested as receiving information from the senses, processing that information, and forming responses. The human nervous system is composed of a variety of specialized cells, including the neurons and other cells of nutritional and structural support. Most neurons are between 5 and 100 micrometers in diameter and are surrounded and nourished by glial cells. Neurons vary greatly in length; while most are very small, some are well over a meter in length. Neurons transmit information along their axon portions, using electrical signals (action potentials) which propagate along the cell membrane surface through a mechanism of biochemically controlled ion concentrations. Transfer of the action potential signal between axons requires crossing a small gap, the synapse, which often utilizes a biochemical (neurotransmitter) to complete the electrical coupling (Figure 9.2).

Neurotransmitters may be excitatory or inhibitory, whereby the signal arriving at a synaptic cleft may be either transmitted or suppressed. Glutamate is the dominant excitatory neurotransmitter in the human brain, being utilized by about 90 percent of the neurons. Once glutamate is released at a synapse, it diffuses across the cleft for temporary binding to receptor proteins in the membrane of the postsynaptic neuron. The glutamate must be quickly cleared from the synapse to prepare for the arrival of the next action potential signal. To accomplish this, glutamate is taken up by nearby glial cells, where it is biochemically converted to glutamine, then released in the vicinity of a presynaptic neuron, where it is collected and packaged for release in a future synaptic transmission. The energy to drive the process of synaptic transmission comes from serum glucose, with ATP as a primary intermediary. The most abundant inhibitory neurotransmitter is γ-aminobutyric acid (GABA), which has a mechanism of metabolism similar to glutamate. A common medication for treating anxiety is diazepam (Valium), which heightens the activity of GABA. The dichotomous nature of neurotransmitter signal transmission is reminiscent of the binary-based mathematics of computer circuitry, where the two basic signals

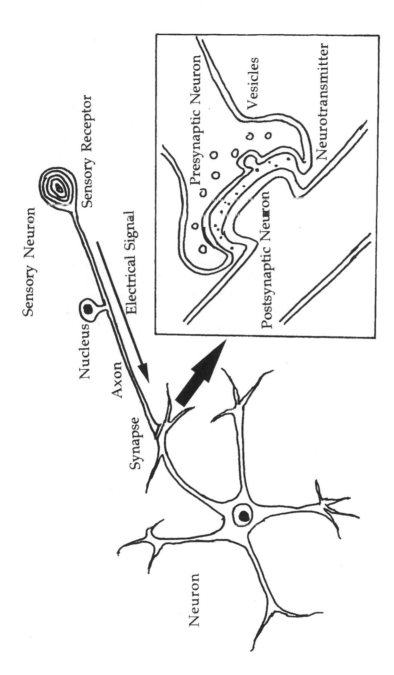

Figure 9.2 The brain's information highway. Electrical signals travel along the neurons propagated by electrochemical properties of the cell membranes. At the synaptic connection, the signal is conveyed across the gap by the release of neurotransmitters from vesicles of the presynaptic neuron. The released neurotransmitter is collected by receptors in the membrane of the postsynaptic neuron where the electrical signal may be resumed.

203

are often denoted as 1 and 0—electrically, the binary digits become voltage or no-voltage.

In retracing the evolutionary steps of human brain development, the paleontological record becomes increasingly sparse the further back in time we proceed. In order to define a place for humans in the natural world, biologists begin by categorizing humans in the kingdom Animalia under the phylum Chordata. As long as 600 million years ago, the phyla Chordata and Echinodermata diverged. On the one hand, echinoderms are marine animals that have a body plan characterized by radial symmetry, such as the starfish and sea urchin. On the other hand, chordates are animals that possess a notochord at some stage in their development; in the case of vertebrates, the notochord becomes the spine. In addition, chordates have a dorsally positioned nerve cord from which the brain and spine are derived in mammals. The first vertebrates were marine and probably arrived during the Cambrian Explosion, but their skeletons were largely cartilaginous, which left little for fossilization—about 30 million years later vertebrates developed the capability to mineralize their tissues to form bones. The earliest chordate forms were small filter-feeding marine animals that lacked a well-formed brain, head, and paired eyes. It would take several million years of evolution to develop these more familiar vertebrate characteristics.

The connection between the vertebrate brain and spine, the brainstem—sometimes referred to as the reptilian brain—first appeared over 500 million years ago and represents a basic coordinating center of the central nervous system. The brainstem is involved with the maintenance of vital functions, including respiration and heart rate. The first vertebrates probably arose from a small marine worm ancestor during the Cambrian—fossil candidates of the earliest such vertebrates include *Pikaia,* found in Canada's Burgess Shale, and *Haikouella,* found recently in China. Study of present-day species of such worms (e.g., lancelets) suggests a relatively more advanced nervous system was already present in the vertebrate ancestors than had been expected. In addition to the basic dorsal nerve cord and segmented pattern, the lancelet nervous system has a collection of pigment cells located at the front of the creature, which may represent a single primitive "eye." Their vision is probably not in the context of images but more akin to the impression of moving shades of light to indicate movement in the environment. The lancelet nervous system may also have the beginnings of a limbic system suggesting behavioral characteristics such as fleeing danger and certain basic social characteristics (e.g., aggression, migrational habits, etc.), which may have been present in our wormlike ancestor. Higher cognitive functions arose later when the rudimentary beginnings of a cerebral cortex appeared about 200 to 300 million years ago.

Pelycosaurs, Permian predecessors to the mammals, roamed the Earth 300 million years ago, living on a diet composed of both plants and meat. There is evidence that some of the pelycosaurs had features of warm-bloodedness. Cold-blooded animals have the benefit of a minimized basal energy expenditure, which requires minimal food intake. Although cold-blooded creatures may be active in a warm environment, once chilled they become lethargic and relatively inactive. Conversely, warm-blooded creatures can remain much more active in colder environments, but that comes at the price of a substantially increased (about ten times) requirement for food. The constancy of body temperature in a warm-blooded creature permits physical activity that is much more independent of external environmental conditions and also provides internal thermal stability for improved efficiency of biochemical and cellular function.

During the Permian, the pelycosaurs gave way to the therapsids—reptiles that walked upright rather than crawled. Among the various groups of therapsids were the cynodonts, which represent the direct ancestors to mammals. Over thousands of millennia, as ever more mammal-like species came into existence, the trend continued toward the high-volume diet of a warm-blooded lifestyle. In fact, the demand to maintain a rich, voluminous diet required significant cognitive capabilities of these ancient creatures. In order to be more effective at acquiring food, the early mammals required skills in recognizing and efficiently securing sources of nutrition. A crucial step includes the development of a comprehensive "mental map" of the environment to recognize and locate potential food sources, as well as tactics that promote effective foraging—the more complex cognitive demands require a corresponding complexity of the creature's brain. These foraging skills were present in the hominid ancestors and were paralleled by changes in their brains. Reviewing the fossil record of human predecessors, one finds that the hominid brain has tripled in size in the last 3 million years and the cerebellum, the portion of the brain involved with coordinated motor function (e.g., walking, running, throwing, etc.), has tripled in size during the last 1 million years. A period of particularly rapid increase in hominid brain size, which occurred between 150,000 and 600,000 years ago, was at a time marked by progressive social and technological advances. In addition, there may have been a concurrent development of spoken language, which would have been advantageous in the cooperative acquisition of food.

In order to process vast amounts of information from its neurons, the human brain has developed a very complex and highly organized anatomical architecture with extensive connections between areas involved with specific functions. The dominant anatomical feature of the human brain is the cerebral hemispheres, which have an outer surface layer, the cerebral cortex that is composed of "gray matter" and an underlying core of "white matter" primarily representing extensive neuronal axonal pathways connecting the different areas of the brain. The gray matter, which makes up about 40 percent of the human brain's weight and varies between one and four millimeters thick, is composed of neurons that produce, receive, interact, and process signals from other neurons. The cerebral cortex is highly convoluted in humans, maximizing the surface area confined to the relatively limited space of the skull—if the cortical surface were spread flat, it would cover an area of approximately 1.5 square meters (about the size of an office desk top). There is a neuroanatomical theory that higher surface-to-volume ratios of the brains (i.e., relatively more cerebral cortex) correlate with increasing intelligence between different mammalian species—the highest ratio is found in humans. The clefts of the cerebral cortex folds (sulci) contain the majority of the brain's gray matter. The sulci separate the upraised convolutions, gyri, which have a relatively consistent pattern among individuals. Utilizing the constancy of gyral patterns and microscopic characteristics of the gray matter cell layers (cytoarchitecture), the human brain has been divided into forty-seven anatomic Brodmann areas. Among the areas of the brain relegated to specific functions are the motor cortex (Brodmann areas 4 and 6), sensory cortex (areas 1 to 3), visual cortex (areas 17 to 19), auditory cortex (areas 41 and 42), and vast regions of association cortices. The white matter of the brain, for the most part, contains complex arrays of bundles of neuronal axons, which transmit information between the different parts of the brain. The brain's two cerebral hemispheres make up the bulk of the brain, and two smaller cerebellar hemispheres lie below and to the back (dorsal) of the cerebral hemispheres (Figure 9.3).

The outermost cell layers that cover the cerebral hemispheres, the gray matter, are composed of about 10 billion neurons and are arranged in six layers, the neocortex. The neocortex is unique to mammals; it came into existence between 160 and 300 million years ago. Reptiles and amphibians lack a neocortex, indicating it originated after the evolutionary di-

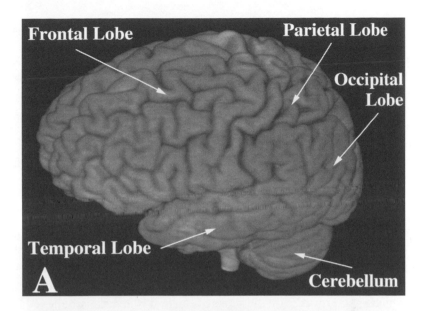

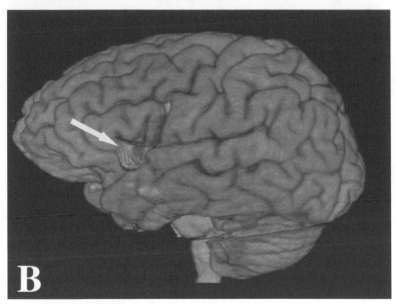

Figure 9.3 A) Surface reconstruction from MRI data of a normal brain with major neuroanatomical divisions labeled. B) Surface reconstruction of a normal brain with superimposed functional MR response (arrow) to language activation delineating its location in Broca's area of the frontal lobe. *Credit: Massachusetts General Hospital, 3-D imaging Service.*

vergence of mammals about 300 million years ago. Furthermore, the presence of the neocortex in all mammals suggests it was present in the common ancestor to mammals, which existed at least 160 million years ago. The human neocortex has approximately three times the surface area of our nearest phylogenetic relatives, the African apes and monkeys. The human cerebral cortex also has some remnants of its ancestral roots, the archicortex, preserved in the single cell layer of the hippocampal gray matter located deep within the temporal lobes.

The human brain has many differences from other primates. In humans, the primary visual cortex of the occipital lobes is relatively small while association areas, which further process the visual information, are relatively increased in size compared to the nonhuman primate brain. Not surprisingly, the areas devoted to speech and language are relatively enlarged in the human brain. In humans the frontal lobes, responsible for thoughts and intentions, are double their relative size compared to other primates. However, there is a relative diminished size of the olfactory areas, which are involved with the sense of smell. It is generally believed that the human brain achieved its modern form after the appearance of *Homo habilis* about 2 million years ago.

The modern era of human neuroanatomy and neurophysiology began near the close of the nineteenth century. An Italian anatomist, Camillo Golgi, developed a pigment using silver salts, which, for unclear reasons, stains about 1 percent of the neurons in each sample of brain tissue. Furthermore, the neurons stained by the silver salts are colored in their entirety, making the complete cell stand out for the microscopist to identify. Cortical brain tissue is densely packed with neurons, making it a confusing tangle of cells, but the selection of relatively few neurons by the Golgi stain greatly clarified the analysis and accelerated the early work in microscopic neuroanatomy. Santiago Ramón y Cajal, a Spanish contemporary of Golgi, made extensive studies of microscopic neuroanatomy and even clarified the concept of individual neurons communicating with others by synapses (Figure 9.4).

The human brain is a formidable organ for anatomic study. With a consistency of half-formed gelatin, the average human brain weighs about fourteen hundred grams, has a volume of 1,450 cubic centimeters, and is composed of approximately 10^{11} neurons, making an estimated 10^{14} (synaptic) connections. The signal processing by a typical postsynaptic neuron is basically an "averaging" of inputs from its synaptic connections with other neurons, which then determines the signal the postsynaptic neuron

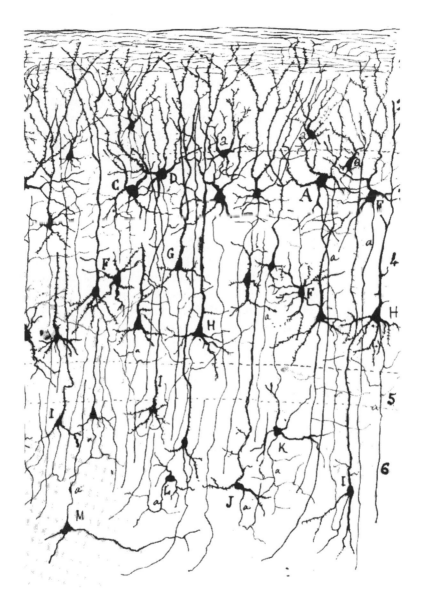

Figure 9.4 One of Cajal's sketches of the visual cortex in a rodent dating to 1888. The numbers denote the cellular layers as one goes deeper into the cortex. The capital letters denote individual neurons. *Credit: Cajal Institute.*

transmits along its axon. The typical signal (action potential) transmitted along the axon of a neuron is about 100 millivolts (one-tenth of a volt) and travels at about 100 meters/second. The averaging process of neuron synapses is more complex than the binary language (1 or 0, "on" or "off," etc.) of digital processing performed by the logic circuits of computers. Instead, in the human brain, neuron networks are extensively crosslinked in series and parallel configurations that operate at a much slower rate than the integrated circuits of computers. This parallel processing of information is particularly evident from studies of how the human visual system sorts and organizes electrical impulses from the optic nerves to create and interpret an image. Furthermore, unlike computers, the neurons of the brain are not rigidly hard-wired—new connections can be made, old ones removed, and even new neurons may develop.

Neurons may be specialized in a variety of ways affecting how they process and transmit their signals. To begin with, there are several different types of neuron cells, which have their own distinct structures and characteristics. The synaptic connections between neurons come in a variety of types, including excitatory and inhibitory. The prototype synaptic connection—where a neurotransmitter is released by the presynaptic axon terminal, diffuses across a synaptic cleft, and is collected by the receptors of the postsynaptic dendrite—may use one of a host of different neurotransmitters (about one hundred have been described so far). Furthermore, certain types of synaptic connections that use specific neurotransmitters may be sequestered in specific areas of the brain. Pharmacology exploits this site specificity in the treatment of certain neurological conditions where medications that exhibit their effect on these specific neurotransmitters can control the disease symptoms. Monoamine oxidase (MAO), an enzyme found in most tissues, with particularly high concentrations in the brain, is important in the chemical formation of the hormones epinephrine, norepinephrine, and serotonin. Serotonin is used by the body as both a neurotransmitter and a hormone. In the brain, it is in particularly high concentrations in the hypothalamus and basal ganglia—one of the many roles of the hypothalamus is the regulation of emotional activity. Decreased levels of serotonin activity are associated with depression; there are a variety of antidepressants (including Prozac) that stimulate the activity of serotonin to overcome the depression. The designer drug Ecstasy (methylenedioxymethamphetamine), used as a stimulant and hallucinogen, causes a massive release of the brain's serotonin, which leads to the sensation of euphoria.

Investigators have found that MAO enzyme levels in individuals with violent behavior (from serum analyses of extremely violent prisoners) are about a third of that found in normal (nonviolent) volunteers. Conversely, some individuals who suffer from phobias and extreme anxiety have abnormally high MAO levels. Investigations of individuals with relatively dangerous careers (e.g., Spanish matadors), who are not necessarily pathological in their behavior, tend to have higher MAO levels than controls. Studies of broader populations have shown that MAO levels are generally lowest in young males and are lower overall in men than women. Thus, as people age, the MAO levels tend to rise but remain lower in males, perhaps providing a physiological basis for decreased aggressiveness as people mature and the tendency for males to have a more aggressive nature than females.

The anatomy of the brain is generally determined by genetic instructions; its complex structure is largely in place early in gestational development. The genetic influence on the development of specific areas within the brain is suggested by many genetic disorders that manifest in the central nervous system. The complete development of the human brain is not accomplished during gestation but requires several more years, with minor changes occurring throughout life. The basic anatomical framework may be in place by the time of birth, but neurons continue to migrate, mature, and form their synaptic connections. In the maturing brain, the fine-tuning of neuron connections may not occur until directed by electrical stimulations or chemical factors secreted within the brain—steroids have been shown to stimulate synaptic formation in particular areas of the brain. The process of axon guidance is driven by the recognition of specific chemical factors in the brain tissue by receptor proteins on the axon surface. The detection of the chemical directs axon growth toward a destination where a synapse may form. The production of the chemicals and axon receptor proteins is ultimately under genetic control. Furthermore, stimulation of the senses (e.g., vision) early in life may assist in directing axons to their appropriate locations. During embryogenesis, some of the brain cells are randomly interconnected, and through development and experience (neural stimulation from the senses) these circuits adapt by modification (e.g., strengthening of synaptic connections, regression of redundant neurons, etc.) to create the functional mature neural networks.

Neuron networks of the brain can modify their signal processing through the plasticity of their synaptic connections. The electrical stimulations of action potentials can affect the metabolic activity of a neuron, which may also alter neuron function. Furthermore, axons have been shown to develop new synaptic endings, while established endings may be altered. Synaptic plasticity plays a role in imprinting information into memory—called engrams in neuroscience. In addition, long-term neuronal changes, such as those required in the storage and stability of permanent memories, may require alterations of gene expression within the involved neurons. As events are witnessed during life and converted to memory, a select set of neurons become activated and modify their interconnections—fundamentally, memories are a pattern of neuron connections. Individual memories may involve thousands of neurons across the cerebral cortex, but not all memories become stored long-term. In fact, the neurons of the cerebral cortex are those most directly responsible for memory, motor function, perception, intellect, and consciousness.

The seventeenth-century French philosopher René Descartes claimed, "Cogito, ergo sum" (I think, therefore I am) to point out that consciousness is essential to existence. The mind, the source of consciousness, is certainly located within the brain, but the identification of a specific anatomical "conscious center" is under heavy investigation, with several candidates proposed. Numerous investigations have shown complex electrical activity of the brain's neurons in subjects undergoing external stimulation (e.g., sounds, lights, etc.). The exact underlying pattern of these neuron signals remains unclear, but they suggest an interaction of multiple centers of the brain to create conscious awareness of the external environment. In general, correlation between messages received from the senses and those from memories is used to create the perception of reality.

Consciousness and what it entails is a subject of controversy. Some experts suggest that consciousness is merely a complex biological neural network that will someday be explained, whereas others argue that it is an entity that lies beyond the pale of human comprehension. Serious challenges are encountered in attempting to formulate a neurobiological basis for such an intangible and highly subjective phenomenon. An anatomical approach includes functional MR imaging investigations in humans to reveal activity in brain areas during states of conscious awareness. Computer algorithms, including those of neural networks (computer models of the interconnections of neurons), have been formulated that are able to identify ("recognize") subtle patterns from a complex mathematical matrix, just as conscious life-forms must recognize food or prey from an environment filled with distractions.

The subjective nature of consciousness raises the question of its variance between individuals (and species). Perhaps the aura of one's own consciousness differs from that perceived by other conscious creatures. The visual perception of a specific color may vary between observers as each individual's interpretation of a color is determined by the complex neuronal connections innate to their brain. Does blue really appear the same to you as it does to someone else? In those with protanopia, the most common form of color blindness, the colors red and green are indistinguishable. Does the tone of middle C sound exactly the same to everyone? Do the sensations of pain, joy, and sadness feel the same to everyone? Since it is difficult to address these highly subjective aspects of consciousness, a more general approach to the subject is useful.

The physiological state of awareness, or consciousness, requires a concerted interaction of multiple areas of the brain, including input from short-term memory. In the awake state, there are characteristic 40 Hz electrical oscillations—the gamma frequency of the electroencephalogram (EEG)—which are apparently involved in this coordinating process. The oscillations may be used to synchronize the activity of various brain areas to establish a coherent state of awareness. The neural basis of consciousness appears to be largely dependent upon connections between the neocortex and the thalamus. There are other areas of the brain, including the reticular activating system of the midbrain and hippocampus of the temporal lobes, which also contribute to the conscious state. Fundamental to cognitive function is the ability to form memories of stimuli received through the senses (sight, sound, etc.) from the external environment. Recent investigations on the formation of memories were performed using functional MR imaging to detect and anatomically localize metabolic activity of the brain in human volunteers undergoing visual or auditory stimuli. The ability to remember visual stimuli seems to be related to activity in the right prefrontal cortex and the parahippocampal cortices (deep within the temporal lobes) bilaterally. Memories of verbal stimuli also involve the temporal lobe cortices but include activity of the left prefrontal cortex. Previous studies of patients with brain injuries substantiate these conclusions. Furthermore, patients who have had portions of their medial temporal lobes removed (e.g., treatment for intractable seizures) may lose the capability to form new memories, although experiences prior to the surgery are recallable.

The awake mind may actually be an intrinsic state of consciousness that is modulated by input from the senses as they comb the environment. The brain exhibits activity regardless of whether the individual is awake or asleep, both of which are actually forms of conscious states. During rapid eye movement (REM) sleep (i.e., the deep sleep in which one is often dreaming), the threshold for awakening is highest; greater efforts are necessary to arouse the individual. In the state of REM sleep, the mind is tuning out sensory input from the external environment and concentrating on internal information such as memories. Once awake, sensory inputs actively modulate the brain's intrinsic conscious mechanism. The thalamus, having extensive connections with the cerebral cortex, may be the source of the mind's intrinsic conscious state, which can either continue without significant external sensory input, as during REM sleep, or be modified by the senses in the awakened state. The synchronous 40 Hz global cerebral oscillatory electrical activity apparently involves the thalamus as well. In fact, it has been suggested that some of the neurons of the thalamus may produce their own intrinsic 40 Hz oscillation, providing a degree of autonomy in establishing a thalamocortical consciousness center.

The different areas of the human brain involved with memory and cognition are highly interconnected. Sensory information is acquired and interpreted by several brain areas, reorganized into individual conceptual bits by the association areas, and delivered to memory pathways for storage and later retrieval. The initial formation of a memory occurs in the hippocampus and is later transferred to the cerebral cortex for long-term storage. A theory of visual recognition, which requires access to memory information, suggests that the human brain reduces objects it sees into a discrete set of shapes, perhaps as few as a couple dozen. According to the psychologist Irv Biederman, these simple shapes (circles, squares, cones, etc.), which he calls geons, are used by the brain to create a matching template of a visualized shape by combining them in various ways. In performing the recognition process, the individual geons may be altered in size, contour, and orientation to construct the brain's internal match of the observed object. Although geon constructions may apply to the recognition of geometric shapes, the ability to recognize a human face is apparently an altogether different matter. Studies of brain injury patients, as well as brain-imaging studies in healthy volunteers, imply that object recognition and face recognition are processed separately.

Information is continually streaming into memory from the senses as they examine the outside world. Basically, a witnessed event is converted to memory as the sensory information is stored in the respective sensory areas. Recall of a memory is organized by the hippocampus through reassembly of the various pieces. Psychological testing suggests that there are

214

several different formats of information storage in memory, including iconic (visual symbols), verbal/auditory, and conceptual algorithms. Furthermore, certain types of information may have specific areas within the brain where they are stored. In humans, functional MR brain imaging of facial images reveals strong activity in the lateral fusiform gyrus of the temporal lobe. Visual recognition of different objects (other than faces) evokes activity in different areas of the temporal lobes. Patterns of neuroanatomic specificity for memory formation have been demonstrated in nonhuman species. In monkeys, the inferior temporal sulcus region contains an area that is specifically concerned with recognition of faces. In lower animals, such as birds, certain brain areas are programmed to recognize a particular song. Although the exact numbers of neurons involved in visual or auditory recognition of specific items are not known, it apparently involves relatively small collections of neurons rather than global activation. Experiments with primates have shown that prior to an intended movement usually fewer than 100 neurons in the motor cortex will fire in a coordinated manner a few milliseconds before any muscle activity takes place.

Thoughts are basically combinations of discrete mental concepts retrieved from memory. The process of forming thoughts involves retrieving information from the long-term memory stores, which is then transferred to the short-term working memory of the prefrontal cortex, where it is available for correlation with inputs from the sensory areas of the brain. This process allows recognition of the sensory information, enabling familiarity with a visual image or the comprehension of verbal or written information. The cognitive disorder schizophrenia involves impairment of the normal thought processes through an internalization of reality, development of delusions, and withdrawal from interaction with the environment—the neurobiological defect of schizophrenia is believed to lie within the prefrontal cortex. Some of the neurons of the prefrontal cortex contain dopamine receptors on their membrane surfaces. These receptors permit modulation of the neuron activity by variations in the concentration of the dopamine neurotransmitter within synaptic clefts. Research on primates has shown the administration of antischizophrenic medications leads to alteration of the number of dopamine receptors of the neurons located within the prefrontal cortex, linking their association with schizophrenia.

Regarding sentience, the mind must construct a mental image in which the individual is included: self-awareness. The ability to function in a conscious state requires simultaneous activity of multiple cortical areas as well as activation of thalamic neuronal pathways, evidently collating both retrieved information from memory and constantly updated sensory input. The next level of function, behavior, is what determines how individuals will interact with the environment of which they are aware. Behavior (formulating and implementing intended activities) largely stems from frontal lobe processing. In fact, patients who have suffered frontal lobe injuries may remain completely conscious but lose their ability to initiate tasks or make decisions.

There are many interesting situations of central nervous system injury where conscious function is strained. Injury to the visual cortex of the occipital lobes can render individuals "cortically blind;" they have no conscious awareness of seeing, although the eyes and optic nerves still gather sight information. In fact, those suffering from cortical blindness may have a "subconscious" awareness of information from their visual system despite their inability to actually see anything. Studies of the optic pathways indicate that up to 10 percent of the visual input from the eyes is directed to areas of the brain other than the visual cortex. A much more devastating form of brain injury is the "locked-in syndrome," where the afflicted individual is unable to physically interact with the outside world. The locked-in syndrome may occur from a stroke or other form of injury to the brain stem (blocking the ability for the brain to send signals to the rest of the body) or from disease such as amyotrophic lateral sclerosis (the neurons to the body lose function as they whither away). The syndrome leaves one conscious but trapped inside his own mind, without the ability to interact with the environment. Among the recent innovative approaches to provide these patients with some ability to communicate is the neurosurgical placement of a "neurotrophic electrode" within the brain tissue of the motor cortex (an area that may be spared in these conditions), enabling direct electrical monitoring of the cortical neurons. Whenever we move a limb or initiate other muscular activities, the motor cortex is intentionally activated: locked-in syndrome patients are conscious and therefore able to purposely initiate activation of their motor cortex. Cortical activation produces electrical impulses in the motor cortex neurons that can be detected by the electrode, which in turn transmits a signal to a computer. Preliminary work has shown that patients can be trained to use their cortical activation to consciously "move" a computer screen's cursor to select displayed icons as a means of communication. A less invasive technique utilizes brainwaves detected by EEG electrodes that are applied to the scalp. With training, the patient can modify his EEG signal to permit communication through a computer intermediary. Thus a blending of neurobiology and technology provides these individuals with a window to communicate with the outside world. In another situation, groundbreaking work in prosthesis technology is enabling amputees to regain some rudimentary sensations. Sensors on the prosthesis that are activated will transmit an electrical signal to electrodes inserted into intact sensory nerve endings located near the amputation stump margin—activation of a prosthesis sensor results in a signal to the brain, giving the illusion that the prosthesis has sensation.

Beyond its cognitive capabilities, the mind can also manipulate the physiology of other organs in the body. The abilities of some individuals to consciously control their heart rate, defy lie detectors, etc., are well known. Individuals with psychosomatic disorders can develop physical symptoms under the influence of the mind or the brain's higher functions (emotions, ambitions, fears, etc.). Many conditions, such as asthma, eczema, or peptic ulcer disease, may become symptomatic when the individual comes under psychological or social stress. The placebo effect, thoroughly studied in medical science, exemplifies the ability of the mind to persuade one's consciousness (ergo, the body) of a fictitious benefit.

Typically the placebo effect occurs when patients receive what they think is a genuine medication, which unbeknownst to them lacks the pharmacologically active ingredient, and they actually derive or truly believe it provides a medicinal effect. In studies of patients with diseases including angina pectoris, asthma, and ulcers, up to 60–90 percent of those receiving placebos may derive clinically proven benefit. A study of patients with angina found that those taking a placebo had an average improvement of 10 percent in their exercise tolerance over those not taking medication, while those taking active medical treatment had a 22 percent improvement. In stroke patients, it was found that although those with severe neurological deficits had no significant benefit from placebo therapy, up to 50 percent of those who suffered mild neurological symptoms did improve to the same degree as those on standard medication. A controversial meta-analysis study of antidepressant drug studies suggested that placebo recipients derive up to 75 percent as much benefit as those taking clinically proven medications. In such cases of placebo benefit, there is an element of expectation (a benefit will come from the presumed "medication"), which influences the outcome perceived by the placebo recipient. The placebo effect appears to be regulated by positive expectations of the patient and care provider and is strengthened by a good relation between them.

Psychological mechanisms of the placebo recipient include the release and activation of endogenous opioids (endorphins) within the nervous system. The release of endorphins is known to affect pain tolerance, mood, and behavior. Interaction between the endogenous opioid system and other neurotransmitter systems may also contribute to the placebo effect. The clinically proven benefit of placebos cannot be ignored, and placebos have even been considered an inexpensive and safe form of treatment (for certain conditions). However, just as with most standard medications, placebos have also been associated with undesired side effects. The side effects suffered by placebo-receiving patients tend to parallel those of the substituted medication—consistent with a psychological expectation.

One of the defining elements of being "human" is the ability to comprehend and make judgments based on learned information: intelligence. The learned information (i.e., past experience) is retrieved by the mind from memory stores in the brain. Within the prefrontal cortex of the frontal lobes lie the various centers for a working memory. In fact, anatomic correlation with those primates most closely related to humans reveals that there is relatively more expansion (development) of the human prefrontal cortex than any other part of the brain.

There is debate as to when ancestral hominids became sentient (i.e., cognizant) of their own existence: self-awareness. Most experts consider the development of language as the major milestone required for a species to be intelligent; it implies sentience, self-awareness, and formulated thought. Impressions on the inner surface of *Homo habilis* skulls suggest that the language center, Broca's Area (found in the left frontal lobe of 97 percent of right-handed individuals and about 60 percent of those who are left-handed), was developing 1.9 million years ago. It is likely that the function of the cortex in Broca's Area was not originally involved with spoken language but perhaps other communication skills. Evidence for the origins of language in the archaeological record is circumstantial and subject to controversy. However, the appearance of meaningful images drawn by ancient man is considered strong evidence for language by some experts, which places the beginnings of language around forty thousand years ago. Some experts argue that participation in cooperative events (e.g., big-game hunting) during the late Pleistocene implies cognizance, as they require decision making and communication among members in a group. Another indirect measure of sentient behavior is the level of social structure in ancient societies as implied by the complexity of their architecture and craftsmanship. Some authorities argue that self-awareness of

218

ancient peoples is implied by the use of jewelry and ceremonial burial practices.

As computational technology progresses, there are ongoing attempts to create a Turing machine–based computer that can electronically mimic the conscious state. In 1950, Alan Turing, an English mathematician, developed the "imitation game" to ascertain if a machine was intelligent. Application of the game, which has become known as the Turing Test, requires the subject—the computer—be questioned for a while and if its responses are indistinguishable from human responses, it passes the test and therefore has a human mind. However, not all cognitive scientists consider the ability to carry on a conversation as the definition of human consciousness. Thus far, software development meant to mimic the synaptic interconnections of the human brain has failed to pass the test but has found applications in commercial markets. Such qualities as plasticity and dynamic behavior may provide an insurmountable barrier to the creation of a truly "conscious machine." Some have proposed a controversial theory where quantum effects play a crucial role in establishing the conscious state. Although computers are becoming ever faster, capable of performing billions of operations every second, and the human brain is relatively slow as it operates at just a few hundred cycles per second, the extensive parallel connections of the brain's neurons which allow rapid dissemination of signals are unlike the sequential arithmetic steps of typical computer algorithms. However, despite the complexity of the brain, the reduction of all the neural elements to a finite set of variables is not impossible in theory, and once it is established, the artificial emulation of the human mind may become a reality.

Sexual Dimorphism

In addition to the innate complexity of the human brain, the brains of the sexes are actually "wired" slightly differently, although they contain the same structures. At the beginning of the twentieth century, the first investigations into the differences between male and female brain development were a combination of scientific study and gender bias. Some investigators aimed to ascertain which sex was "superior" according to gender-specific characteristics of the brain that differed between the sexes. The claims of such investigators were inconclusive at best. The notion of a "superior" brain according to sex never gained valid scientific support and has long since fallen into disfavor.

Anatomical studies including both pathological specimens and imaging studies in living subjects have revealed a variety of gender-specific differences. Male brains tend to be larger and contain about 16 percent more neocortical neurons (23 billion in males versus 19 billion in females), the preoptic area of the hypothalamus is slightly larger in males, the neuron density in the posterior temporal cortex is greater in females, and there are differences in the shape of the corpus callosum between the sexes; the list goes on. Recent imaging studies in healthy volunteers reveals a relatively higher percentage of gray matter in females. However, the absolute amount of gray matter is higher in males, as their brains are larger on average. Gray matter is evenly distributed between the cerebral hemispheres in females, whereas in males there is an asymmetry of gray matter, which is slightly increased in the left hemisphere. However, despite the asymmetry in males, females still have relatively more gray matter in the left hemisphere. In males there is a corresponding higher percentage of white matter and CSF (cerebrospinal fluid, a clear fluid that surrounds the brain and fills its cavities, providing protection and nutritional support). There is suggestion that the relatively smaller crania and the relatively lower percentage of white matter in females may reflect a more efficient use of available white matter than in males.

Differences between male and female brains have been found not on only anatomical and physiological bases but also on behavioral levels. Men tend to have a cognitive advantage with tasks involving orientation of objects in space, while women are more adept at verbal and nonverbal communication comprehension. A relatively increased number of intrahemispheric neural connections in males, with relatively higher amounts of white matter (in males), may in part explain the better spatial orientation skills of males. The relative increase in gray matter of the left hemisphere (which contains the language center) in females may be related to their typically more astute language skills. Functional MR studies of verbal tasks suggest men have a lateralized response with activation of the left inferior frontal gyrus, while women showed activation of this region bilaterally. These variations may be attributed to more cross-connection between the cerebral hemispheres in women, which allows a more global input for comprehension. In the clinical setting, relatively increased interhemispheric and/or more efficient distribution of neural connections of females may substantiate their generally better recovery from unilateral (one-sided) strokes than found with similar lesions in men. In a stroke patient where a particular region of one hemisphere is damaged, substantial compensation may be achieved by the corresponding area of the undamaged hemisphere through interhemispheric neural connections—the corpus callosum, which is composed of connections between the cerebral hemispheres, tends to be relatively larger in females. The teleology of the structural differences between the sexes is unclear, but genetics and the different exposures to sex hormones of the *in utero* brain may contribute to the dimorphism. Studies on rodents have shown that sex hormones and some steroids affect neuroanatomical and biochemical properties of certain brain tissues. The morphological differences between males and females are relatively minor, and the differences between individuals, regardless of their sex, can overshadow these sex-linked subtleties.

In humans, neuroanatomical differences are not limited to sexual dimorphism but also include variations observed between heterosexual and homosexual males and between heterosexual and transsexual males. Some investigators suggest the neuroanatomical differences are related to behavior and are determined in early (embryonic or fetal) stages of development. The plasticity of the brain throughout life raises the possibility that such changes may not necessarily arise at the earliest developmental stages. In fact, transsexuals receiving cross-sex hormones may reverse their sexual roles, and sex offenders receiving anti-androgens may become more sexually passive.

Language and Behavior

It was early in the nineteenth century when the doctrine of phrenology came into vogue. Its premise was a presumed relation between the shape and contours of the skull and the degree of development of the underlying brain. Phrenology assumed that more-developed mental faculties were related to extra growth of the brain tissue responsible for that quality. Proponents believed that expanded areas of the brain, representing sites of advanced development, would produce matching contours on the overlying bone of the skull—the skull deformity was readily amenable to the touch of a skilled phrenologist's hand. An extensive map of skull contours was developed to assure consistent assignment of behavioral characteristics based on interpretation of a patient's scalp. In 1834, investigators eager to apply the technique exhumed the body of the notorious Marquis de Sade—a man repeatedly imprisoned for acts of debauchery and atheistic beliefs—for analysis. Despite the incongruent phrenological conclusions of a nearly perfect human skull with particularly prominent contours suggesting benevolence and religious faith, the discipline continued to grow in popularity. The theory eventually fell into disfavor as scientific investigations repeatedly failed to confirm a correlation between cranial contours and behavior. However, nineteenth-century naturalists were curious about other aspects of the human mind and behavior as well.

Alfred R. Wallace, contemporary of Charles Darwin, was a biologist credited as having independently discovered natural selection. Wallace constructed an interesting paradox of the developmental status of the human mind. He proposed that the capabilities of the human brain lie beyond that achieved by natural selection alone. Observations of contemporary primitive cultures suggested simple languages and lifestyles were sufficient for survival and were probably similar to those developed by ancient man, who was subject to natural selection on the evolutionary pathway to modern man. Wallace correctly surmised that the people of contemporary primitive cultures were anatomically equivalent to those living in "modern" industrialized societies of his day. Furthermore, he noted that people from primitive societies who were placed and educated in more advanced cultures could fully adapt to their new environment. Wallace went on to propose that this supposed excessive intellectual capability of man was the result of intervention by a "superior intelligence" for some "special purpose." To counter Wallace's paradox, evolutionary biology theory suggests that the intrinsic complexity of the brain enables it to perform functions outside perceived evolutionary constraints. The origins of language and complex behavior represent similar processes. The levels of advancement in knowledge and the technological success modern man enjoys may seem beyond the intellectual capabilities of an ancient human, but our society has preserved vast collections of knowledge and adopted extensive educational programs, which have allowed our cultural progress. One must remember that human knowledge is the result of human understanding and discovery: placing it within that context makes the argument of cultural aspects lying beyond the natural capability of the human mind less clear.

The road to human dominance over other primates relied heavily on the capability of inventive behavior. Studies suggest that a significant por-

222

tion (possibly the majority) of the features of human behavior originates from hereditary (genetic) factors. The archaeological record suggests inventiveness (e.g., elaborate tools and artwork) was present at least 50,000 years ago. In addition, the development of spoken languages may have occurred at a similar time. It is perhaps the codevelopment of language and resourceful inventive behavior that gave man undisputed dominance over his fellow creatures.

The rise of human intelligence incorporates a variety of cognitive functions requiring sufficient development of different areas of the brain. The elements of memory and learning are critical in forming intelligent be havior. Behavioral genetics studies of human intelligence, based on IQ scores, suggest that intelligence is relatively stable throughout an individual's life and is about half inherited (genetically determined). Many physical qualities, such as height and weight, are closer to 90 percent inherited. Studies of rodent learning models indicate that enhanced function of the N-methyl D-aspartate (NMDA) receptor—triggered by glutamate released by nearby neurons in signal transmission—heightens function of memory-forming neurons found in the hippocampus. Genetic manipulation to enhance expression of the NMDA receptor produces mice that exhibit more rapid learning capabilities, suggesting a relation between genetics and cognitive function.

It is generally believed that through learning synaptic connections are reinforced and new ones created. Learning implies a change must be made to store the new information (for later retrieval), which is believed to occur at the level of the synapse. Hebb's rule states that a synaptic connection may be strengthened when the involved neurons are activated at the same time. A form of synaptic plasticity, called long-term potentiation (LTP), occurs in many parts of the brain, including the hippocampus (involved with memory formation). In the case of the transgenic mice with enhanced NMDA receptor expression, they also express LTP phenomenon, assumed to be related with improved memory function. Investigations of rodent behavior have shown altered activity of neuron groups in the cerebral cortex. It was found that as a motor skill was learned, neuron ensembles—groups of about two dozen neurons—in the motor cortex expressed changes in their average rate of signal firing, pattern of signal firing over time, and developed coordinated signal firing of the neurons within an ensemble.

As enhanced function of synaptic transmission is related to memory formation, the loss of (former) synaptic transmission implies loss of stored information. During the normal aging process of humans, about 25 percent of the synapses deteriorate between the ages of twenty-five and fifty-five years—as does the ability to recall memories. Memory loss begins at about the age of 30 years in normal individuals. Much pharmacological research is dedicated to preventing age-related cognitive decline. Various treatments of questionable efficacy have been promoted: megadoses of antioxidants (vitamins C and E), flavonoids (pigments with antioxidant activity found in blueberries), ginkgo biloba extract (may increase blood flow in some tissues), various anti-inflammatory medications (may reduce risk of developing Alzheimer's disease), lecithin, vitamin B_{12}, folic acid, and aerobic exercise. Medications designed to block the degradation of the neurotransmitter acetylcholine show evidence of slowing the cognitive decline in Alzheimer's disease.

One of the defining characteristics of the human mind is the ability to communicate through verbal language. The development of the capacity for language during hominid evolution has an unclear history. Arguments continue as to whether it was a relatively recent development of modern humans or had its origin in more distant ancestral hominids. Creatures using language must fulfill certain minimum requirements: sufficient cognitive capabilities and appropriate anatomical structures. The ability to communicate requires a relatively advanced central nervous system capable of the cognitive and comprehensive skills necessary for language. In addition, there are also functional anatomical components for the delivery of spoken language through precisely controlled movements and articulations of the face, tongue, neck, and airway. Through meticulous examination of the remains of ancient hominids, anthropologists attempt to identify key anatomic structures that may be directly related to the capability of language. Much of that work has focused on the hominid skull.

Major characteristics used in determining ancestral relationships in human evolutionary biology are the size and configuration of the brain. Although the tissues of the nervous system do not survive fossilization, the size of the brain that once occupied a skull can be estimated easily: material may be poured into the empty cranial vault to estimate its volume. Historically, water displacement and millet seed techniques have provided braincase volume estimates—millet seeds may be used to fill the skull, then re-collected and their total volume measured to determine the cranial vault volume. Such techniques provide an estimate of the brain volume but say nothing about its shape (morphology).

The majority of the space in the cranial vault is occupied by the brain such that the inner contours of the skull roughly form a mirror image cast of the surface of the brain it once contained. Some investigators claim contours found on the inner layer of some hominid skulls indicate development of the speech centers of the association cortex—complex higher functions of the brain are coordinated through multiple regions of the cortex known as "association areas." However, examination of the inner skull in many fossilized specimens is difficult, as it is often hidden from view. An artificial representation of the cranial vault is an "endocast"—in turn, the endocast is a volume re-creation of the intracranial contents, chiefly the brain (as the majority of the cranial vault is filled by the brain). CT scan imaging technology can produce virtual endocasts without damaging the fossil specimens. The accuracy of previously accepted methods of fossilized hominid brain volume determination is being challenged with the development of virtual endocast imaging. The computer-generated endocasts accurately determine the volume and shape of the cranial vault, and the imaging data is amenable to further processing by investigators. In addition, such data can be easily shared for research without damaging the artifacts, as direct handling of the specimens runs the risk of irreparable damage. Many archaeological specimens are incomplete or were damaged prior to or during the fossilization process. In some cases, missing portions of specimens can be "replaced" in the virtual rendering. On the assumption of symmetry, the computer models can easily "fill in" the missing portions—providing there are mirror-image remnants of the other side of the body to use as templates. Despite these advanced techniques, the task of reconstructing the ancestral hominid brain leaves much of the picture incomplete.

In addition to the brain, other organs are involved in the process of verbal language. The ability to speak requires phonation, which utilizes the larynx (voice box), where air from the lungs is channeled to create the range of sounds used in speech. Investigations of the evolutionary development of the throat indicate the necessary changes may have taken place as long as 500,000 years ago. Another feature of the hominid anatomy that has been investigated is the hypoglossal canal of the skull base, through which the hypoglossal nerve of the tongue passes. Certain investigators propose that the (cross-sectional) size of the hypoglossal canal is relevant because highly coordinated tongue motion is required to articulate effective speech, which in turn requires sufficient neural connections via a correspondingly large nerve. It was found that the nerve's canal is relatively larger in humans than in other primates, and it was enlarged in hominid ancestors dating back as far as 400,000 years ago. Study of the remains of several *Australopithecus africanus* specimens indicates the hypoglossal canal was less prominent than in more recent hominid forms, which may represent limited spoken language capabilities. However, some investigators have found that the human hypoglossal canal–to–oral cavity ratio is exceeded in speechless species, including some prehuman australopithecines, as well as other contemporary primates. Cadaver studies raise questions as to the validity of assuming a direct relation between the canal cross-sectional area of the hypoglossal canal and size of the hypoglossal nerve that traverses it. Some of the disagreement of the hypoglossal canal ratio measurements between studies may reflect a wide range of interspeciate variability; resolving the discrepancy may require examination of larger numbers of specimens to favor one conclusion over the other.

Although controversy persists, review of the hominid fossil record suggests that the capability of speech may have arisen between 400,000 and 500,000 years ago. Immediately preceding and during this period, the hominid brain underwent significant proportionate increase in size. It has been proposed that the increase in brain size reflected evolutionary pressures favoring intelligence and skills of manual dexterity. Perhaps the ability to conceptualize language became possible through the more advanced capabilities of a larger brain as an indirect consequence of unrelated evolutionary achievements. The cognitive and motor skills required for language utilize many areas of the human brain such that the capability of language required adequate development of multiple areas of the brain, consistent with a global increase in brain size as occurred in the fossil record.

Under the pressures of natural selection, the human mind (along with the body) was altered over generations. Behavioral and physical characteristics play critical roles in survival and are therefore subject to selection pressures. According to some cognitive scientists, the appropriate structures of the brain required for language were gradually developed during the course of evolution. It is improbable that Broca's and Wernicke's Areas, along with other regions involved with language and communication, were the result of a predetermined plan, but rather they had a different original function, which was later readapted for language. Study of endocasts in the areas responsible for communication of australopithecines indicates a pongid (apelike) configuration, while those of habilines (*Homo habilis*) are more humanlike. Thus, with the first *Homo* lineage, the neuroanatomical structures for language may have been present but not necessarily used for verbal communication. In addition, the ability to use language is a highly specialized form of communication, which probably was continually refined under natural selection. The strong association of "handedness" (hemisphere dominance) with the speech areas has suggested a possible link whereby these areas may have been previously involved with function of the hand.

The brain receives information about the external environment through the five major senses: vision, hearing, touch, taste, and smell. Functional anatomical studies of the neocortex have revealed distinctive topographical maps of the senses; the sense of smell is an exception, as it is organized differently in the paleocortex. These sensory maps not only are present in humans but also exist in much more primitive mammals, suggesting these maps represent an early evolutionary manifestation in mammals. The closest species to humans is the chimpanzee, which shares a 99 percent genetic concordance with man. Among the various types of chimpanzees, the pygmy (bonobo) chimpanzee displays some of the most similar behavioral characteristics to humans. Chimpanzees have been taught to use sign language and other systems of pictorial communications to exchange information and can perform certain complex tasks that are requested through language. Controversy exists in the interpretation of psychological studies (which date back to the 1960s) of the cognitive abilities of chimpanzees: skeptics argue that the communications initiated by chimps do not represent cognitive communication but are more akin to a meaningless list of words.

Comparative anatomical studies between humans and other primates have revealed some brain structures that are shared. Of particular interest is similarity found in the region of the auditory association cortex in the temporal lobe. In humans this area has a slight asymmetry between its right and left counterparts—believed to reflect development for specialized function. A similar asymmetry has been found in autopsy studies of chimpanzees and in MRI scans of living specimens. Furthermore, the association cortex asymmetry is not apparent in monkeys (which are less closely related to humans than are chimpanzees), implying that it is not a feature common to all primate brains. If the anatomical correlation truly reflects a language feature common to chimpanzees and humans, this structure was probably present in the common ancestor millions of years ago. Therefore, the capability for primitive language skills in the ancestral hominids is not unreasonable. However, the role of these association cortices may not have been dedicated to verbal language in ancestral species; it may have been involved with other forms of communicative skills. As humans and

227

chimpanzees diversified through evolutionary isolation, the auditory association cortex of the two different species may have persisted but developed different capabilities. In the chimpanzee, the function of the auditory-association cortex may have become devoted to the interpretation of their own form of communication, i.e., verbal calls, gestures, and expressions.

Hints of a relationship between neuroanatomic structure and sites where specific elements of behavior originate may be found in patients who have suffered a brain injury. Studies of brain injury patients indicate that damage to the prefrontal cortex alters judgment, including social conduct, decision making, and the ability to estimate quantities. Injury to Broca's Area of the dominant cerebral hemisphere's inferior frontal gyrus affects language production, while an injury to the nearby posterior temporal lobe in Wernicke's area affects language comprehension. Although certain areas of cortex have specific roles, language is a complex skill that requires activity from multiple centers in the brain—ranging from identification of a written character by the visual system to the comprehension and generation of language through activity of other areas. Intraoperative studies on humans using direct electrical stimulation of the language association cortex have shown some variability in the exact anatomic locations of these areas between different individuals. Despite individual variability in the location of specific functional sites within the association cortex, each individual still has cortical representation of all of the sites. It has been suggested that the variable locations of these areas among individuals may reflect a relatively recent origin of the language association cortex.

Behavior has both determined and malleable components. The determined component is biological (including *instinct*) and is largely a consequence of genetic determinants. The genetic contribution to behavior is difficult to quantify but is certainly polygenic, requiring hundreds to thousands of genes. While in some diseases a single gene defect can be cited as the culprit (i.e., sickle cell disease), behavior is a complex quality that has multiple contributing factors. Genes code for proteins, some of which may affect neurological processes, but a gene doesn't necessarily code for a specific behavior. Genetic contribution has been associated with some diseases having behavioral components, such as Alzheimer's disease and schizophrenia. In the case of schizophrenia, if genetic predisposition is present, the individual has approximately a 50 percent chance of developing the syndrome—indicating that other factors, such as environmental stimuli, are necessary to trigger expression of the disease.

228

The malleable component of behavior is subject to influence through environmental exposure. Classic experiments on humans have shown that fear can be a learned response to previously nonthreatening stimuli simply by association with stimuli that typically induce fear. Thus the environment can manipulate behavior and thereby affect the expression of some behavior-associated genes. From studies in psychology, experiences at an early age can affect behavioral traits much later in life; e.g., the lack of adequate nurturing in an infant may lead to an inability to form normal social relationships as an adult. Studies of twins suggest that about half (30 to 70 percent) of personality traits are a result of hereditary (genetic) factors, leaving the rest to the social environment. It is estimated that only about 5 percent of personality differences are related to being raised by one family versus another (assuming exposure to a relatively "normal" environment with sufficient nurturing and appropriate care). The determinants of the remainder of personality, approximately 50 percent, are unclear and may include brain development prior to birth, exposure to a variety of experiences, and social interactions throughout life.

The appreciation of physical attractiveness has both genetic and malleable components. Although many social scientists argue that what constitutes beauty is shaped by the advertising industry, there are instinctive levels of appreciation. The basic components of appreciation include such things as body symmetry and proportion. In seeking out a mate, those who meet such basic characteristics should be less likely to have certain health problems or diseases—such unfavorable qualities may make an individual less successful in terms of future offspring and stability as a long-term partner. Those who carry instinctive (genetically based) appreciation for an "attractive" mate (i.e., implying a mate that is more likely to be healthy) will be at a reproductive advantage, as will their progeny, assuming they inherit the same behavior in choosing a mate. Studies suggest that attractive qualities are cross-cultural. Particular qualities favor certain advantages; women with full lips and short, narrow jaws tend to have a relatively fertile hormone balance. However, there are certainly cultural influences that attempt to define beauty.

The degree of expression of "deterministic" genetic traits can be influenced by the environment. In developed nations, height is considered 90 percent inherited. Historically, the heritable contribution to height was reduced because populations often experienced periods of poor nutrition, which did not permit expression of the trait to its fullest potential. Behavioral traits are also under some degree of control by the environment, where expression can be altered or suppressed in response to the individual's experience. Interestingly, a recent (and controversial) claim is that the trait of homosexual behavior is a composite of a series of genes on the X chromosome. Studies of homosexuality have suggested a 50 percent heritable component, but a recent study of Australian identical twins reveals only a 20 to 24 percent inheritable element. Investigations of adopted children and identical twins raised separately suggest that the social environment of the family in which one matures provides only a limited contribution toward future behavioral development. Apparently, as long as a family offers at least a level of minimal support, the behavioral outcome of their offspring will be predominantly independent of the social environment.

The genetic contribution to behavior is indirect, as it is actually made through intermediaries: proteins and other biochemicals. Certain hormones, a class of proteins, affect various parts of the nervous system associated with specific behaviors or emotions. In situations that create anxiety or depression, those individuals more prone to a certain behavior may have a lower threshold to an environmental stressor that incites the particular behavior through abnormal levels of a key protein or biochemical. However, biochemical influences generally create behavioral trends rather than absolute behavioral responses.

The activity of the brain is much more than just a consequence of the complex connection pattern of its neurons; it is also strongly influenced by biochemical agents. The cells of the nervous system, as well as the rest of the body, are bathed in a biochemical environment, which must be constrained to a narrow range for normal function. Alterations of the biochemical environment through either endogenous or exogenous means can upset the natural balance and manifest as abnormal function. Behavior may be affected through altered biochemistry in the tissues of the central nervous system. From neuroanatomic manifestations of behavior, the malfunction of neurons in specific areas of the brain produces the behavioral symptoms. In some types of schizophrenia (e.g., where aberrant behavior is present), patients may get relief from their psychosis with medical management that alters the brain's biochemistry. One agent of treatment for schizophrenic patients is chlorpromazine, a medication that affects the levels of the neurotransmitter dopamine in the brain.

Chemical substance abuse has plagued man since ancient times. Many ingested chemicals are known to affect the function of various regions of the brain. Many of these chemicals are natural agents derived from plants—the creation of synthetic agents is a relatively recent practice. Recent investigations of addictive behavior have brought the role of the neurotransmitter dopamine to center stage. Dopamine has the effect of regulating pleasure through its interaction in the limbic system, particularly in the area of the nucleus accumbens. There are natural levels of stimulation induced by "routine" pleasing events, e.g., the taste of a favorite food. A multitude of drugs (cocaine, heroin, nicotine, etc.) are known to artificially activate the same regions of the brain, but with much more rapid and intense results. Over extended periods of exposure to these agents, the brain loses the ability to produce its own dopamine because the stimulus for its production has been replaced by the exogenous source (i.e., the drug). After chronic use of the stimulant methamphetamine hydrochloride (which raises the brain's levels of dopamine and norepinephrine), abrupt discontinuation of the stimulating agent leaves the natural dopamine levels deficient; the individual is void of the normal sensation of pleasure, leaving him feeling miserable—part of the withdrawal syndrome. In order to feel "normal," the individual must resume using the stimulant, which leads to addictive behavior. There is some evidence that recreational drug addicts with naturally decreased levels of dopamine tend to be depressed and the circumstance makes them more susceptible to drug usage as an emotionally satisfying means of escape. Other neurotransmitters with behavioral associations include serotonin and acetylcholine.

The roles of genetics and environment have been an area of intensive study in the investigation of the causes of alcoholism. Its long history, prevalence, and destructive nature make alcoholism a focus of much attention in hopes of deriving effective means of treatment and prevention. Research has shown that there are risk factors that point to both genetic and environmental influences. The mere fact that alcoholism runs in families suggests a genetic predisposition. Furthermore, studies of separated identical twins, where one twin is an alcoholic, found an increased risk of alcoholism in the other twin. Children of alcoholics, when adopted by nonalcoholic families, maintain an increased risk of developing alcoholism. Individuals raised in an environment where alcohol is consumed in excess will be at an increased risk for developing alcoholism. However, a more objective criterion has been identified: individuals with a higher level of tolerance to alcohol were found to be more likely to become alcoholics than those who are more easily affected by the drug. Although extensive analyses of the data points to a nearly equal contribution, there may be a slight predominance of environmental over genetic influence. One must recognize that there are many variables that confound research regarding behavior, but as with many behavioral traits, there seems to be significant contribution from both genetic and environmental factors in the development of alcoholism.

Behavior clearly plays a major role in survival. Instinctive behavior for self-preservation is essential for a species to continue. Intimately associated with self-preservation is the fear response, which is ingrained at multiple levels. Activities resulting in painful stimuli are feared, which minimizes future exposure through memory of the unpleasant event. Furthermore, the nervous system has developed a "fight or flight" mechanism where the sensation of fear or danger activates a specialized part of the nervous system, the sympathetic nervous system, which has direct connections to various organs throughout the body and also liberates biochemicals (chiefly norepinephrine). Activation of the sympathetic nervous system rapidly affects the body's physiology in a variety of ways, increasing heart rate, blood flow to the muscles, and blood pressure, overall producing a state of anxiety—the individual becomes alert and heightened physically, in preparation for either confrontation or escape.

The expression of aggression (or violent behavior) likely has its origin teleologically as an instrument of survival. Such behavior, particularly in a primitive setting, can be used as an agent to secure essentials from others as well as to ward off danger. The root of aggression appears to be buried in the substance of the frontal and parietal lobes of the brain's cerebral hemispheres. In addition, abnormally elevated levels of serotonin and dopamine are associated with violent behavior; research with monkeys has shown that increased levels of serotonin are associated with more aggressive behavior patterns. In humans, a study of college males revealed an average of 25 percent higher serotonin levels in those occupying a position of authority over their peers. Elevated serotonin levels can induce other mood disturbances, including compulsive behavior disorders. Some dietary pills will artificially elevate serotonin levels in the brain to create a feeling of satiety, which reduces the appetite. Decreased serotonin levels can also affect mood, where it has been implicated in depression, suicidal tendencies, premenstrual syndrome, and affective disorders. However, serotonin does not create aggressive behavior but rather activates biochemical restraint mechanisms, which, at low levels of serotonin, are minimized or ineffective.

The instinctive drive of aggressive behavior is one of the major elements of human behavior. In ancient times, aggressive behavior was a ne-

232

cessity for survival, and its instinctive nature remains imprinted on the human psyche. However, aggression is not necessarily expressed in only violent terms. In contemporary civilized society, nondestructive release of aggressive behavior may be manifested through various competitive physical activities (e.g., organized sports) or through technological means such as immersing oneself in violent cinema or engaging in challenging video games.

Senescence

The integrity of the highly organized and complex physiology and biochemistry of life inevitably "fails" the organism. In the case of humans, the body is at its peak of health at about the age of eleven years, when the effects of aging have not yet taken any significant effect. In fact, from the age of eleven years onward, the risk of death doubles about every eight years. The average human life span has significantly increased in the last century through improved hygiene, nutrition, and medical care. At the beginning of the twentieth century, the average life span in developed nations was forty-six years; a century later, that number increased to nearly eighty years.

The implications of the end of corporeal existence have intrigued mankind for hundreds of centuries. It was difficult to accept that the arrival of death marked an absolute endpoint of existence, and therefore various forms of eternal destinations were proposed. A hundred thousand years ago, stone age people would paint and bury their deceased with ornaments for use in an afterlife. Such practices continued into civilized times. Hadrian, who ruled the Roman Empire during the second century A.D., built an extensive villa at Tivoli that included an amphitheater and subterranean system of tunnels (the "underground") where spectators could experience death and the afterlife as a means of preparation for the actual event. The fascination continues into modern times in the realms of religion, cults, science fiction, and medical pseudo-science.

Research suggests that humans manifest the characteristics of aging through lack of self-repair of the body's tissues. Many of the tissues of the body are constantly being repaired or replenished throughout life. The gradual decline in these processes allows the body's cells and tissues to degrade. The dedication of an organism to maintaining self-repair has a

233

heavy cost, as it requires additional physiology and energy to carry out the process. During life there are other "responsibilities" one's body assumes that are considered vital and therefore divert resources, e.g., maintenance of normal metabolism and physiology, childbearing, physical exertion, etc. The mechanisms behind the process of tissue self-repair are not entirely clear, but genetic influences must play a role. Observation of elderly people indicates that individuals manifest the changes of aging differently, and it has been shown that longevity often has a familial association. It is likely that the genetic influence on aging involves hundreds of genes, but there may be a small group of genes that actually have the broadest influence. A study of the DNA of 137 families with some siblings aged beyond ninety-one years, suggests a region on chromosome 4 may harbor a longevity gene (or genes). Stemming from such genetic research, there are predictions that through genetic manipulation the average human life span could be tripled. However, experts believe that only about half of the differences in aging between individuals are related to genetic influences, the balance made by environmental factors such as diet, lifestyle, and certain vices.

There are a variety of specific processes thought to be related to the effects of aging. As cells go about their metabolic business, their biochemistry occasionally produces oxygen-free radicals, which are highly reactive chemical groups. These toxic radicals can chemically injure the structures of a cell, which in turn degrades the cell's ability to function normally. Biochemical treatment with antioxidants can neutralize these radicals, converting them into harmless molecules. Protection against free radical damage to a cell's lipid membrane components is offered by the presence of the fatty soluble antioxidant α-tocopherol (vitamin E) and protection of the cell interior by the antioxidant properties of ascorbic acid (vitamin C). Another form of cell damage arises with increased levels of glucose, the six-carbon sugar that provides us with an instant source of energy. Glucose will form chemical cross-links between certain proteins in the body, which can disturb the normal function of certain tissues, such as blood vessels. The manifestations of glucose damage are found in diabetics, where many tissues of the body are prematurely injured.

A common manifestation of aging is a relative increase in body fat and atrophy of the muscles. During aging, the levels of growth hormone manufactured and released by the anterior pituitary gland decrease. Studies have shown that individuals receiving growth hormone injections undergo some reversal of the body fat to muscle mass ratio. However, its

234

actual effect on the process of tissue aging has yet to be established. The levels of growth hormone are critical, particularly in a growing individual, as low levels lead to dwarfism while elevated levels cause gigantism; both conditions are associated with relatively shortened life spans.

As few things in life seem predictable, the undeniable certainty of death makes a grim exception. The life tables generated by insurance companies are used to make accurate predictions of a client's life expectancy. Although the time of death for a particular individual may be stochastic, as the population sample size increases the life table predictions of the sample become increasingly precise. What is it about the human body that determines the appropriate time to expire? There are general characteristics associated with the aging process—graying and thinning of the hair, decrease in body water content, laxity of the skin, etc.—that in themselves are not lethal. Further insight may be obtained by investigation of the life cycle of the individual cells.

The human body contains about 100 trillion cells, with each cell having an intrinsic life span. The reason for a deterministic period of healthy function is unclear, but it may be a form of protection from developing cancerous (abnormal) cells. In the laboratory, cells from a living person may be grown artificially and sustained in petri dish colonies. Cultures of healthy cells, under ideal conditions, will continue to grow and divide for about fifty generations—then the process comes to an abrupt halt. Furthermore, cell specimens obtained from older individuals produce fewer generations of cell division before they cease to divide. These experiments imply that the mechanism for programmed cell death is not dependent upon chronological or metabolic time but depends on the number of divisions (mitoses)—implying an internal "mitotic clock." In addition, as people age, their cells tend to divide more slowly, which may contribute to the coincident progressive deterioration of organ function that is typically associated with the aging process.

Although growing cells may stop dividing after a certain number of generations, they do not necessarily make a final division and immediately call it quits. Upon their final division, many cells continue to live without dividing, having become "senescent." Typically senescent cells become larger than their forebears and can assume distorted abnormal shapes. There is further evidence that these cells may produce toxic substances as their function gradually declines. These toxic substances can leak out of an aging cell to injure nearby healthy cells, and may contribute to the changes tissues undergo that we commonly associate with the aging process. In fact, there are indeed an increased number of senescent cells found in the tissues of older individuals. Thus if the deterioration of the senescent cells can be controlled (e.g., curbing the release or noxious effects of liberated toxins), the manifestations of aging may be controlled as well.

The process of programmed cell death may in fact be a result of injury to the cell's DNA over its lifetime. In aging cells, the repair of defective DNA is less effective and leads to altered protein synthesis, with consequential declining function. It has been shown that the telomeric region found at the ends of chromosomes may determine when a growing cell stops dividing. In human germ cells, the telomeric region of the DNA consists of tandem repeats of a specific sequence that is about fifteen thousand base pairs long. Each time a somatic cell divides, the telomeres are shortened by about one hundred base pairs. After multiple cycles of divisions, the telomeres are nibbled down to stubs when the process of division ceases—the cycle of telomeric reduction may be reversible.

In the interest of longevity, experiments have shown that adding an enzyme, telomerase, which actively rebuilds the shortened telomeres, leads to an increased life span of the cell. In fact, some stem cell lines and germ cells naturally contain their own telomerase, and many of these cells have extremely long or indefinite life spans. Application of these genetic modifications to multicellular organisms is potentially disastrous—cell longevity is one of the key factors in the formation of cancerous cells. However, the presence of telomerase is not synonymous with cancer, as some normal cells naturally carry and express the enzyme. In addition, ongoing research indicates that genes themselves may produce proteins affecting the process of cellular longevity in ways other than telomerase.

The brain is not immune to the effects of aging. At approximately the age of five years, the human brain has its maximum number of neurons. As we age, there is a gradual loss of neurons in the brain—thousands each day in healthy individuals. Normal aging will bring a relatively increased loss of gray matter to white matter and a generally more pronounced loss of brain tissue in males. There are a variety of pathological processes that will cause accelerated death of neurons, including Alzheimer's disease. Although there is a long list of substances known to be toxic to neurons, it has been shown that merely the effects of stress can be injurious to the brain. Other substances can be neuroprotective, including the hormone estrogen. Some investigators have found that growth hormone may preserve the function of neurons, and studies are under way to determine if injections of the hormone will have a real benefit. Medical technology is working on ways to repair or replace brain cells damaged by Alzheimer's disease, or other diseases, by placing healthy cells in the damaged tissues. Preliminary research utilizes fetal and stem cell lines, which are cells that are "undifferentiated" and have potential to produce offspring cells with specific characteristics (e.g., a neuron) able to replace lost cells.

From an evolutionary standpoint, extended longevity of the individual is not necessarily a goal of natural selection. The process of species propagation requires individuals to survive to reproduce—life of the parent beyond the production of viable offspring is less directly associated with their genetic contribution to future generations. Certainly, in the short term, the survival of the parent is beneficial primarily through nurture and protection. More abstract long-period benefits may include education of offspring to improve their chances of survival. In fact, primates require extended nurturing, as their offspring are born in an immature state. In situations where parental support is effective during development of their progeny, and assuming these traits have a genetic component, these qualities would tend to be passed on to future generations. However, some biologists propose that in aging beyond the period of reproduction the individual enters a phase outside the constraints of natural selection, where bodily deterioration and senility may be the consequences of evolutionary anarchy.

For thousands of years, there has been interest in forms of existence that follow death of the body. A hundred thousand years ago, Neanderthals buried their dead with items for use in an afterlife. Many tribal and modern religions celebrate an afterlife where the deceased enjoy a blissful existence in an alternative reality. Beyond archaic dogmatic accounts, the majority of the information gathered in recent times comes from individuals who have been considered clinically dead but were successfully resuscitated and claim to have had a recollection of a "near-death" experience during their breach with their own mortality. The first recorded near-death experience was documented by the ancient philosopher Plato when he described a soldier who awoke on his funeral pyre—the soldier remembered leaving his body and was confronted by a bright figure. More recently, a modern-day European study of patients who were resuscitated from a cardiac arrest found that about 18 percent described a near-death experience. A recent study in London investigating thousands of near-death experience cases found that about 30 percent had a relatively pleasant experience while about 4 percent found it profoundly unpleasant. Those claiming a near-death experience typically describe leaving their body to float

237

through a long dark tunnel that has a bright light or beautiful landscape at the other end. During their journey, they may encounter people or a spiritual figure and eventually reach a point where a decision to continue on the journey (i.e., to die) or return to their body (i.e., continue living) is encountered.

One must keep in mind that during a near-death experience the brain is typically deprived of its nutrition and normal sensory input. Under these conditions, the brain may enter a period of dissociation where it can internally activate various areas, bringing up memories and creating sensations; the individual may sense he is having an experience that is actually independent of his surroundings. This fabrication may be constructed from subconscious desires and expectations, creating an experience that may conform to what is anticipated. The argument is not airtight, as some who have had a typical near-death experience state no previous belief in an afterlife. A study conducted in the United States of children resuscitated from clinical death found that about a third described a near-death experience. In some of these cases, the children's descriptions of their experiences seem too detailed to simply write off as recollections of ingrained teachings or expectations. Prospective investigations have been designed where specific objects (e.g., a sign displaying certain words) are strategically placed in hospital rooms where a patient may have a near-death experience. The test object is placed out of view, requiring an elevated vantage point for its detection—near-death experience patients who claim they were outside their body looking down into the room have failed to accurately describe these test objects. Many less innocuous conditions have been known to create similar sensations as those described in near-death experiences: severe emotional trauma, loss of consciousness, hallucinogenic drugs, etc.

The Ideal Human

The desire to accurately predict (or control) the characteristics of unborn offspring has been around throughout civilized human history. One aspect of particular interest has been the gender of an expected child. Historically, most cultures feature some form of a patriarchy, where social influence and inheritance are along male lines, thus giving male offspring advantages in terms of social status. The men of ancient Greece would perform sex lying on their right side to improve the odds that a male child would result. Although the practice of superstitious methods has been historically unreliable, solace may be found in modern technology that is introducing techniques that significantly reduce the random nature of gender selection. One recent method borrows from techniques used in animal breeding, where it has been shown that when sperm are doused with a fluorescent dye, exposure to laser light will provide an estimate of the amount of DNA they contain. Since a sperm carrying a Y chromosome (destined to produce a male) has about 2.8 percent less DNA than one carrying an X chromosome, a distinction between the two sperm types is possible. Once a particular sperm's genotype is identified, the sperm can be isolated, collected, and stored. A female artificially inseminated with a sample containing a high concentration of one sperm genotype will have the gender of her offspring influenced accordingly. The relative simplicity of the technology may make gender selection affordable for broad usage by the mainstream population. However, as a form of genetic manipulation combined with the potential for widespread utilization, ethical concerns are raised. In the United States, many consider the ideal family as one that has one child of each sex, implying that given the choice, an overall even balance of males and females would be maintained. In families that contain known genetic diseases that are expressed in a particular sex (sex-linked), the ability to choose gender can directly determine survival of the child.

In the early twentieth century, the notion that human characteristics were purely a product of genetics became increasingly popular. Proponents argued that most qualities are inherited from the parents and previous generations. This theory was put into practice as the fashionable eugenics movement. The term *eugenics* was coined by the nineteenth-century anthropologist Francis Galton, a cousin of Charles Darwin. Galton was interested in analyzing the impact of genetics on human development through studying twins. He concluded that genetic inheritance overwhelmed the effects of social influences in determining the qualities of an individual. In fact, he recommended "breeding quotas" as a means to avoid adding individuals considered "genetically inferior" to the population.

The motivation behind the eugenics movement was to improve the quality of the human race through selective breeding. In the United States in the early twentieth century, the movement gave recognition to families who displayed the qualities of "superiority," with the hopes of creating a better working class. Resentment in the mainstream population had been growing with the constant inflow of immigrants, which was felt to contribute to a decline in genetic quality through introduction of an inferior gene pool to the North American population. Also during the early twentieth century, but on the other side of the Atlantic, the growing Nazi party in Germany provided financial incentives to "healthy" families to produce more children. Individuals considered as having an inferior genetic constitution were subject to sterilization, segregation, or extermination. After World War II, as a philosophy of optimistic social engineering prevailed, opinion swayed, giving more attention to environmental influences rather than purely genetic factors. The changing political climate brought reforms directed toward improving the social environmental condition of the population.

It is unfortunate that despite these changes in the global attitude, ethnic purification continues to be a priority in some political regimes. The mass executions of 300,000 Chinese civilians and military personnel over a six-week period during the Japanese campaign against Nanking in December 1937 were horrible war crimes carried out by thousands of soldiers and their officers. Although some historians claim the atrocity was an act of revenge for Japanese losses incurred from fighting at Shanghai earlier that year, most consider the slaughter was aimed at eliminating a people and culture deemed inferior. Perhaps the most extensive such campaign since the industrialized death camps of Nazi Germany was that of Cambodia, where between 1975 and 1979 up to 2 million people were exterminated in the name of "ethnic cleansing" under the reign of terror of the Khmer Rouge. Pol Pot's regime systematically exterminated nearly one-fourth of Cambodia's population as a means to replace the modern culture with an idealized agrarian society to restore former glory the nation had enjoyed several centuries ago. In some cases, the lack of stable governmental control allows subdued ethnic hatred to erupt. During the decline of Soviet control in 1991, ethnic conflict resumed in the former Yugoslav states, which led to over a hundred thousand deaths in the years that followed. In 1994, when the Belgian government withdrew its control in central Africa, 800,000 Rwandans were brutally slaughtered by their own people in just a few weeks. During the twentieth century, over 100 million people were killed through warfare that had its roots in ideological principles, typically including political, economic, or ethnic motivations. In fact, more people have died in such conflicts during the twentieth century than have died in all wars combined since the beginning of recorded history.

The differential contribution of genetic and environmental factors in human development remains controversial. In 1969, Arthur Jensen, professor of educational psychology at the University of California, Berkeley, proposed that an individual's IQ is essentially a product of heredity and, therefore, money spent on educational assistance programs for underprivileged children was of no potential value. In the years that followed, he bolstered his argument with research suggesting that raising IQs through focused educational tactics did not provide any substantial or lasting benefit. However, learning—the acquisition of knowledge—is not equivalent to intelligence. Intelligence can be thought of as a more comprehensive quality, which includes many processes, such as memory function and capacity, ability to compare and discriminate information, judgment, etc. In 1994, *The Bell Curve,* by Charles Murray and Richard Herrnstein, presented research along hereditary lines and contended that individuals prefer to marry those with similar backgrounds, creating intelligence gaps between the races that would continue to diverge with time. The premise of the book has been highly criticized as a product of pseudo-scientific analyses, but it does bring to light differences in IQ results among ethnic subpopulations, which may actually reflect significant social environmental influences.

Attempts at separating the influence of environmental from genetic contributions in molding human behavior have proven a difficult challenge wrought with controversy. A readily available living biological laboratory was found in the study of identical twins. If a pair of twins is separated immediately after birth, in theory they share identical genetic inheritance but will have different experiences as they are exposed to different environments. Thus, the differential contributions of inheritance and environment become amenable to investigation. Recent studies with large collections of identical and fraternal twins suggest about 75 percent (69-78 percent) of one's intelligence is attributed to hereditary mechanisms. Other studies based on first- and second-degree relations produce figures between 30 and 50 percent. However, studies of separated identical twins, critics argue, are flawed in that they don't actually represent a complete isolation between nature and nurture.

Although identical twins separated at birth may have identical genes and may go on to mature in different social environments, the first part of their development (the nine months of gestation) was spent together in their mother's womb. Indeed, the brain undergoes significant development during the gestational period, a time when environmental factors can influence development. There are numerous risk factors to the normal *in utero* development, including maternal disease, poor nutrition, and exposure to environmental toxins, as well as poor prenatal care. This point of view is substantiated by investigations of non–genetically identical twins—as twins they will generally share the same gestational exposures. Although fraternal twins are as genetically different from each other as any nontwin siblings, studies have found that they tend to have IQs more similar to each other than to their nontwin siblings. In addition, twins who are separated may spend some time in the same environment before the separation takes place—the long-term influence of early life experience on future behavioral development is well known.

As techniques for early (prenatal) diagnosis of genetic diseases continue to improve, the potential for a new wave of eugenics mounts. Whereas historically the practice of eugenics focused on selective breeding with the goal of healthier children, early genetic testing may help couples decide on the "genetic acceptability" of their embryo. Currently genetic testing is largely performed in pregnancies at risk for genetic diseases (e.g., Down's syndrome, Tay Sachs disease, etc.), where it provides useful information regarding the presence of disease and affords an opportunity to either act on the test findings or disregard the results. As knowledge of the human genome grows and the link between genetic material and its expression is better understood, the role of prenatal genetic testing may become significantly broadened. The directive to reduce the incidence of genetic disease may be replaced by parents' expectations for their unborn child's physical (or mental) qualities.

The Chinese government has been criticized for its 1994 infant mortality law as an apparent eugenic facade. The law was designed to reduce infant mortality by impeding couples carrying certain genetic diseases from having children. The obstacles include the inability for such couples to marry unless they agree to sterilization or long-term contraception, and abortion is encouraged for any abnormal pregnancies. A survey of Chinese genetic counselors found this measure as a means to reduce harmful genes in the gene pool. However, critics claim the Chinese infant mortality law is ineffectively written and is not being seriously enforced. Ethical implementation of genetic testing and counseling must be adopted to avoid abuse.

Gene insertion technology and other developing methodologies may provide ways to purposefully modify genetic material. By using an altered virus, genetic material may be added to the genome of a living human. The

242

genetically engineered viruses are injected into an individual, where they spread throughout the body and insert their (altered) genetic material into cells, which in turn incorporate the new genetic material. There are limitations, as human cells do not always adopt or express the new genetic material. Clinical trials are under way that use gene therapy methods of inserting new genes to override abnormal genes or replace missing genes in the treatment of various disorders, including cystic fibrosis, cancer, and AIDS. Currently many of these techniques are relatively crude, but as further refinement will certainly improve efficacy, it will likely raise eugenic issues as well. Eventually society will have to formulate ethical policy concerning the implications of such interventions for the future of humanity.

Progress in the field of genetics is at the precipice of deciphering the entire human genome. Perhaps, a few years thereafter, the ability to rapidly sequence the genetic code of an individual will become a reality. Mapping of small segments of DNA (DNA fingerprints) is already being used as a means of identification of biological evidence (e.g., blood, semen, etc.) in criminal investigations. Theoretically, applying a rapid sequencing technique to human embryos early in development would open the opportunity for early intervention—perhaps permitting assimilation of any genetic modification of the embryo's DNA desired. If genetic material is successfully incorporated into the first cells from which the embryo develops, the modification will be present in all of the cells that follow, including those destined to be germ cells. Thus artificially introduced genes would become part of the embryo's DNA and inheritable by future generations. The feasibility of such modification is currently questionable; genetic sequencing of the early embryo and subsequent genetic interventions must be performed very early in a pregnancy, making it logistically difficult to accomplish. Some of these obstacles may be overcome through artificial fertilization techniques (i.e., *in vitro*) or measures to slow the process of cellular division, allowing more time to examine and modify the genetic material. Although an intriguing capability, in reality the cost would likely be prohibitive. Indeed, these genetic manipulations place man in the role of directing natural selection, a role historically relegated to deities.

It is generally accepted that knowledge benefits mankind. The ability exists to identify genetic diseases that may directly affect one's mortality, including some diseases not yet manifested in those tested, who are otherwise in good health and may show no symptoms to raise suspicion. Unfortunately, in many cases a genetic disease may be diagnosed, but there is no effective cure. Is society benefited when the individual can obtain knowledge of her own mortality? Is medical science advancing beyond the limits of acceptable ethical constraint? There is the story of the psychology professor who began his lecture by handing an envelope to each student: "Within your envelope is a number that is the age at which you will die. It is your choice whether to open the envelope or not." Most students left their envelopes sealed. Although they may be nearly three thousand years old, the insightful words of the preacher (allegedly King Solomon) still hold meaning: "For in much wisdom is much vexation, and he who increaseth knowledge increaseth sorrow." (Ecclesiastes 1:18)

10

Culture

The seeds of the first cultures were planted once ancient man was released from the constant struggle to survive. Once a community becomes sufficiently organized, its members can develop a long-term social infrastructure. Over generations a community may evolve its own elements of customary beliefs, social structure, and material trades that define its culture. In fact, the social environment of those living in some ancient societies (e.g., Mesopotamia several thousand years ago) was similar to that of modern society—citizens lived in homes, worked for wages, had families, attended religious ceremonies, etc. Although many amenities were certainly more primitive in ancient times, personal ambitions and the social structure of the population were not unlike those found today.

The first civilizations began several thousand years ago, and much of their stability relied on the dependability of agriculture, which in turn was strongly influenced by the climate. Examination of ice core samples from the Greenland ice sheet has provided a reliable history of temperatures in that region for the past 15,000 years. In addition, the ice sheet data can be extrapolated to establish global trends in climate during the same period. Analyses of these cores indicate the last ice age ended about 11,700 years ago, which marks the recent epoch characterized by a steady relatively warm climate: the Holocene. The warming trend from the last ice age ended about 7,000 years ago, with a rather stable temperature ever since. It is during this period of a stable climate that human civilization began. Although the formation and stability of the ancient civilizations was a multifactorial process, a stable climate was a major contributing factor.

Society

Social bonds are deeply ingrained in the human psyche. A human infant, like many other primates, is born in a relatively premature and fragile state, requiring a significant period of nurturing to ensure survival. It is well documented that even if one survives to adulthood, lack of appropriate social interaction early in life can have dire consequences. The social bonds are not limited to interaction only with the parents but extend to other members in the social environment. It is a logical extension of this primal social behavior that groups of individuals would naturally congeal to form communities. In the evolution of a society, as it develops a more complex social structure—division of labor, tiers of social class, practice of religion, etc.—the necessity for organization and control becomes increasingly important. In order to protect and maintain the integrity of a society, a political body of some form is created. Prior to the twentieth century, with few exceptions, it was popular opinion that "primitive" societies lacked social order and their people lived in a state of anarchy. To the contrary, archaeological study of primitive societies has proven that stable society was likely in existence long before such political governing institutions were ever conceived, much less formally created. Furthermore, the development of such institutions involved groups of people rather than a few inspired individuals. Adam Ferguson, an eighteenth-century writer, studied the work of philosophers as well as accounts of missionaries and explorers describing their experiences among primitive societies. He noted that through conflict and competition the individuals of a group are encouraged to develop an organized political system. In addition, he found that the propensity for division of labor leads to the increased prosperity, size, and complexity of a society.

The development of society abides by the principles of human behavior and is a beneficial instrument for survival. Creatures forming complex social communities are not limited to humans, as numerous examples abound—from nonhuman primates to ants, bees, and elephants. Clear advantages include protection and safety in numbers, a behavioral survival technique used by many species. The collective force, rather than the individual, is better able to overwhelm a potential enemy threat. In addition, in the event of the loss of an individual (to a predator or natural disaster), a large group will endure relatively less adversity than a small group consisting of just a few individuals. Larger communities permit individuals to share responsibilities that enhance survival (e.g., obtaining food, building shelter, quelling enemy threats, etc.). The division of labor also enables individuals to find ways to improve their particular skill through personal dedication. In addition, useful skills may be taught to others of future generations. In a growing society, there are a greater number of individuals to contribute through knowledge and experience, which further accelerates progress. Thus in a society providing freedom for individuals to excel, cultural development can flourish.

The earliest cultures centered around a variety of survival-related tasks, including the building of tools and shelter, and primitive religious customs. Various cultures developed as the ancients settled in different geographic and climactic regions. Many of the early communities had seasonal migration patterns, responding to climactic variations and the migrations of the beasts they hunted. Although the conditions under which ancient man lived were harsh by our standards, some of the basic amenities have been in use for millennia. Excavations in Europe indicate that fire was used for cooking as long as 500,000 years ago. There is suggestion of animal skin clothing being worn throughout Europe, the Middle East, and central Asia as early as 200,000 years ago. As the Paleolithic hunter-gatherer groups gave way to the Neolithic settlers, the initial elements of social and political structure began to develop.

The popular notion of ancient man living in peaceful commune with nature has been contested by extensive research in the studies of both contemporary and historical primitive societies. As primitive man found himself at the mercy of the environment, his ability to understand nature and desire to appease it demanded considerable attention and effort. Some of the patterns of nature lend themselves to prediction, as they are cyclic (i.e., the change of seasons); however, natural forces can also be unpredictable and devastating. The desire to predict or at least achieve some level of control over these mysterious natural forces led to the development of superstitious practices. It was found that the simultaneous development of superstitions and religion in primitive societies may lead to a plethora of ritualistic responsibilities for the individual. It is not uncommon in primitive societies for individuals to participate in various elaborately complex social practices, as well as to avoid numerous taboos, in order to maintain harmony for themselves and the community.

The progression of society from the primitive hunting and gathering lifestyles of the Paleolithic to industrialized societies often follows a general pattern. As diverse as society appears, upon closer inspection of its historical development there are certain general characteristics expressed in its evolution. Ancient civilizations arose from hunter-gatherer tribal communities that drifted toward more stable agricultural and domesticated patterns. These changes in lifestyle led to increased central political structure, which often gave way to theocratic chiefdoms. The early primitive societies were largely divided into kin groups, some of which developed into institutions governed by a central authoritative body. In these simple societies, the governing body is neither large nor constant in size. In small primitive societies there may be no clear leading individual or group and only in times of necessity was a leader identified. In studies of contemporary primitive societies, the leaders are usually well-respected elders thought to have important qualities such as experience and intelligence. In a tribal society, the leader often has a temporary role and his "power" is more accurately described as an influence over the community. It has been suggested that when a community has a desirable leader, the community anticipates their leader's progeny will have similar qualities of leadership. It is here we find the origins of a ruling class in primitive society, where a subset of the population inherits an elevated status of authority over the remainder of the community. These small segmental tribal communities likely constituted the vast majority of societal structures throughout human history.

The vast majority of the historic tribal communities were dead ends; however, a few managed to develop a more centralized and stable governing body as a chiefdom. Evidence of the development of chiefdoms can be gleaned from the archaeological record, which reveals burial sites indicating different levels of social status, specialization of crafts, theocratic themes of public monuments, and evidence of trade—the practice of these focused community activities would be difficult in the less-organized societies of segmental tribal or hunter-gatherer communities of prior times.

Although the governing of a tribal community may seem limited by its inherent ephemeral form of rule, it does offer interesting ways of distributing authority. In situations where multiple tribal groups become involved with a common outside antagonist, they can easily form a confederation to strengthen their position. Furthermore, under appropriate conditions the confederations may rapidly fragment into constituent tribes, as in the conflicts between the American Indians and the European and early American military during the settlement of the North American West. However, societies that are governed by a chiefdom are more easily exploited by invaders; the cohesive quality of a central authority offers easier control by an outside aggressor.

The formation of the early classic civilizations was generally consolidation of smaller chiefdoms that were unified under a bureaucracy. In the interest of assuring a peaceful succession of leadership in ancient civilizations, it was common practice to create a hereditary aristocracy and a bureaucratic hierarchy—the beginning of a true government. In cases where a dominant group incorporates other bureaucracies and territories, often using militaristic power, an empire is formed. The larger heterogeneous population of an empire enables significant and rapid cultural developments (e.g., written language, scientific, artistic, and philosophical advances). In addition, when different cultures are brought together in forming an empire, elements from each may combine to form new philosophies and religions. In the case of the rise of civilized societies in the modern era, as with the populating of the Americas, the progression to a civilized community is rapid, as the culture and structure of society (political system, language, religion, etc.) are borrowed from the migrating populations' ideologies. However, some cultures are dramatically changed in a matter of a few decades. Throughout the world are countless examples where relatively primitive cultures rapidly lost their identity as they were exposed to the cultures of industrialized nations—from the islanders of the Pacific and the Aborigines of Australia to the indigenous tribal cultures of Asia, Africa, and the Americas.

Scholars had originally reserved the definition of a civilization for those societies that developed a written language. More recently, V. Gordon Childe, an archaeologist and scholar, proposed a different approach with consideration of numerous specific qualities. He defines civilizations as those societies that possess urban centers, a class of inhabitants involved in various trades (merchants, craftsmen, food producers, etc.), authorities (civil, military, religious), purposeful activity in the development of the sciences/arts, and a governing political structure. Contrary to what one might assume, urbanization of a civilization does not necessarily imply that the population lives in large cities. In cases where external conflict with competitive neighbors develops, defense fortification with a central migration of the peripheral community would lead to a more densely populated urban center. In addition, fluctuations in the production of food would tend to create fluctuations in the population, which may arise in times of environmental hardship or warfare. It was under these conditions that the Mesopotamian and Teotihuacán civilizations developed, with their relatively densely populated central regions. Other civilizations maintained a relatively dispersed population with a central ceremonial district, as did the Mayans and residents of ancient India. It was during various occasions, such as religious or political activities, that the population would concentrate in the central, more "urbanized" area. The cohesive nature of a civilized society relies on the ability of its governing body to use an administrative influence to maintain ceremonial, economic and military functions.

Many ancient civilizations arose independently and flourished at different times in human history. In addition, particularly within the Mediterranean basin, there was a series of longitudinal civilizations as one dominant civilization gave way to the next. The geographical extent of a civilization, particularly in ancient times, depended heavily on environmental barriers such as topography (rivers, lakes, and mountains), productivity of the land, abundance of wildlife, and climate. In ancient Egypt and the early civilization of coastal Peru, natural boundaries concentrated the populations and prevented avenues of escape in the event of warfare, promoting the development of powerful integrated societies. In ancient Egypt, survival was feasible only in settlements along the banks of the Nile, because the surrounding desert was inhospitable. In the case of coastal Peru, the habitable land is situated along a narrow strip between the Pacific Ocean and the steep Andes Mountains. In both cases, their civilizations were relatively peaceful.

Throughout recorded history, perhaps the most common and devastating pressure on population centers was the threat of attack from outside enemies, which generally is one of two forms: similar cultures that develop geographically close to one another or nomadic raiders. In the first case, stability can be sought politically through agreements made by the leaders of the respective communities. In the second case, the defense against raiding cultures is less amenable to negotiation, as aggressor groups tend to be less politically stable. With a persistent threat, communities would establish a form of defense including forts and barriers. In fact, one of the most ancient cities known, Jericho, had a stone wall fortification, suggesting that conflict was a major concern 10,000 years ago. In addition, ancient civilizations that were under constant threat of invasion generally actively developed their urbanization and agricultural aspects. Perhaps the most crucial element in maintaining integrity and stability of a civilized population center is its political structure.

Political power is basically applied to society on three levels: persuasion, leadership, and arbitration. To remain in control, the political body must gain support from the population through some form of persuasion. Once a political action has been decided, by whatever means, applying it effectively requires a form of leadership. Arbitration is concerned with the mediation of disputes to define and enforce the rules of society. In early primitive societies, there were no formal institutions handling these functions. However, societies that developed into the chiefdom phase would often have a hierarchical aristocracy providing guidelines without a developed institution of arbitration. Instead, control was usually relegated to a religious body, and in some cases leadership assumed the role of a priest-chief Law in tribal societies typically arose from sanctioned customs, where breach could bring severe punishment. Eventually these rules would become standardized, enabling legal authority to enforce them through either persuasion or threat of force.

Maintenance of society requires civil control on multiple levels. For the individual, social etiquette is learned through lessons in morality. Incongruous behavior may be addressed by society through sanctions (rewards or punishment) imposed by a designated authority. The implementation of sanctions often implies force; however, in some societies (particularly in primitive African societies) the hierarchical authority is deemed "absolute," where the wishes of the leadership were followed without question for fear of occult reprisal. However, these mystical powers may also be used for peaceful tasks such as improving crop yields and providing safety from an enemy. Scholars suggest that early forms of law were developed in societies with a central chiefdom governing body.

In cases where accepted social etiquette is breached in tribal cultures, the punishment may be determined by the leader or by consensus, as no governing body would exist to make such decisions. More serious conflicts (e.g., feuds or warring within or between tribes) were rare, as these communities tend to be small, and the potential loss of some members could be devastating. In addition, leaders were often transient, which limited their ability to persuade young warriors to commit to a dangerous cause. In fact, serious disputes would more likely be resolved through physical contests. In the case of the Eskimos of North America, duels of singing were conducted to settle conflicts. In addition, warfare among primitive communities was generally not the highly lethal confrontations found in the military engagements of recent times. Physical confrontations would often be limited to a few individuals, and the endpoint would be the first wound that brought forth blood. More organized forms of warfare (i.e., killing opponents quickly and in numbers) were developed about twenty-five hundred years ago by the hoplite soldiers of the feuding city-states of ancient Greece. The methods of warfare have become far more efficient over the millennia.

The survival of governments, particularly during times of conquest, is reminiscent of "Darwinian" selection; the better political system (at conducting and surviving warfare) survives. In times of conflict, systems with a more central control tend to remain more stable than loose tribal associations. In fact, warfare became more common in many of the ancient civilizations as they developed. A common theme in the history of civilization is for developing societies to conquer others to secure resources (mineral wealth, strategic advantages, increased manpower, etc.). Although it is historically common for civilizations to fall into decline through conflict with other groups, the role of natural disasters cannot be ignored. Some investigators claim that the harsh climactic conditions of centuries of drought may have adversely affected the great ancient civilizations of Mesopotamia, Egypt, and the Indus Valley around 2200 B.C. The strong dependence of these civilizations on agriculture may have made them particularly susceptible to climactic change.

Philosophy and Science

The earliest elements of human engineering can be traced to the Paleolithic, over 2 million years ago, when stone implements were first fashioned into tools. As communities became more sedentary and populations grew largely through improved agriculture, the great ancient civilizations arose and brought the first elements of genuine philosophical and scientific investigations. The invention of written language allowed rapid dissemination of information, which bolstered the progress of these intellectual pursuits. Philosophy was nurtured as man's relation with the community and nature was coming into focus. Initially, the combination of these interests led to superstitious conjectures, as there was lack of understanding of the true nature of the forces at work.

Classic philosophy can trace its origin to the ancient Greek civilization where Thales, during the sixth century B.C., speculated on the nature of matter and the Universe. By the fourth century B.C., Athens had become the center of philosophy and early science, with contributions by Plato,

Socrates, and Aristotle. The study of science and philosophy had a common origin in Greece, where the fundamental principles of Western tradition were laid. The Greeks approached science differently from their predecessors in that they attempted to explain the properties of the world through material influences rather than incorporating the interventions of deities. Concurrent with the flowering of philosophy and science, an appreciation of the visual and written arts and architecture flourished. It must be remembered that the early philosophers pondered the workings of the natural world without the luxury of modern-day technology to assist in their investigations. However, despite such limitations, amazing insight into principles of science was derived—ancient Greek philosophy rationalized the existence of atoms as tiny (too small to be seen) individual particles that combined to form matter.

In the Hellenistic Era of Greece, a three-hundred-year period beginning with the death of Alexander the Great in 323 B.C., philosophy entered a trend to focus on one's search for happiness. Among the Hellenistic philosophies was Stoicism, which held that individual happiness lay in avoiding dependence on items that were ultimately under the control of others. The pacifist ideology and high ethical standards of the Stoics were adopted by the Romans and continued until the arrival of Christianity. A great library and museum were built in Alexandria, where intellectuals gathered to study such fields as mathematics, astronomy, medicine, and literature. As the Mediterranean basin fell under the influence of the Romans, a period of stagnation ensued, since the empire was not inclined to pursue such intellectual endeavors—it was more concerned with material conquests. With the fall of the Roman Empire and domination of Western civilization by religious convictions and ideals, the importance of the individual was considered secondary to his obligation of service to the Church. However, the philosophies of Plato and Aristotle influenced the origins of Christianity and dominated Western thought for centuries.

The Renaissance period of fourteenth- and fifteenth-century Europe returned trust to human thought and reason, despite resistance by the Church. The Italian humanists of the fourteenth century reexamined the teachings of Christian theology with a classical motivation that stressed more tangible qualities than pure theological and metaphysical concepts. By the end of the fourteenth century, the humanist movement began to press for political service to spread scholarship and mold society appropriately. In the late fifteenth century, the humanist movement produced the semireligious Neoplatonism, which returned to the teachings of Plato, with an emphasis on contemplation of life and God. Predominantly located in Florence, the Neoplatonists rejected political aspirations and focused on enlightenment of the individual. The ideals of humanism spread beyond the borders of Italy to gradually influence Europe. The question of religious goals and authority led to the Reformation of the last half of the sixteenth century, when Protestant and Catholic ideals were bitterly tested. Eventually religious authority waned as academic achievement gained credibility through advances in agriculture, science, medicine, and literature. It was finally accepted that man could benefit from his own intellect and abilities. As the period of Enlightenment arrived during the eighteenth century, European society blossomed and intellectual curiosity entered an unprecedented period of philosophical and scientific exploration.

The philosophy of social order was perhaps best exemplified by those in Asia. In the fifth century B.C., Lao Tzu, a man from western China, is credited with founding Taoism. Legend has it that he was displeased with his people's refusal to seek the indigenous happiness of humanity that he advocated, and (atop a water buffalo) he ventured out of civilization. Before his departure, he left the *Tao Te Ching,* which summarized his placement of humanity in the Universe and became the basis of Taoism. Taoism is practiced in various forms and may be considered either a philosophy or a religion. It encourages utilizing life's vitality without wasting it. Instead of merely accepting it, some Taoists will attempt enhancing vitality through various means, including diet and meditation. Around the fourth century B.C., K'ung Fu-tze (Confucius) traveled through China to tutor local rulers in his philosophy of appropriate moral conduct as a means to obtain social order. He believed people were born virtuous but required discipline to maintain purity through example by their superiors. He argued that the implementation of force or punishment indicated failure of the political system. While Confucianism encourages appropriate ethical behavior, Taoism advocates freedom and earthly blessings. Although different philosophies, Taoism and Confucianism exemplify the dichotomy of the yin and yang philosophy of Eastern thought.

Philosophy is by no means a stagnant discipline. The principles of philosophy are timeless and as valid today as they were thousands of years ago. During the Vedic Period of ancient India, between 1000 and 500 B.C., a philosophy developed that taught of an underlying order in nature that determines all phenomena. This assumption of constancy of physical laws is central to the principles of modern science. In modern society, philosophy has applications in countless fields, including ethics, artificial intelligence, law, and other social issues.

Religion

The human psyche has an underlying desire for a benevolent mentor—96 percent of the U.S. population believes in a supreme being (i.e., God). Human life begins in a dependent state where survival demands parental nurture and protection. This situation stems from the relatively frail condition in which a human infant is born—an evolutionary compromise for the immature state of development of the brain at birth. As the brain matures, the infant naturally forges social and emotional bonds with her supportive family. Even as independence is gained, the comfort and respect for protective authoritative figures remain. In situations where there is lack of such authority, insecurity and anxiety develop—unpleasant emotions that may be alleviated by the auspices of a dominating figure. During periods of subjugation (e.g., during the European Middle Ages), in which the ruling authority dictates the individual's role in society, the individual gains a level of psychological relief by the freedom from making many major decisions (e.g., occupation, marital decisions, etc.). The feudal tripartite model of Middle Ages society supported basically three occupations, generally without the luxury of choice: nobility, clergy and land-tilling peasantry. The development of mythology and religion provided followers with the security of an order to their world, typically overseen by a deity, who often physically resembled a human, but with special qualities. The strong prevalence of such a conviction in most cultures implies that the basis for such desires lies deeply rooted in the human conscience.

Ancient man, living at the mercy of a world governed by mysterious forces, longed for a rational underlying authority. As an attempt to create a (perhaps false) sense of understanding, superstitions and rituals arose. As irrational as some primitive superstitions may seem, analysis of many such beliefs reveals underlying threads of rationality. Primitive superstitious beliefs often associate events with material items in the environment and that these items can be used to control future events. However, in some cases the factors that control events were thought to be the wishes of other beings: deities. In order not to displease these powerful beings, one must show respect through some form of worship (e.g., celebration, offering of food, or even sacrifice). Following ritual traditions to appease the deities afforded the ancients a sense of security. In addition, they assumed that treating deities with respect could bring benefits, such as plentiful food and good health.

As a stabilizing element in society, delegated religious roles may define different levels of social status for individuals in the community. In some societies religion was applied as an instrument of control over its population; priests or the ruling individual maintained authority through supernatural means. All of the major ancient civilizations either developed or adopted forms of religious or philosophical traditions. Furthermore, the integrity of a populace is strengthened by the social unity promoted by common convictions. Most religions include a moral code, which would further maintain social stability in a population. The codevelopment of religions within the cultures of the ancient civilizations suggests a critical role in the cohesiveness and stability of society.

Human history predates the origins of the dominant world religions of today, all of which originated in Asia within the past four thousand years. Archaeological studies suggest the earliest forms of religion were practiced by at least the time of the Neanderthals, 100,000 years ago. Prehistoric man lacked a system of writing, but studies of burial sites clearly suggest a belief in an afterlife. Some cave wall paintings are believed to depict a spiritual element to their society, perhaps used to ensure successful hunting, as well as other prehistoric concerns of which we can only guess. During the Neolithic period, settled communities formed, which typically ranged from groups of about ten individuals to villages of up to a few hundred people. The increased social interaction of this community life led to the development of the Neolithic cultures and religious beliefs. At some sites, excavations have found evidence of human and animal sacrifice, as well as miniature figurines imprinted with designs of unclear significance. The primal religions, ubiquitous in their forms, have been present in human history much longer than the major religions: their contribution to human society was crucial, and many forms are still practiced today.

Most primitive religions center around elements of nature, which clearly have a major impact on daily life. Primal religions tend to promote order in the community and nature, with facets for securing benefits such as food, rain, fertility, and protection. Unlike the major religions, the drive to obtain a state of eternal purification is absent, and the concept of an afterlife is usually that of a mystical existence in a congenial place of natural abundance. The early primal religions were ubiquitous, and the vast majority will never be known, as they have been lost over the millennia, forever buried in the past with the ancients who created and practiced them.

The oldest world religion is Hinduism, which began through oral tradition by the people of the ancient Indus civilization. The central god of Hinduism, Shiva, originated from a cult of the ancient Indus region people, with additional deities borrowed from the religious practices of invading nomads from the north (Aryans) and other neighboring influential cultures. The religion eventually took form as sacred hymns and poems, called Vedas, which were recorded in written form as long ago as 1500 B.C. In the centuries that followed, additional Vedas were added. The *Upanishads,* the last Veda, focus on the concepts of morality and obligation, which many consider the central theme of Hinduism. Later writings include the *Bhagavad Gita,* which emphasizes one's societal discipline and responsibility to the caste system. The philosophy of Hinduism includes karma, a form of spiritual destiny that is determined by an adherence to moral obligations. Jainism and later Sikhism have their origins in Hinduism but separated as monotheistic revisions.

As Hinduism became more ritualized and embedded in the caste system, Buddhism was founded by Prince Siddartha Gautama of India at the close of the fifth century B.C. Legend claims the prince was discontented with his life of earthly pleasures and was consumed with the sufferings of mortal life. He left his privileged life to wander the country for several years as a beggar in search of a philosophical solution. After forty-nine days of intense meditation beneath a tree, he was able to find truth and became enlightened, taking the name Buddha. The philosophy of Buddhism centers on the "Four Noble Truths," which state that the difficulties of life stem from desires, which can be resolved by following the path of "right conduct." The path is essentially a code of morality, which if adhered to leads to *nirvana,* a state of spiritual bliss. Partial assimilation of Hindu values into Buddhism led to the conviction that individuals who fail to adequately follow the righteous path will be reincarnated as a different life-form rather than reaching nirvana, in accord with their inadequacies. Buddhism spread from India to other parts of Asia, while its presence in India dwindled with a gradual assimilation into Hinduism.

In ancient Mesopotamia, the Sumerians worshiped anthropomorphic gods representing natural forces (heaven and water). Life was a harsh struggle against the ravages of the climate, and their gods were equally harsh, placing humans in the role of servants. In addition to the four gods of creation, there were many lesser gods, some of which could be called upon to help in times of need. In contrast to most religions, the afterlife was depicted as an unpleasant existence, which may have been a reflection of the hardships of life for those living in Mesopotamia. However, there were similarities to the later religions of Mesoamerica, where many religions recognized a central jaguar god to which sacrifices were made at large temples.

Judaism originated in Mesopotamia but was unlike the other contemporary ancient religions, as it was monotheistic. There are interesting parallels between Judaism and the Mesopotamian religion, including the story of an ancient flood sent as punishment to mankind. Although such stories were clearly borrowed from other cultures, the tales were modified to fit the different roles of man and his gods in the different religions. Abraham, the celebrated founder of Judaism, was a resident of the ancient Mesopotamian city of Ur, from which he fled around 2000 B.C. Over the following centuries, his followers carried on his teachings and eventually migrated to Egypt. Around 1200 B.C., the Hebrew followers of Abraham's teachings left Egypt in the exodus led by Moses, with divine guidance from their God, Yahweh. Their migration eventually took them to Palestine. At the time of early Judaism, the ancient religions of the Mediterranean were pagan—followers worshiped many deities, which had specific attributes. Often the pagan deities had little concern for mortals, and their actions were generally amoral. Part of the appeal of Judaism was its promotion of a righteous god with a genuine concern for everyone, not just reserved for the aristocracy or privileged. The visions of Abraham and Moses were that the responsibility to teach the world about their religion lay with the chosen people, the Hebrews, and being a member of this group required adherence to various rituals and restrictions.

Near the northwest shore of the Dead Sea, a religious sect was established at Qumran. The community had its beginning about 200 B.C., and shortly afterward they began preserving their religious tradition through writings on animal skins. These writings were eventually stored in pottery vessels in about a dozen caves scattered around the landscape. The manuscripts have become known as the Dead Sea Scrolls, and more than eight hundred fragments have been found thus far. Among the writings are excerpts from the early Hebrew Bible, predating the prior record holder by 1,000 years. The remainder of the scrolls includes excerpts of the contemporary literature and the teachings of the scroll authors, the "Sons of Light." The community lived in the desert for about 150 years, practicing a strict religious tradition that predicted a catastrophe that would end the Earth, at which time they expected to be delivered to a better place. The Qumran community was discovered by the Romans around 70 B.C. and completely annihilated—a graphic fulfillment of their prophecy. There is controversy regarding the contents of the scrolls, as some discrepancies have been found between the modern Hebrew Bible and the translations of this ancient source of Hebrew tradition.

Christianity can trace its beginning to the ancient region of Judea, at roughly the location of modern-day Palestine. Historians place the birth of Jesus at approximately 4 B.C. in the small Judean town of Nazareth and believe he spent the majority of his life in Judea. Despite the tradition of a birth in Bethlehem and a peasant heritage, historians suggest Jesus was probably of a higher social class than previously thought and may have been multilingual. It was after his baptism by John the Baptist that Jesus began teaching his interpretations of Judaism. He largely followed Judaic tradition but disagreed with some of its restrictions of social conduct—he considered the rigid practices of Judaic tradition inconsistent with a compassionate God. Similar to the contemporary practice of Judaism, Christianity was active in helping the destitute; such conduct was instrumental in gaining membership support, since the Roman Empire showed little concern for the welfare of the poor. The division between Christianity and Judaism was not clearly defined until long after the execution of Jesus. During this early period, many regional forms of Christianity were practiced. In the second century, a consensus of Christian authorities adopted a collection of religious writings as the New Testament in order to unify Christianity.

There was religious rivalry in the Mediterranean basin during the early years of Christianity. Mithraism, originating in Persia around the fourth century B.C., was a paganistic practice that was popular in the Roman Empire. By the third century A.D. it was competing with Christianity for followers. Mithraistic tradition held a popular annual celebration on the twenty-fifth of December, the winter solstice by the Julian calendar (which represented the day Mithras was born from a rock as the god of light). The early Christian doctrine did not assign a date for the celebration of the Nativity. However, Christians in Egypt began to celebrate the birth of Jesus on January 6. The Western faction tactfully adopted December 25, coinciding with the popular festival of Mithras, as the day for celebration of the birth of Jesus. The practice became widely accepted by the late fourth century. The celebration of the Epiphany, the arrival of the Magi, was then held on January 6. Such politically ingenious maneuvers enabled Christianity to gain popularity in the Roman Empire, through its compassionate philosophy and preservation of popular festivals. After centuries of Christian persecution, in 313 the Emperor Constantine enacted the Edict of Milan, which provided religious freedom to Christians.

During the twilight years of the Roman Empire, in the early part of the fifth century, Augustine, Bishop of Hippo, published *The City of God.* Augustine was a prolific writer of Christian theology and Greco-Roman philosophy, which he jointly applied to explain the nature and destiny of humanity. In fact, he promoted the attitude that the decline of the empire was punishment delivered by God for the sinful ways of Roman society. Furthermore, he argued that man's immorality and unrelenting desire for material pleasure require the establishment of a political state to maintain order. In other words, if mankind were morally perfect, the necessity of a political state is eliminated. Augustine's contributions to Christian philosophy remained influential for centuries after his death.

Islam is the most recent world religion and is currently practiced by approximately one-fifth of the world's population. While praying in a cave in A.D. 610, Muhammad, a wealthy Arab, had a revelation from Allah to repudiate pagan idolatry. Muhammad lived in Mecca, an active trade center with a grand religious sanctuary that housed hundreds of religious idols for worship by a variety of pilgrims. Although initially a religion of three gods, Muhammad revised his original doctrine to a monotheistic form. Combining elements of Judaism, Christianity, and Arab pagan traditions, the Koran (the holy book of Islam) was written. In order to be a Muslim, one who practices Islam, one must adhere to the Five Pillars of Islam. The Pillars are a collection of rules requiring submission to one God, payment of alms, and instructions regarding prayer, fasting, and pilgrimage. The jihad, a spiritual conflict, is sometimes considered the sixth Pillar and may be in the form of an individual's spiritual plight or an aggression against enemies of the religion. Muhammad was a soldier, and his doctrine included a call to fight those who resist accepting Allah as the only god. The jihads and conquests that followed led to a vast Islam empire a century after Muhammad's death. The history of early Islam is plagued with inner conflicts, and eventually the religion split into two factions, the Sunni and the Shi'ites. Most Muslims are Sunni and continue their worship as decreed by Muhammad and the Koran. The Shi'ites believe that Muhammad's son-in-law, Ali, was his first divine successor, followed by eleven other prophetic successors. The Shi'ites follow the teachings of the Koran and the twelve successors.

The original religions of China were essentially in the form of philosophies. The philosophy of Confucianism was often combined with elements of Buddhism or Taoism to create a form of religious philosophy. Under the Shang dynasty, the leaders considered themselves divine and had their subjects perform human and animal sacrifice to ancestors and to various gods of nature.

Religion has played a crucial role in pacifying the human psyche, an objective that science strives to accomplish in its own right. For uncounted millennia, religion has played a central role in providing answers to some of the most difficult philosophical questions: Where did our universe originate? What is the purpose of life? And what happens when the life force has left our body? Great strides in science have been made, particularly in recent history, that have provided some insight into many of the fundamental questions of humanity. The comforts provided by religion and the answers by science are not necessarily mutually exclusive.

Religious influence has not always been benevolent. Perhaps the best-known examples of abuse of religious authority came in the form of the infamous Inquisitions. The first Inquisition, the Albigensian Crusade, was ordered by Pope Innocent III in 1208 to eliminate the Albigensians, a sect in southern France that promoted a practice of extensive abnegation, including rejection of the Church and its practices. In 1231 Pope Gregory IX formally created the Inquisition to convert heretics and stop the spread of heresy. A band of Inquisitors was formed to carry out the task using principles of Roman law and torture. The accused had no right to counsel, to question witnesses, or even to know of the crime with which they were charged. Although some victims died in horrible ways, including being burned alive at the stake, the majority managed to evade punishment. During the late fifteenth century, the Spanish Inquisition, particularly concerned with the religious conversion of Muslims and Jews, was especially notorious in its use of torture. Published in 1486, the *Malleus Maleficarum* (Hammer of the Wicked), authored by a former Inquisitor, described the archetypal witch. Many editions were published, and for hundreds of years it was widely used as a handbook for the detection and interrogation of suspected witches. One of the most famous Inquisition victims was Giordano Bruno, an Italian philosopher, who in 1584 wrote *The Infinite Universe and Worlds.* The book was a blend of mysticism and Copernican astronomical theory depicting an infinite universe containing many suns and worlds like our own—it even entertained the possibility of intelligent life elsewhere. The Roman Inquisition declared him a heretic and had him burned alive in 1600. A little over thirty years later, Galileo Galilei faced the Roman Inquisition for charges of heresy and was forced to recant his scientific findings. The authority of the Inquisitions eventually waned, and in 1820 the Spanish Inquisition was outlawed, bringing an end to a dark era in history.

From the governments of nations to radical cult groups, history is teaming with examples of attempts at creating a perfect social order—no attempt has been without its critics. Religions and philosophies that are designed for moral and social conduct are generally not sufficient to govern a population. Despite the benevolence of religion and science in their purest forms, they have been used on many occasions for destructive purposes, ranging from the Islamic jihads and Christian crusades of the past to the lethal technologies unleashed in modern warfare.

The strong bonds among members in a population sharing common cultural interests (including religion) form a solidarity that should not be underestimated. In the aftermath of World War I, the treaty made at Versailles included the reorganization of geographic boundaries of several countries in eastern Europe—divisions made without respect of cultural boundaries. The oversight led to repeated episodes of bloodshed throughout the twentieth century as various ethnic populations sought independence.

Technology

In the late nineteenth and early twentieth centuries, popular opinion held that mankind armed with technology would champion adversity to pave the way to a utopian society. The accomplishments rapidly changed society, as they brought about the industrialized era. However, as the years passed, there were some disappointments in the melding between technology and human nature. George Orwell, in 1948, wrote a vivid account of an overpowering technological society with loss of individuality and freedom in his novel *1984*. The pendulum of popular opinion reflects the volatility of the future of mankind and technology.

Agricultural technology allowed Stone Age man to form stable sedentary communities. The improving efficiency of agriculture over the centuries that followed fostered the phenomenal growth of the global human population. A new era of agricultural technology began in the twentieth century with the application of genetic engineering techniques. Historically, agricultural biotechnology focused on protection from pests and on efforts to improve crop yield, nutrition, and taste. Recent efforts in molecular farming are creating genetically altered plants that produce medications used in the treatment of human diseases—vaccine- and antibody-producing plants are undergoing clinical trials. The conventional use of genetically altered bacteria to produce medicines is relatively expensive, while molecular farming techniques may significantly reduce cost. In some cases, a medication may be ingested directly by eating the genetically altered plant—giving new meaning to *hospital food.*

The social role of technology has been debated and frequently cited as a threat to replace the working individual. Historically, technology has actually freed laborers to explore less manually tedious occupations. Admittedly, there is a displacement of some labor-intensive jobs, but other opportunities are created by technology. In fact, during the last thirty-five years of the twentieth century the average American manufacturing employee works approximately one hour less per day, largely from the technological infiltration of the workforce. Furthermore, studies suggest that this additional free time has been utilized, at least in part, for the pursuit of intellectual or cultural interests.

The impact of technology can be a complicated interplay of multiple apparently unrelated factors. In the early years of the twentieth century, the automobile was notorious for being expensive and untrustworthy. Henry Ford redesigned the automobile into what he thought Americans wanted: affordable, simple, and reliable. In doing so, he created a massive industry that reshaped the future of America. The impact of automobile technology was far more widespread than just the development of a device for travel. Ford pioneered a new manufacturing technique for assembly line mass production, which ultimately changed labor economics. Furthermore, society was directly affected by the liberties offered by the automobile—it was no longer necessary for workers to live within city limits, and families no longer had to remain geographically sedentary. Thus development of affordable transportation for the masses not only brought the suburbs but also affected the nation's economic and social structure.

Historically, communication was clearly one of the landmark accomplishments of mankind—it secured our position as the dominant species of the animal kingdom. Technology has been extensively applied as a tool to enhance the ability to share information. Several recent national polls of the most influential individuals in human history placed Johannes Gutenberg at the top. His invention of the printing press in the fifteenth century heralded a new era of written communication, which until that time had been laborious, slow, and expensive. His innovation was unprecedented in bringing information to the general populace—at least to those who were literate. Nineteenth- and twentieth-century innovations of the electronic mediums of the telephone, radio, television, and computers have further revolutionized communications. In fact, communications technology has provided a means to easily overcome geographic barriers that have isolated populations throughout history.

The Internet was initially developed as a means for scientists to exchange information and papers, but after incorporation of the World Wide Web (WWW) network widespread access and versatility provided a new medium of communication for the masses. The Internet has already proven a useful tool for business, where rapid access to updated information and communication across time zones and national boundaries are nearly effortless. A recent survey by the American Management Association found 53 percent of those in executive and management positions spend up to four hours a week on the Internet and one-quarter of those anticipate continued increase in Internet utilization. However, the ability to easily reach such a vast audience invites abuse. Unfortunately, this service is subject to the same sorts of exploitation that have occurred in other facets of the communications industry, which bring consternation from special interest groups and the advocates and opponents of censorship.

The transfer of information and knowledge, including formal education, is a major priority of society. Computer-based communication networks have already begun to assume roles as sources for education, research, and entertainment. In fact, it has been predicted that the traditional form of higher education will be replaced by virtual forms within a few decades. Several major institutions of higher education already offer courses on the Internet. Proponents of a virtual educational system cite advantages such as information access to wider populations affording increased diversity across economic and ethnic lines. However, studies have shown that access to Internet technology is dependent upon household income and lower-income residences have more limited access. Critics argue that a purely virtual education deprives the student of the personal interactions with classmates and other social events—activities that are a major component of a university education environment.

The differences in quality between a virtual education and a traditional one are under investigation. In favor of remote learning through virtual university programs is the historical success of well-established college correspondence programs. In addition, educational forums for periodic updates of material, continuing education, and other brief topics are a very efficacious educational application of Internet technology. However, there are limitations of certain transactions that may be performed by such technology; humans are social creatures accustomed to interacting with one another. Despite many decades of telecommunications technology, plenty of business deals and other negotiations are still done face-to-face.

The union of computer and machine technology has had a major impact on industry. A subset of software technology used to run machines has come to be known as artificial intelligence. The spectrum of artificial intelligence technology can be divided into three principal categories: robotics, expert systems, and neural net technology. Robotics technology has been extensively exploited and is perhaps best known for its application in the automotive industry. Robotics technology generally utilizes machines designed to perform repetitive humanlike tasks with a high degree of speed and accuracy, leading to increased efficiency. Expert systems software combines a highly developed software algorithm and database and is designed to formulate conclusions based on the data provided, as an expert provides useful advice for a consultation. One application of expert system software includes medical decisions, where a patient's symptoms and physical exam findings represent input data from which the program derives a diagnosis or a list of probable diagnoses (with different degrees of likelihood). Neural net computing, the most complex form of artificial intelligence, has a biological basis in its design, capitalizing on the billions of years of evolutionary development of the central nervous system.

The human brain, which generates only ten watts of power—equivalent to a very dim light bulb—is the result of a long, arduous evolutionary pathway culminating in the creation of an extremely complex organ consisting of a hundred billion neurons, with interconnections (synapses) numbering many times more. The processing complexity of the brain not only lies in the number of its functional cells but also includes the multiplicity and plasticity of its interconnections. Nearly all of the neurons of the central nervous system are multipolar; they interconnect with more than one other neuron. In addition, the interconnections may cause different effects; stimulation or inhibition of the activity of the other neuron(s).

Further neuron signal complexity includes variability in the levels of output response by the rate of input signals received from other neurons. The many ways in which neurons process and transfer their signals differ from the relatively simplistic manner of a digital computer circuit (either on or off). The signal manipulations of neurons imply an element of signal processing, rather than assuming that the neuron acts as a simple transmission link or relay station restricted purely to data transfer. Indeed, the multiple levels of complex signal processing in the brain involve both anatomic and physiologic components. Some investigators propose that there is yet another degree of neuronal signal modification by quantum effects—they argue that quantum effects of the brain's neural networks are essential in creating sentience and, further, that these ultramicroscopic effects preclude the ability to create sentience in nonbiological models.

Anatomic organization of the neurons of the cerebral cortex is probably best exemplified by the visual cortices of the occipital lobes. It has been shown that activity of the neurons of the visual cortex is not a direct map of signal information from the retina and optic nerve but involves complex signal processing: specific visual field stimuli are required to elicit coordinated cortical activation. This pattern of organization is different from that used in the production of an image in photography or that displayed by the phosphorus of the cathode ray tube in a television monitor. In these technological cases, the individual elements of the image data correspond to specific positions on the image, with a relatively direct relation between the two. Conversely, the human brain manipulates the image projected onto the retina through a biological convolution creating a complex network of pattern-sensitive neurons: pattern recognition. Thus when a particular pattern (such as a vertical line moving from left to right) is detected by the retina of the eye, a subset of neurons in the visual cortex will react as they "recognize" their specific pattern. The various elements of what we see are broken into many patterns by the brain, and the recognition of the various patterns forms conscious awareness of the visual world. The neural network approach of artificial intelligence is rooted in the concept of pattern recognition and is implemented through mimicking the interconnections of the neurons of the brain. State-of-the-art software algorithms are far from producing a "sentient" machine, but the technology has many applications in industry.

"Directed evolution" software has cleverly borrowed the principles Charles Darwin introduced in evolutionary biology and applied them to design new algorithms. The technique, which was first proposed in the

1960s, combines data sets composed of parameters describing critical design aspects of the subject under investigation (analogous to the genes of chromosomes) to yield new designs through modifications of parameter combinations. The resulting parameter-modified "progeny" is analyzed to determine if it is an improvement (analogous to natural selection), and if so, its "chromosomes" are placed into the gene pool, where a new combination is made and the next "generation" model tested. In addition, "mutations" may be introduced representing random alterations in design parameters to further diversify the range of models created in subsequent generations. After perhaps a hundred generations, a virtual model of the product begins to take shape that approaches an optimized form based on the desired characteristics of the directed evolution program. The applications are nearly infinite, ranging from the optimization of product design to employee work schedules.

As the ability to construct and implement more complex computer algorithms escalates, it is conceivable that an accurate representation of the "human" decision process will someday be achieved. Albeit an extremely complicated organ, the human brain is composed of a finite collection of interacting cells, which implies that the possibility exists to create models that precisely mimic the neuron interaction, perhaps in the form of a computer program. Artificial intelligence utilizes the concept of basing computer algorithm design on models of decision making. Extension of this technology could be used to create a basic algorithm designed to remove (human) bias from the artificial intelligence decision process. Thus a consistent and "fair" decision matrix in theory may be achieved whereby rational decisions could be formulated purely by computational methods—a role previously considered appropriate for human intelligence alone. The applications of such technology are myriad. Business, medical, legal, political, military and countless other situations would be amenable to an artificial intelligence source of decision. Obviously, many of these facets of modern society already utilize computer technology to some degree.

Projections of continued progress in computer technology suggest that computers will be able to match the computational complexity of the human brain by the year 2030. A decade later, the capability will lie within a device that can be placed on a desktop. How much information does a human brain contain? Some experts claim the human brain contains roughly 100 million memories. If one compares the processing capability of a human brain (on the level of information processing by the neurons) with that of a computer, the brain performs approximately 100 trillion instructions per second—about 1 million times faster than a typical desktop personal computer of today. Like the human brain, much of the capacity of a computer is concerned with the gathering and processing of data, which in itself is crucial in the decision-making processes. Therein lays the potential for society to delegate its political and social decision process to a resource utilizing artificial intelligence. An optimal computational decision algorithm could be designed that would faithfully reproduce the desirable outcomes as determined by the opinion of society, ensuring public support. Parameters, such as economic and social indicators, could be systematically evaluated, and analysis would supply data for application in the design of new policies. Furthermore, proposed policies could be tested using directed evolutionary techniques, to arrive at an optimal political solution. Although the ability to create a completely nonbiased decision-making political system may be possible in theory, it is doubtful that human nature would permit man to relinquish control of the future of society to a technological device utilizing basic biological principles. However, in the meantime, the dependence on computer technology continues to rise.

Computer technology, a very recent product of human innovation, has already permeated nearly all facets of society in developed nations. In fact, modern philosophers suggest that computer technology is the latest form of enslavement. The implication of an inanimate beast of burden is a curiosity. Although a topic of science fiction writers, perhaps a reversal of roles between man and technology is worthy of contemplation as technology has a major influence on the direction of industrialized society.

Part IV

Ad Infinitum

But for man, no rest and no end. He must go on.
                                    —H. G. Wells, *Things to Come*

11

Immutable Legacy

Although humanity's future may be unpredictable, it will undoubtedly flow from a legacy of incessant striving for innovation and progress. The path from the dawn of sentience to modern civilization was not a trail of predictable sequential achievements but more akin to the complex interaction of man with his ambitions, his nature, and the environment. Throughout history, innumerable civilizations and political and social establishments arose, but fate has left just a few as tales of historical lore while most have been forgotten.

Beyond intrigue, is the future of humanity worthy of serious investigation? There are discernible adverse changes in our environment directly attributable to the influences of man and society. Although the magnitude of some of these alterations is in question and their ramifications the subject of speculation, the implications of serious consequences affecting the future survival of humanity clearly deserve attention. It is unfortunate that it is not uncommon to ignore important issues, particularly when they seem unsolvable, a convenient method of avoiding the displeasure of presumably pointless anxiety. Perhaps the most relevant issues of immediate danger to humanity are those of a self-inflicted nature, particularly warfare and terrorism. As these confrontations put large populations at risk and are based on ideological concerns, their resolution requires political intervention. A recent study shows that over half of Americans surveyed fear that civilization will be exterminated by a man-made disaster sometime during the twenty-first century. The ability to avoid devastation from many of the possible global disaster scenarios, both natural and otherwise, is actually within the control of humanity.

The doctrine of determinism claims future events are predictable, as they are the direct consequences of the past. It encompasses all events of nature, including the formation of the Universe, our own evolution, actions, thoughts, and dreams. Furthermore, the principles of determinism can be extended to include predestination—the future is completely predictable. Through the work of great minds, such as Sir Isaac Newton, the mystical workings of the Universe gradually entered the realm of scientific investigation, where phenomena could be rationally explained and predictions made. Fueled by a rapidly growing scientific knowledge base, the opinion arose that through intensive mathematical investigation of the properties encompassing all matter, space, and time reliable predictions of all future events was inevitable. Thus if at the beginning of time the relevant physical properties of all matter were known, then their future interactions could be calculated with scientific precision. Furthermore, not only would cosmic events be predictable (i.e., formation of galaxies, stars, planets, life) but also the actions of man (as thoughts are produced by the brain, which is composed of matter and therefore obeys its properties). The very foundation of this "mathematically precise and predictable" Universe came into question as the understanding of the properties of nature's ultra-microscopic structure began to take shape.

It was in the early twentieth century when the properties of quantum mechanics were first uncovered. Unlike the teachings of classical physics, such as those followed for centuries based on Newtonian principles, the principles of quantum mechanics are intimately associated with probabilities and therefore "blur" calculation results—the quantum mechanical realm is notorious for lack of absolute precision when considering a single parameter or particle (Heisenberg Uncertainty) but is extremely accurate when describing behavior of large collections of particles. Thus it is not possible to exactly predict the interactions of individual particles of matter (and energy). Consequently, the absolute predictability of determinism is lost—at a fundamental level, unpredictability prevails.

The inability to predict future events may leave some with a feeling of dissatisfaction. Einstein found the statistical properties of the quantum world disturbing—in a letter to a friend he included the famous statement: "I am convinced that God does not play dice." Upon learning of Einstein's comment, Heisenberg purportedly questioned Einstein's intimate knowledge of God's activities.

12

Evolution

Since its arrival, terrestrial life has been subjected to the whims of natural selection, which produced the biodiversity of our world. Models of the process of evolutionary change describe it as one of sporadic periods of large-scale species radiations followed by long-term survival of relatively few distinct species types. When Charles Darwin was proposing the processes responsible for evolutionary change, he believed the possibilities of creature variation were unbounded. Despite a theoretically limitless range of phenotypical variations, generally only a few distinct morphological forms are expressed. This pattern of limited phenotypes is a consequence of restrictions imposed by constraints on an evolving species. Evolutionary biology theory suggests that evolutionary constraints favor long-term stability of species' structure over tens of millions of years. Analysis of families of vertebrates reveals many situations where retention of similar anatomic structure has persisted for periods lasting over a hundred million years.

Evolutionary constraints come in a variety of forms, including environmental, developmental, structural, and biochemical. The environment in which a creature lives will tend to favor (i.e., afford an improved chance of survival) certain species characteristics, which in turn limits the spectrum of phenotypical traits expressed in surviving offspring. Species living in aquatic habitats clearly have different environmental constraints from those in an arid region. Under the boundaries imposed by environmental constraints, nature has many examples of independently evolving species occupying the same environment yielding very similar structural forms in their evolutionary pursuit of optimization. The gradual isolation of the continents with the breakup of Pangea provided many similar environments that were populated by local species. Isolation of the Australian continent led to extensive development of parallel phenotypes. The Australian marsupials have diversified to occupy many niches, with some creatures that are strikingly similar to different species found on other continents occupying similar niches. Developmental constraints refer to the limited amount of biological substrate a species has from which to call upon to produce modification. Evolution generally works through the modification of existing structures (ultimately defined by genetics) to cope with the pressures of natural selection. Examples of developmental constraints include the modification of mammalian forelimbs into wings (in bats) or fins (in dolphins). There is a point of irreversibility along certain pathways, as in cases where evolutionary modification eliminates expression of a particular structure, thereby eliminating it for use in descendants. However, under certain circumstances, new sections of DNA may be added, or others that were dormant may again be expressed, providing the opportunity for expression of new structures. Structural and biochemical constraints refer to limitations imposed by the types of body tissues and their biochemistry. Tissues have inherent qualities that place limits on their properties, such as strength and elasticity. The biochemistry of organic life was determined billions of years ago, and such fundamental qualities cannot be altered easily without the high probability of incompatible (lethal) results. In the event of "relaxing" of evolutionary constraints, the potential for major adaptive radiations occurs and new species may result.

In the rise to the present form of *Homo sapiens,* characteristics that were particularly influential were intelligence and dexterity, which are evident in the advancement of the human brain and body. The intellectually gifted hominid tended to survive and had offspring that tended to be intelligent (as they resembled the parent and received a similar genetic inheritance). To the contrary, in civilized society individual survival (to produce offspring) is less dependent on the primal necessities of ancient man but more closely linked to social aspects. Currently issues of health (e.g., nutrition, disease), which are often a consequence of economics, public health standards, and political establishments, are the influential elements in human survival on a global scale, particularly in the case of less-developed nations. In developed nations, as in the United States, young adult mortality is more likely from trauma (e.g., car accident, firearms, etc.), often a consequence of urbanization and other social pressures. Despite environmental differences from those of early hominids, the forces of natural selection may seem less heavily applied in the present but are not entirely absent. In terms of survival, those with genetic resistance to various diseases (e.g., the AIDS virus), other infectious agents, and cancers—have an advantage. Some forms of natural selection are actually more cultural than natural, including cultural bias, genetic counseling, and eugenics. Furthermore, the possibility of technologically introducing new genes into the human genome invites the opportunity to circumvent natural selection entirely.

Generally, evolutionary processes are effective through isolated subpopulations that develop favorable traits that enable them to either replace or, after many generations, coexist as new species. The gradual decline of cultural barriers through education, increased communication, and world travel leads to ethnic intermixing. In its ultimate form, the production of a single race, the combination of the individual factions, is a possibility: an undoing of racial identity achieved during the previous 100,000 years, to return to a single *Homo sapien* form. However, various forms of voluntary segregation, such as those prompted by cultural and religious motivations, will likely prevent complete ethnic intermixing.

Will humans continue to evolve? During the past 2 million years of human evolution, the brain has doubled in size, language developed, and in the past several hundred thousand years the body size and body hair have decreased. Thus should we expect these trends to continue, with humans becoming hairless, frail, large-headed creatures? Furthermore, the rise of technological comforts has brought (comical) predictions of regression of the arms and legs, as they will be unnecessary in the distant future. Such predictions should not be taken seriously, as they ignore the principles of evolution and, in part, are based on the outdated Lamarckism theory. The subduing of the effects of natural selection largely through technology, combined with a relatively homogenous world population, argues against any significant alteration in the human species in the foreseeable future. In addition, for evolution to have a significant impact, it usually involves relatively small groups where newly introduced genes can be more effectively distributed among the population and into future generations—the large and mobile populations of today blur the prospects for change, as new genes are likely to be "lost" in the masses. However, future evolution of the human species may be driven by biotechnology through direct alteration of the human genome.

Some authorities argue that "de-selection" is being applied to the human population on a global scale. Advances in medical science allow the successful treatment of ailments that until recently carried high mortality rates. Certain infections, cancers, and even genetic diseases that had been uniformly fatal are now treated with nearly miraculous results. Although the treated individuals clearly benefit, should they have progeny, those offspring may inherit the same genetic susceptibility of the parent (which otherwise may have been eliminated had the parent not been treated). However, the majority of modern medical care treats those beyond child-bearing, making inheritance issues less direct. Another source of broad "artificial" selection pressure is manifest during times of war, when a largely healthy sector of the population is subjected to increased mortality.

Are evolutionary pressures still in effect? Are certain traits being selected over others? Individuals may prefer expression of certain characteristics in their children, which may in turn influence their choice of mates. Perhaps, through genetic manipulations in the future, it will be possible to alter the DNA of a child early enough so that the desired characteristic may be expressed independent of the parental genetic contribution. If the change were truly an alteration to the DNA (and included in the chromosomes of the offspring's germ cells), a transmittable genetic trait would be artificially created. This concept may be expanded to allow creation of new traits (new genes). Under such a scenario, the course of genetic change in the human evolutionary path would be a form of *unnatural* selection.

The rise of humans from sentience to cognitive intelligence is perhaps the greatest achievement of terrestrial evolution. In addition to the genetic recipe for the structural components, the development of the human brain is also affected by extragenetic influences of the environment. General health, nutrition, and appropriate social stimulation are crucial elements for normal development of brain function. According to the fossil record, the basic anatomical design of the human brain has remained generally unaltered for a hundred thousand years or longer. As society continues to advance technologically, one may wonder if human intelligence will remain adequate to function in an increasingly complex environment. In 1982, Arthur Jensen, professor of educational psychology at the University of California, Berkeley, proposed that in a complicated technological society (i.e., where intelligence is at a premium for societal stability and function) the population must be able to assimilate an adequate knowledge base. He contends that innate human intelligence has limitations imposed upon it and that the ability to improve intellectual prowess will require eugenic manipulations to enhance intellectual capabilities to cope with the advanced societies of the future. However, one can argue that since the society in which man lives is fundamentally his own construct, its properties are ultimately derived from human intellect.

The twentieth-century American philosopher George Santayana proclaimed "Those who cannot remember the past are condemned to repeat it" as a paraphrase of human nature. Although there may be changes in the physical and social environment—technological, governmental, cultural, etc.—the qualities of human nature will remain and continue to shape humanity's pathway. Indeed, there hasn't been significant change in humans for the past hundred thousand years, and probably the same will be true for the next hundred thousand. Although there is no clear evidence to suggest future speciation of *Homo sapiens,* the path is made somewhat uncertain with (new and possible future) advances in genetics. The evolutionary consequences of these techniques, should they be applied (if deemed ethical by those in the future), is only speculation at this time.

13

Technological Prowess

As man began to investigate the workings of the natural world, the discipline of science was born. Technology applies the principles of science to provide means to improve the human condition. The ubiquitous technologies of today are the results of thousands of years of human progress—future technologies will stem from concerns of times yet to come of which we can only guess.

Many of the recent advances in technology have focused on the development and improvement of computer-related devices. Miniaturization and increased computational speed have revolutionized the capabilities and applications of such technology. It has been predicted that by the year 2020 the miniaturization of the components of microchips will be to the point where a wire in those circuits will be the width of an atom; the limit of miniaturization will have been reached. First introduced in 1965 by Gordon Moore, a cofounder of Intel, was "Moore's Law," which states (in its updated version) that the number of transistors that can be placed on a silicon chip will double every eighteen months. The concept has held during the three decades since it was proposed. In order for the Moore's Law principle to remain true, the next generation of semiconductor chips should arrive by 2005. However, this next generation will be at the limit of current technology, where the individual transistors are composed of fewer than 100 atoms. At that point, the physical qualities of the materials will approach thermodynamic limits and even quantum mechanical effects will begin to interfere with transistor function. Thus the microchip industry is approaching a physical barrier that will require fundamental changes to its technology if it is to continue to progress at the same rate.

The next major phase in the computer innovation industry may be the quantum computer. First proposed in the 1980s, quantum computers were thought to be inevitable, as the microchip miniaturization cycle would eventually lead to computer components so small that their properties

278

would fall into the realm of quantum mechanics. In the years that followed, theoretical advantages to the computational capabilities of quantum logic arose. The language of conventional computers (Boolean logic) is a binary mathematical code where an individual datum is represented by a 1 or 0 (electronically either a charged or uncharged capacitor) and denoted as a bit. A quantum computer stores data by quantum phenomena (ground or excited states of an atom, photon polarizations, nuclear spin states, etc.). Although two quantum states may be used to define a 1 or 0 (as in binary systems), superposition of quantum states allows both (1 and 0) to simultaneously exist. Thus a conventional two-bit computer can exist in only one of four states at a time (00, 01, 10, or 11), while a two-qubit quantum computer can exist in all four—the same concept can be applied to larger quantum computers, where the difference between the two computers becomes exponentially more significant. Furthermore, the quantum computer can theoretically perform calculations simultaneously in each of its states (quantum parallelism), with a single value occurring once the output is measured.

The parallel computational capabilities of the quantum computer have practical applications. In 1994, Peter Shor, a researcher at AT&T, proposed that quantum computation would be highly efficient in determining prime number factors of large numbers, a calculation that is tedious by conventional computing. Shor's algorithm is useful in cryptography, where the security of a code relies on the complexity of the factoring of its base. A conventional supercomputer may take 10 billion years to factor some of today's encryptions, whereas a quantum computer of several thousand qubits may solve the problem in less than a minute. In addition to code breaking, the quantum computer may be useful in searching large sets of data—a capability that becomes more useful as the information age continues.

The construction of a quantum computer is challenging, as their operation requires isolation from outside influence—once a qubit is disturbed, it will lose quantum coherence and assume a definite state. As long as the system can be put into quantum superposition and the qubits interact without significant decoherence, it may function as a quantum computer. To achieve computational practicality, the quantum bits must be suspended in their indeterminate state long enough to perform about a million operations—approximately a second. Currently the state may be maintained for about one hundred microseconds (ten thousand times too brief). Theorists anticipate the multiple states of the qubits of quantum computers will require quantum error correction and calculation reliability algorithms to compensate for decoherence. Although quantum computers are in their infantile stage, there are several methods under investigation—particularly promising techniques include quantum optical systems and nuclear magnetic resonance, NMR (the same principle behind MRI scan technology used for medical diagnosis). Qubits have been made in the laboratory using interactions of the photons of polarized light as they pass through an array of cesium atoms. Nuclear magnetic resonance can be used to "program" the spins of hydrogen and carbon nuclei (in a chloroform fluid) for quantum calculations where the result affects the magnetic field (from the nuclear spin states), which is detected by an RF antenna. Some experts predict that quantum computers able to perform certain types of calculations about a billion times faster than the fastest microprocessors of today will be available by the year 2030.

Historically, the future consequences of technology have often been unpredictable. The influence of a given technology can have ramifications outside the original concept, including extensive economic and social implications. In the case of computer technology, the very intimate association it is developing with society has potential for widespread consternation. The benefits of computer technology have allowed it to flourish; however, beneficial aspects can also provide opportunities for exploitation. Software innovations (e.g., smart viruses) repeatedly demonstrate their potential for serious interference or destruction of information, programming, and communication capabilities. In addition, as computers and their programmers are not infallible, there is always a chance of malfunction without intent or malice. The infamous Y2K (Year 2000) software glitch earned widespread media attention at the close of the twentieth century—raising concern from the home computer owner to the heads of national and international businesses and governments. Although the Y2K impact was essentially negligible, some of those who had predicted major technological disasters (which did not occur) claim the billions of dollars spent analyzing and rewriting software beforehand prevented catastrophe. Yet countries that did little or no preparation for the Y2K software glitch had little consequence. However, the dependence of so many facets of modern society on computer technology must not be underestimated. As the pillars of society continue to lean more upon computers and related technology, the consequences of damage from a failure of this technology become increasingly serious.

The consequences of technology on civilized society have

well-known industrial, political, and economic implications; its effects on the social structure are more complicated. There is evidence that modern communications technology may have negative effects on social interactions. A recent study on the effect of those spending time with the electronic social mediums of E-mail and the Internet has indicated a relative decrease in time spent interacting with family members and friends. In addition, individuals had decreased their frequency of social contacts and had slightly, but statistically significant, increased levels of loneliness and depression. These findings suggest a trend toward deterioration in the social and psychological status of individuals spending time on electronic social networks. Previous studies on the effects of television viewing also indicate a reduction in social involvement. Furthermore, it has been suggested that humans find virtual social relationships less psychologically satisfying than those entailing conventional social interaction. Robert Putnam, professor of political science at Harvard University, points out a trend toward individual isolationism among Americans since the 1960s, which he attributes to the developments in communications technology. The lack of psychological satisfaction from virtual (or remote) relationships stems from a fundamental need of humans to be able to directly interact with fellow beings. The absence of adequate physical contact in newborns has been shown to lead to inappropriate behavioral development that may last a lifetime, and in extreme cases it may have fatal consequences. Thus there may be limitations to the extent to which humans and future technology may peacefully coexist.

The late twentieth century saw the dawn of the era of integrated biotechnology. Among the great advances in medicine is the ability to successfully transplant organs, which has given extended life to many patients who had little hope of survival only a few decades before. Temporary biological support through biotechnology has become commonplace—e.g., cardiopulmonary bypass used in cardiac surgery, dialysis machines for kidney failure, etc. Limited success has been achieved with the implantation of artificial organs to replace or assist ones that are diseased. The applications for artificial life-sustaining devices implanted *in situ* have become increasingly diverse. Prosthetic limbs have been used as structural support for centuries; however, it has become possible to give these devices mechanical function guided by conscious control of the prosthesis recipient. Individuals with certain forms of deafness can regain some hearing through the implantation of a small cochlear implant device, which directly stimulates the auditory nerves of the brain according to sounds its

281

microphone detects. Experimental work through implantation of tiny integrated circuit chips into the retina has given people with certain forms of blindness a rudimentary sense of sight. Thus technological assistance of biological function can be used to sustain or even improve the quality of an individual's life. Currently such biotechnology is generally reserved for those under severe circumstances. Investigators studying other aspects of human health, such as the aging process, seek ways in which to artificially extend life, even in the absence of known disease. The application of biotechnology to enhance the function of the human body beyond that considered normal (e.g., increasing strength, improving vision, expanding intellectual capacity, etc.) is an intriguing venue of future technology.

Historical review of theories on the development of technology divides the models into two major categories: *revolutionary* and *evolutional.* By the late nineteenth century, the first credible attempts at explaining the progress of technology were based on the revolutionary model. This view suggests that advancements in technology occurred as a series of relatively discontinuous innovations. The evolutionary model followed, which has since become the more popular theory. It considers technological advancement as a consequence of accumulated knowledge from the results of a series of innovations. On a more individual basis, technological advances are the results of investigation utilizing contemporary means to achieve a goal. As the tools of investigation and goals are continually evolving, the overall scheme of technological development favors a continuum more than a discontinuous series of events. The "selection pressures" for the products of innovation may be political, such as economic or military, but may be culturally or socially driven as well. Thus the direction of future technology will be largely influenced by the course of future society. As in the present, future large-scale roles of technology will be controlled to some degree by financial feasibility and practicality. The bounds of technology may seem nearly infinite, but without a clear benefit or market value many inventions fall into obscurity. Unfortunately, the use of technology for perverted political motivations can lead to great devastation, as evidenced repeatedly in human history.

Despite all of the humanitarian aspects of technology and its marvelous advances, one must be cognizant of intentionally destructive applications of technology: the resolution of conflict. Perhaps stemming from retained aggressive behaviors that gave modern humanity's predecessors a survival edge, these characteristics breed the development of technological methods to incapacitate an enemy. The earliest forms of organized conflict among humans probably occurred during the hunter-gatherer stages, with the raiding of small unprotected settlements by hunting parties—spoils obtained by the threat of fist, club, or spear. As the technology of hunting instruments progressed, raids could be performed more effectively. As time passed, political bodies would muster armies, which could be sent to subdue an enemy or protect their own community. Organized warfare was perhaps first clearly developed by the hoplite soldiers of the ancient Greek provinces, where fighting was typically over disputed territories. Alexander the Great modified the weapons and strategies of Greek warfare on his successful campaigns of conquest throughout the Mediterranean basin and Asia. It wasn't long before Roman engineering and military tactics further improved and organized military capabilities, which enabled the creation and maintenance of their empire. In the centuries that followed, interval discoveries—primarily ordnance materials (e.g., explosive materials)—brought much more devastating weapons.

The twentieth century brought the construction of weapons that harness the same method of energy production used by the sun: nuclear fusion. In the wars fought during the twentieth century alone, there have been over 100 million people killed, a toll greater than all of the dead from all previous wars since the beginning of human history. The dedication of time and resources to this industry is staggering, but one must agree that the "progress" made is equally staggering. The atomic detonation over the city of Hiroshima killed 80,000 people instantly, with about 200,000 dying afterward: one weapon, smaller than a sports car, can kill a quarter of a million people. The hydrogen bomb, which uses nuclear fusion (rather than the fission process of the Hiroshima and Nagasaki atomic bombs), releases about one thousand times the energy of an atomic bomb. Ironically, despite abundant technology, more coalition forces were killed by friendly fire than by their less high-tech enemy during the Gulf War. As effective as these weapons of advanced technology have become, the much more mundane weapons are used in the bulk of conflicts. The international terrorist attacks on the U.S. World Trade Center and Pentagon buildings on September 11, 2001, using commercial aircraft and the (less costly but equally alarming) subsequent distribution of anthrax spores as a biological weapon through the postal mail system killed several thousand American civilians. In 1994, within just a few weeks, 800,000 Tutsi were killed by Hutu factions in central Africa, spurred by the revival of cultural tensions that arose with the withdrawal of governmental control by Belgium—the genocide was conducted with little more than guns, knives, and axes.

Technology is often sought for solutions to many of the great problems that plague mankind: pollution, population growth, disease, etc. However, at the core of these major concerns are sociopolitical forces, which likewise require sociopolitical solutions. Although technology may provide some solace, one should proceed with caution, as technology may not give a lasting remedy but, rather, postpone inevitable consequences. Technology has enjoyed widespread use in making our lives more comfortable, improving health, communication, and transportation, but has also been touted as a solution to aggressive human conflict with inventions such as the machine gun (designed by a Civil War physician, Dr. Richard Gatling) and nuclear weapons—devices ironically labeled as technological solutions to *end* warfare when they were first developed.

A theoretical application of technology in the distant future (about 10^{30} years or so) would be to sustain the physical existence of mankind itself. The very matter of which the Universe is composed may be unstable and, if so, will eventually decompose (according to the Grand Unified Theory). Thus the atomic building blocks of everything in the Universe, including ourselves, will gradually become scarcer and scarcer. Speculation as to how humans could survive in an era of disintegrating matter has been explored by Freeman Dyson, professor of physics at the Institute for Advanced Study. He proposes that, in the distant future our descendants may be able to cope by converting the matter of their bodies to those forms of matter that will be available. His argument is based on the premise that the human body is a physical structure that allows what we consider *human* to exist within it. Thus, if the function of the body can be exactly reproduced (albeit even if constructed of different material building blocks), it would be able to support "life." This form of existence would be nonhuman in its physical form, an evolution into a radically new species that is driven by technology itself!

Analysts of technological accomplishments suggest that it is progressing at an exponential rate—a rate of growth that is unparalleled throughout human history. One must exercise modesty before predicting an impending decline in this escalation, as the limits of technological achievement have been seriously underestimated historically. In fact, in 1898 Charles Duell, the director of the U.S. Patent Office in Washington, D.C., recommended permanently closing the office with his famous claim: "Everything that can be invented has been invented."

14

Future Civilization

In *Things to Come*, written by the English novelist and historian Herbert George Wells during the early twentieth century, we find a future marked by global devastation resulting from extended warfare and disease. Wells proposed that science and technology were the means to bring civilization back from the brink of extinction, but not without the price of civil discontent. Fortunately, human history has provided a long record on which to base conjectures of future trends. Although society may seem entirely different in the developed nations of today compared to what it must have been like on the savannas of central Africa a million years ago, there are underlying similarities. The basic qualities of human nature persist through the ages and in part shape the societies in which we live, as they did for the ancients.

The basic drives of "human nature" were responsible for the survival of our ancient predecessors—these primordial ambitions can also interfere with the stability of civilized communities. The disciplines of intellect and rationality must prevail to control primitive socially obstructive drives, which are a natural component of the human psyche. Perverted ambitions have plagued society since antiquity, from the emperors of ancient Rome to notorious leaders of modern-era nations. Romantic predictions of future society as a congenial utopian community are comforting but not realistic, as they ignore the history of humanity and would basically require a fundamental alteration of human nature itself. As long as society is populated by humans, it must acknowledge the consequences of human nature. One can review the works of philosophers, such as Saint Augustine of ancient Rome, and find that the dregs of his society were not unlike those found in modern civilized society. One should not despair—despite the counterproductive nature of some elements of human behavior, great accomplishments have been and will undoubtedly continue to be made in the future. Since the earliest civilizations, humanity has struggled with many forms of adversity—including disease, malnutrition, poverty, and illiteracy. Although science and technology are often turned to as a source of solutions, solving these great maladies will require sociopolitical interventions on a global scale to have a significant chance of decisive eradication.

The continual fragmentation of boundaries between cultures—largely through improved education, communication technology, and increased travel—may lead to a new era of cultural assimilation. Perhaps the ultimate result will be a dominant uniform world culture, an eclectic blend of elements borrowed from the many individual cultures that preceded it. On a smaller scale, this type of assimilation has played out in previous civilizations. The ancient Greek civilization adopted characteristics of the neighboring Mycenaean, Mesopotamian, and Egyptian civilizations, which were then incorporated into Roman culture when Greece was conquered in the second century B.C. The cultural elements of language, art, philosophy, and religion were significantly affected through such historic assimilations, often leading to subsequent hybrid cultures.

The dynamics of human civilization involve a balance between environment, population density, culture, and sociopolitical and economic forces. The environment has constraints; it can support up to a certain maximum population density, determined by limited resources of raw materials. As a community's population increases, the rising demand on resources required to sustain it become increasingly more difficult to secure. The cultural, political, and economic influences serve to define the lifestyle of a populace and also maintain stability of the civilization. Predictions of future communities often anticipate large densely populated cities surrounded by zones of relatively sparse population density that are largely reserved for agriculture and natural interests. Either the lifestyles and culture of individuals within these planned cities may reflect a diverse mix of subcultures or perhaps each city may favor a homogenous culture to serve a community with more focused interests. A third option, a combination of the previous two, may be the more likely, as it more closely mirrors the majority of civilizations of the present and past.

The role humans will play in civilized communities of the future may become redefined by technology. The industrial revolution and subsequent development of mechanized industries have increased the importance of machinery in relieving man of laborious work and redefined the social environment of labor. Just as man's role as a laborer has been changing, the role of intellectual contribution may also change. As artificial intelligence technology evolves, a new era of applications is on the horizon.

Perhaps many of the decision-making tasks of society will be allocated to technology (perhaps in the form of software algorithms): from medical evaluation and business investments to political policy. Although such applications may seem outlandish, the process is under way. Currently Computer Aided Design (CAD) technology is widely utilized, where computers play an active role in device design, including the design of other computers.

Just as humans have undergone evolutionary change through adaptive mechanisms for survival, so, too, must culture and civilization. Historically, adaptation by civilization has involved a balance of relations with other neighboring civilizations, adjusting to changes in climate and accommodating to geographical constraints. A centralized bureaucracy is an efficient way to rapidly and effectively direct efforts to these ends. However, in some cases such adaptive influences may lead to a static stability, resulting in a stagnant civilization. Theocratic societies tend to fail through their underlying inflexibility when encountering pressures of cultural change. Severe societal restrictions may create an environment of stagnancy, which prevents society from further advancement, as in the Middle Ages of Europe. Under these conditions a civilization that is unable to adapt may collapse, followed by sociopolitical reformation.

Failures of civilizations of the past can often be traced to an ineffective bureaucracy that failed to maintain the stability of society or protect it from internal or external sources of decay. In civilizations undergoing growth and expansion, the core bureaucracy will incorporate newly assimilated societies in its jurisdiction. In order to assure stability and avoid insurrection, the core bureaucracy may compensate through political adaptation, which aligns new populations with the collective of the remainder of the civilization. Eventually, through repeated cycles of adaptation, the malleability of the core political system becomes strained. Eventually the core civilization may lose the ability to effectively adapt any further, while the dominated societies are relatively free to undergo rapid change. Many examples in history of expanding civilizations illustrate how their demise was brought about indirectly through interaction with assimilated peripheral populations—exposure to the dominating civilization enables newly acquired communities to rapidly borrow capabilities (i.e., tactics, weapons, etc.), which may be used in retribution. In addition, newly incorporated territories may be politically dynamic, since they would not have the bureaucratic constraints of the core civilization. In the case of the Roman Empire, the frontier societies were able to take advantage of the stagnation of Roman civilization to overthrow it. Historically, expanding civilizations tend to regress as they mature and adapt.

The collapse of many previous civilizations has been attributed to a variety of causes, including military defeats, environmental disasters, etc. However, the decline of some of the great ancient civilizations, including Egypt and the Indus Valley, was through stagnant or declining political economies. Competition for trade seems to have been the primary mechanism in the decline of these ancient civilizations. Applying these theories to modern civilizations to predict future trends is difficult. In the case of the United States and many other industrialized nations, certain crucial components of civilization have been established and the ability to implement change is limited. A contemporary example is the dependence of many industrialized nations on fossil fuels. Reserves of these natural resources are limited, and eventual depletion is inevitable—which places pressure on these societies to seek alternative fuel sources to avoid devastating consequences. In theory, societies that have not developed a fossil fuel dependence are able to more easily adapt to alternate sources of energy.

The role of the environment in civilization may seem subliminal, yet it can lead to widespread devastation. The annihilation of population groups from depletion of natural resources is a recurrent theme in relatively localized populations or groups. In some cases a critical resource such as food becomes scarce, leading to mass starvation and death, as with a World War II Japanese garrison stranded on Wake Island in the North Pacific or Josef Stalin's famine campaign in Ukraine. In the majority of cases, the decline of natural resources was more gradual. Easter Island, a small isolated island 3,200 kilometers west of the shores of Chile, was first inhabited in A.D. 400 by Polynesians who sailed almost 4000 kilometers southeast from the Marquesas Islands. After centuries of prosperous habitation, in about 1680 conflict developed among the islanders as their primary resources (land and trees) were being depleted. Captain Cook visited Easter Island in 1774 and noted that several of the famous *moai* statues had already been toppled and the meager resources of freshwater and crops were nearly unpalatable. During the next century, sporadic visits to the island confirmed the degrading conditions, including near-complete elimination of trees, wildlife, freshwater, and other natural resources. Exposure to smallpox and leprosy had reduced the island population to only 110 people by 1870. Even worse fates occurred on the small, remote islands of Henderson and Pitcairn in the South Pacific. On these islands, the small populations depended on imports from neighboring islands. When trade ceased, survival became increasingly difficult, to the point that cannibalism developed and the original population subsequently vanished entirely. Examples as recent as the twentieth century include the Indian population of San Nicolas island off the Southern California coast, which was reduced to one woman who survived in complete isolation for eighteen years.

Environmental vulnerability is not restricted to small island populations. In A.D. 982, Erik the Red journeyed to southern Greenland, where he established a thriving Viking civilization. Eventually two Greenland settlements were formed (the eastern and western settlements), with a total population of about six thousand. During the early years of the settlements, there was a pleasant climate in southern Greenland, as the Earth was entering the relatively warm period of medieval times. However, cultivation was challenging, and as the climate began to cool over the centuries that followed, the ability to sustain crops became increasingly difficult. Furthermore, the settlements were heavily dependent on supplies from Iceland and Norway. Visiting ships that brought supplies became less frequent. By the mid–fourteenth century the western settlement had been mysteriously abandoned: a visitor to the settlement found the livestock wandering around the village without any sign of people. However, the eastern settlement would continue to confront the hardship for more than a century longer but eventually failed. In about 1540, a German merchant ship found the eastern colony deserted with the exception of a single corpse—a clothed man facedown in the dirt with a knife at his side. Forensic studies of buried remains indicated that the health and life expectancy of the settlers declined in the later years. The Vikings failed to adapt to the mini–ice age conditions, unlike the native Inuit (who survived the hardship and still inhabit Greenland to this day). Apparently the Vikings were reluctant to change their customs of diet and clothing, perhaps through intolerance of their Christian religious tradition, and by the end of the fifteenth century had eaten their breeding stock, sealing their fate. Human remains are notably absent from the living quarters at the Greenland archaeological sites—one would expect to find some human remains in their dwellings representing the last of the colony to die. Historians suggest that as the population dwindled, the last few survivors probably left by sea in hopes of finding a better life; their destiny remains a mystery. As the last traces of the Greenland colony faded, Europeans were making their first explorations of the New World, a continent abandoned by the Vikings 500 years earlier.

The extinctions of isolated communities and cultural groups may seem insignificant in comparison with the world population, but the examples do provide insight. Generally, these populations were isolated and their dependence on outside sources of materials made them vulnerable. Although the apparent abundance of natural resources may be comforting, the first settlers of Easter Island must have had similar thoughts. Likewise, the resources of the entire Earth seem vast, but they are not infinite and therefore cannot indefinitely support growing civilizations. Eventually a balance must be achieved between the consumption and replenishment of natural resources—the price of failure is too high.

Ever since ancient man ventured from northern Africa 100,000 years ago, migrations across the globe have continued, and until only recently have settlements been established on all of the world's continents. However, the idea of no further territorial expansion may be premature. Successful habitation in certain hostile environments can be achieved with the aid of technology, including places that were the subject of science fiction just a few decades ago. In fact, the feasibility of human occupation of the Moon has been raised with the recent evidence of water at the polar regions, as suggested by the Lunar Prospector spacecraft's neutron spectrometer. There may be up to 11 million tons of water beneath the lunar soil in an area up to 50,000 square kilometers—enough water to support a population of 2,000 for more than a century. This figure could be significantly extended with recycling. The lunar water could be used for drinking, as a source of oxygen for air mixtures, and as a source of hydrogen for fuel. However, the long-term stability of lunar colonies, to some extent, will be vulnerable to the same strains historically endured by the isolated island nations of the Pacific.

Projections of human survival in the distant future have been considered should the Earth be no longer habitable for various reasons, e.g., overpopulation, environmental inadequacies, or other catastrophes. Princeton physicist Gerard O'Neill, in 1974, proposed future populations could live in huge spinning cylinders orbiting the Sun. Rotation of the cylinder on its axis would simulate gravity along its inner surface, where inhabitants would live—the inner surface of the cylinder would be designed to approximate the Earth's terrain, complete with residential communities and parks. Such communities might relieve the stress of a large population on the Earth's surface, but the construction and maintenance of the orbiting communities would be colossal projects. There has been a recent growing interest in human habitation of the planet Mars. The logistics are far more demanding than those of lunar voyages, but Mars does have a major difference: an atmosphere. Some of the proposals include utilizing the Martian atmosphere as a resource to manufacture the fuel for the return trip to Earth. Furthermore, long-term plans of terraforming have been outlined where terrestrial organic biochemistry is introduced on Mars to convert it into a more Earth-like planet (although much colder).

Will human intellect and ingenuity lead to an ultimate civilization of a purified political, economic, and social environment where everyone enjoys a fulfilling and healthy life? Throughout thousands of years of human history, no lasting "perfect society" has been developed. In fact, the history of civilization has been a series of rises and declines of many societies. The pattern is likely to continue, as the nature of man is constant and civilization is a construct of humanity. It may be that the primitive behavioral drives of humans preclude an ultimate congenial social existence. In fact, the primordial drive for survival that brought humanity to its current state may deny the voluntary leap to a higher social order. Unless solutions are found, it is unreasonable to expect the present technologically advanced civilizations to continue on the same course and achieve long-term stability as insatiable exploitation plunders irreplaceable natural resources, the human population expands beyond control, the biosphere becomes increasingly polluted, and the resolution of ideological conflicts promotes development of more efficient means of extermination.

In the present, mankind is enjoying a continuously increasing privilege of technological comforts. As these commodities continue to proliferate and become more widespread, will the quality of life really improve? Futurists predict that despite the anticipated advances of future technology and extended healthy lives from future medicine, the human psyche will remain the same. Although science and technology have been crucial in uncovering the story behind many of the mysteries that we ponder, some futurists predict that they will ultimately fail to provide answers to deep philosophical questions such as the ultimate purpose or goal of humanity—if the solutions even exist. Placing such demands on science and technology may be unfair, but futurists such as Herb London (president of the public-policy think tank the Hudson Institute) suggest that from this social dismay new religions will develop, which will be tolerant of diverse philosophies but maintain a high moral standard. Thus the technologically driven society of today may be replaced by one governed by philosophy during the next millennium.

15

Environmental Hostilities

The relationship between man and the environment, forged by evolutionary processes over billions of years, has developed into a strong bond that, if disturbed, can lead to serious consequences. Ecologists and naturalists have been attempting to define key environmental indicators to identify any trends that may put the future of humanity in jeopardy. Some concerning trends are not yet completely defined (greenhouse effect and global warming), while others are more convincing (natural resource depletion).

Natural disasters fill the news on a daily basis: massive storms, floods, earthquakes, etc. In October 1999, the U.S. Geological Survey announced a 70 percent chance of at least a 6.7 magnitude earthquake in the San Francisco Bay Area by the year 2030—numerous fault lines traverse this region, which supports a population of nearly 7 million. The 1989 Loma Prieta quake, which caused cancellation of the third game of the World Series, tipped the Richter scale at magnitude 6.9 but was centered 100 kilometers south of San Francisco. An earthquake of similar magnitude centered closer to more densely populated regions of the San Francisco area would be far more devastating—some scenarios predict greater damage than the 8.3 magnitude earthquake of 1906, which nearly wiped out the city entirely. Other less frequent natural disaster events (e.g., volcanic eruptions) may be less predictable yet have the potential for rapid and severe devastation. The former inhabitants of Pompeii and Herculaneum would attest to this, as Mount Vesuvius erupted and quickly annihilated the two prosperous Roman cities in August A.D. 79. In fact, the effects of volcanic eruptions are not necessarily limited to the reach of the lava flows. The 1815 eruption of Mount Tambora in Indonesia, the most powerful volcanic eruption in 10,000 years, affected the entire global climate with unusually low temperatures from the dust and ash that were injected into the atmosphere. Approximately eighty thousand people starved worldwide from crop failures brought by the climate change. It is estimated that over three hundred thousand people have died as the result of volcanic eruptions over the past millennium. Although these natural disasters can cause severe damage, they tend to be limited to local communities. However, there are many other environmental catastrophes that can wreak havoc on a global scale.

Global Human Population

Be fruitful and multiply.

—Genesis 1:22

Of the multitude of crises facing humans, many scientists agree that excessive population growth is the most pressing. It is estimated that with the anticipated continued global population increase, the ability to simply feed the world's population fifty years from now will strain the very limits of our technology. Although overpopulation won't bring mankind to extinction, it could harbor unprecedented catastrophe with global hardship, suffering, disease, starvation, and death.

Throughout human history, society has generally benefited from a growing population. Increased numbers provides a larger workforce and generally improved stability of a community. Only a few had questioned the dangers of uncontrolled population growth previous to the latter half of the eighteenth century, when it was seriously debated by European intellectuals. A little over two centuries ago, Robert Malthus, an English mathematician and Anglican priest, championed the notion of "biological determinism," with particular emphasis on the limitations of food production. In the years that followed, other limiting factors gained recognition, including sources of energy and other raw materials. In his paper "Essay on the Principle of Population," Malthus argued that the population was expanding at an exponential rate. Exponential growth implies at least a "doubling time," whereby a population rapidly expands by repeatedly doubling in size with each successive doubling time period. Exponential growth is an unstable situation, as over the period of only a handful of doubling times the population will become so large that it can no longer support itself by outgrowing its resources. In fact, until recently the rate of doubling of the human population was an *accelerated* exponential growth—the doubling time was actually becoming shorter with each subsequent generation. However, around 1965 the world population growth

293

rate actually declined. At the close of the twentieth century, the global population was over 6 billion, with an annual population increase of nearly 80 million.

It is generally held that, historically, the dynamics of human population size have been strongly influenced by agriculture. The simultaneous rapid increase in human population with the development of agriculture (and later with technology) invites two interpretations: the "push" and "pull" theories. In the push model, an increase in population brings pressure to develop agriculture for support. However, the pull model states that advances in agriculture led to the population increase. A third approach is "coevolutionary," where components of both mechanisms contribute.

Without question, the population of the Earth has increased throughout history. Reliable estimates of world population have been determined for the past 2,000 years (Figure 15.1). Estimates of world populations suggest ancient humans were in an abundance of a magnitude similar to their contemporary Pleistocene mammals. Population growth was maintained through tool making and the beginnings of culture, which led to a global population of a few million. The arrival of agriculture, domestication, and early civilization brought roughly a hundredfold boost in numbers over the next several millennia. The world's population fluctuated around several hundred million people from the early days of the Roman Empire until medieval times. Until about 1650, the global population increased by about 0.04 percent annually. There was a brief period of sharp decline in the fourteenth century from the ravages of the bubonic plague, war, and famine. Since the middle of the seventeenth century, global population growth increased to a peak of 2.1 percent annually in the late 1960s. The rate has since decreased to about 1.6 percent. The decrease is largely from voluntary reductions in fertility, which initially occurred in the industrialized nations but was followed by decreased birthrates in the third world nations. The decreased fertility rates of the poorer nations are critical in controlling global population, as it is in these nations where most of the world's population reside. Despite these measures the global population passed the 6 billion mark in October 1999. The ultimate limit, the Earth's carrying capacity of human population, depends on several parameters—the major deterministic factors include both natural and social constraints.

One of the simplest methods to determine the Earth's human carrying capacity is to estimate the maximum population density per representative region and multiply that amount by corresponding areas across the planet. Other methods use mathematical models based on population growth

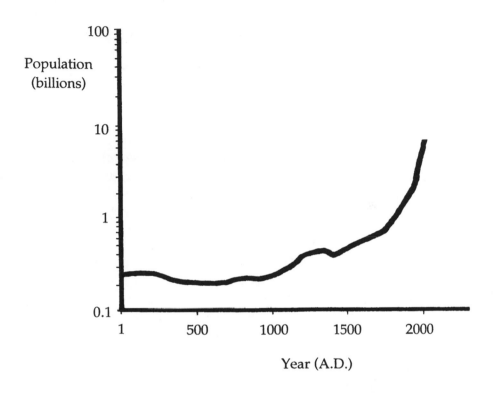

Figure 15.1 The Earth's human population from A.D. 1 to the present.

trends from historical data or by quantity and availability of key constraining factors or combinations of such factors (e.g., food or water). In addition to these natural constraints, there are social constraints on the achievable human population densities. Among these social parameters are technology, environmental quality, economic distribution, and family and cultural values. The results of such analyses estimate that the Earth's capacity falls between 7.7 and 12 billion. Some estimates of global population growth suggest the Earth's human carrying capacity could be reached as soon as the year 2050—at that time, the Earth will not be able to support continued growth in the human population. In 1998, the United Nations' long-term world population estimate placed the global population between 7.3 and 10.7 billion by 2050, with about 8.9 billion as their best guess. Some analysts suggest that the declining rates of population growth will become more significant—averting the global population from ever reaching the carrying capacity. In fact, they argue that the trend will become more obvious during the mid-twenty-first century, curbing the global population to about 9.8 billion by the year 2150. Along similar lines, some experts predict that if the current decline in growth rates continues, the Earth's population will stabilize at about 10 billion by the year 2200.

Contributing to the growing population is the substantial increase in life expectancy, largely through improved health care and policy. The graying of the population does raise concern, as it occurs in the face of a dwindling labor force, which in turn must support the growing pool of retirees. An interesting dichotomy of populations is also emerging. In Asia and Africa the population is growing and relatively young, but exposure to AIDS has slowed the rate of growth and reduced life expectancy. To the contrary, the populations of the Americas and Europe are declining and aging. There are predictions that some developed nations will gradually lose substantial numbers in their labor force, which may put pressure on immigration to supplement anticipated losses.

There are several approaches to address the threat of overpopulation. Some experts advocate pressure to develop technology, to remedy the logistical problems of maintaining a larger population. Ironically, it was advances in technology that were key in creating the population burden in the first case. Another approach is to decrease the rate of global population growth by lowering fertility through contraception, education, family-planning programs, and promotion of economic development. The individual contributions of various methods of lowering fertility are not entirely understood, but in developing nations (such as Indonesia) a combination of techniques has yielded measurable benefit. China has demonstrated significant reduction in population growth control through bureaucratic and social pressures on birth control.

Disease

Disease has been a burden of terrestrial creatures since shortly after life first appeared. Bacteria, the culprit behind many of the infectious diseases that afflict man, are also one of the most successful species, having been in existence billions of years before humans arose. Paleopathology, the study of disease-related abnormalities of human and animal remains, verifies that prehistoric man had to cope with diseases not unlike those of today. The remains of ancient Egyptian mummies show evidence of common ailments such as infections, pneumonia, and even atherosclerosis. Even dinosaur skeletons have evidence of the same infections and arthritic joint disease that affect modern humans.

The art of medicine, which strives to alleviate suffering, has its roots in a time when there was no distinction between religion, science, and superstition. Perhaps the earliest record defining the profession of medicine dates back to ancient Mesopotamia, where rules regarding the practice of medicine are found in the Code of Hammurabi (circa 1700 B.C.). Among the 282 provisions of the Code, there are ten that relate to the practice of medicine:

If a doctor has treated a freeman with a metal knife for a severe wound, and has cured the freeman, or has opened a freeman's tumor with a metal knife, and cured a freeman's eye, then he shall receive ten shekels of silver.

If the son of a plebeian, he shall receive five shekels of silver.

If a man's slave, the owner of the slave shall give two shekels of silver to the doctor.

If a doctor has treated a man with a metal knife for a severe wound, and has caused the man to die, or has opened a man's tumor with a knife and destroyed the man's eye, his hands shall be cut off.

If a doctor has treated the slave of a plebeian with a metal knife for a severe wound and caused him to die, he shall render a slave for a slave.

If he has opened his tumor with a metal knife and destroyed his eye, he shall pay half his price in silver.

If a doctor has healed a freeman's broken bone or has restored diseased flesh, the patient shall give the doctor five shekels of silver.

If he be the son of a plebeian, he shall give three shekels of silver.

If a man's slave, the owner of the slave shall give two shekels of silver to the doctor.

If a doctor of oxen or asses has treated either ox or ass for a severe wound, and cured it, the owner of the ox or ass shall give to the doctor one sixth of a shekel of silver as his fee.

The physicians' fees were substantial in Mesopotamian society; a typical craftsman would expect to make about six shekels of silver annually. However, the penalties for a treatment failure were severe.

Medicine has clearly undergone a long evolution into its current form, with recent advances continuing at a phenomenal rate. By the time of Hippocrates (credited as the father of medicine) in the fourth century B.C., the foundations were already being laid based on knowledge and influence from the various regions of the Mediterranean basin. The Greek physician Galen brought great advances as the most influential writer of medical physiology and anatomical knowledge during the second century A.D.—his works remained unsurpassed for nearly fifteen hundred years. However, his beliefs were influenced by the philosophy of Aristotle ("Nature does nothing without purpose"), which led Galen to some erroneous conclusions.

In ancient times those practicing medicine largely focused on the treatment of infectious diseases. The medieval bubonic plague pandemic (Black Death) struck Europe in 1348–50 and took the lives of nearly half of the population. The plague was later thought to have been brought to Europe by soldiers returning from the Crusades in the East, where it was already in progress.

Ring around the rosy,
a pocket full of posies,
ashes, ashes, we all fall down.

This popular rhyme traces its origin to Europe during the time of the plague. A few days after inoculation (by the bite of a flea harboring the bacteria), there is swelling of lymph nodes (the bubo) and development of a skin rash, which sometimes has blackened margins ("ring around the rosy"). As the skin lesions began to fester, concealed fragrant flowers were used to mask the repugnant odors ("a pocket full of posies"). The dead were often burned in an attempt to avert further spread of the disease ("ashes, ashes, we all fall down"). Genetic evidence recovered from exhumed victims of the plague confirms the infection was from *Yersinia pestis,* a bacterium that is now easily treated with antibiotics. Recurrent epidemics of the plague during a hundred-year period claimed about 75 percent of the medieval European population.

Historically there have been countless major infectious disease outbreaks—the ravages of tuberculosis, typhoid, and smallpox are perhaps the most notorious. The bubonic plague epidemic of the mid–fourteenth century devastated much of the civilized world and occurred at a time when superstition dominated the practice of medicine. Current practice in the control of a dangerous contagion outbreak is one of containment and identification of the agent. In 1918 a pandemic of an influenza A strain (subtype H1N1), which may have started among doughboy recruits at Fort Riley in Kansas, eventually killed about a half-million Americans and up to 30 million people worldwide as it was spread by troop migrations during World War I—more people died from the flu epidemic than were killed in combat. There have been other pandemics of the influenza virus, but perhaps the most widely known ongoing viral pandemic is caused by the AIDS virus.

The tendency of a virus to kill its host is highly variable among different viral strains. With the viruses that cause the common cold, the human mortality rate is extremely low. In the case of the AIDS virus, the mortality is essentially 100 percent; once an individual is infected, the outlook is certain death (unless a cure is found). The Ebola virus outbreak in central Africa in the late twentieth century had a 90 percent mortality for those infected. In fact, many historians believe that it was a similar virus that caused the "Great Plague in Athens" between 432 and 425 B.C. The Hanta virus outbreak of the American Southwest in the 1990s had a mortality rate of 70 percent. Fortunately, exposure to these high-mortality viral outbreaks were limited to local populations, as more widespread exposure would have brought much higher tolls. In the case of the AIDS virus, educating the general public about the disease and measures to limit transmission are used to control its spread. The constant battle of medicine with disease is ongoing; as effective treatments are developed against certain viral or bacterial strains, new strains may develop (via evolutionary mechanisms) that are resistant to conventional treatments.

As the global population expands, people are getting exposed to pathogens that had formerly resided in relatively remote regions. Increased global travel and dense populating of urban centers further promote exposure to pathogens and rapid spread of these infections. As man continues tampering with the environment, exposure to previously unknown virulent agents is a constant threat. In fact, the biological dangers encountered by humans are no longer the notorious carnivorous beasts of prehistory but lie in the realm of microscopic viruses and bacteria.

Not all disease is acquired from the external environment, as many ailments are genetic and thus genetic techniques will prove crucial in their diagnosis and treatment. The promise of gene therapy includes eradication of inheritable diseases, some types of cancer, and other genetic disorders. There is great potential for modification of the human genome to improve the genetic constitution, such as creating a more effective immune system with better resistance to disease and diminished predisposition to developing genetically related ailments. However, genetic manipulation also opens the door to eugenic motivations on a scale far greater than those of the selective breeding policies of the early twentieth century.

Pollution

The production of waste is fundamental to all living organisms and is one of the major factors that limit the size of a growing population. Likewise, pollution is an inevitable by-product of human society and comes in several forms: organic, inorganic, and nuclear. In ancient times, waste was generally biodegradable; when left exposed to the elements it would eventually decompose completely—ancient people quickly recognized the benefits of removing and burying waste. Other forms of disposal popular throughout history included dumping waste into streams as well as incinerating it. The majority of waste is handled in similar ways today, but with its management assisted by technological efficiency. Our rapidly expanding world population with its increasing production of waste will undoubtedly produce greater strain on resources and techniques of disposal.

Urbanization invites serious environmental consequences, including production of waste products, which are often released directly into the environment. A recent example is Nakuru, a Kenyan town known for its beautiful lake supporting millions of flamingos and other wildlife. As the town industrialized, its population grew from 30,000 to 360,000 in forty years, with parallel growth in its production of waste. Despite numerous safeguards, toxic materials are now found in the lake, and the once-abundant wildlife has nearly vanished. Global implications of wildlife loss, the majority from habitat destruction, may lead to extinctions resulting in a significant loss of species diversity by the middle of the twenty-first century. It has been suggested that should the loss of so many species occur, mankind's fate will be sealed. The complex interdependence of life-forms that has evolved over billions of years is a fragile balance, which may become irreparably disturbed (or collapse entirely) should sufficient diversity be lost. Fortunately, in the past there have been enough species surviving period extinctions to maintain the terrestrial web of life.

The first major contributions to environmental contamination came with the arrival of the manufacturing processes of the industrial revolution, which led to a plethora of pollutants. Among the contributors to air pollution are fine particulate substances, which can become embedded deep within lung tissue. The Environmental Protection Agency (EPA) has investigated air pollutants and claims these fine particles contribute to 60,000 deaths annually in the United States and are an aggravating factor in millions of cases of asthma and lung disease in children. Furthermore, studies have shown an increase in premature deaths among those living in areas of higher air pollutant levels. However, there is opposition to proposed controls on the release of such pollutants by industry factions, including the American Petroleum Institute.

A by-product of the Second World War was the introduction of nuclear power. Although it has the advantage of a clean fuel (i.e., not producing carbon dioxide, etc.), the products of nuclear fission remain radioactive for extended periods, which requires long-term containment and storage—they emit their damaging rays and particles for thousands of years. Unfortunately, these materials cannot be incinerated but must be shielded in special containers and allowed to decay with time. These radioactive nuclear waste materials have been packaged and buried underground or dumped into oceans. Nuclear power, which generates about 20 percent of the electricity used in the United States, produces about 3,000 tons of radioactive waste. However, if this electrical energy was produced by the conventional burning of fossil fuels, nearly 200 million tons of greenhouse gases would be produced instead. Although the quantity of radioactive by-products is not excessive, the thought of surrounding ourselves with toxic nuclear waste or submerging it in the depths of the oceans is not an attractive one. However, the same argument applies to other forms of waste.

The United States produces over 200 million tons of municipal garbage annually, with recycling efforts salvaging approximately 25 percent of the total. Although the active programs for recycling of certain materials (e.g., glass, plastics, metals, etc.) have been in effect for many years, there are a limited number of times that some materials may be reused. Thus it is prudent to minimize the amount of material used in the first place—plastic bottles and aluminum cans have become gradually thinner over the years. Biotechnology has been able to contribute to the recycling efforts by producing microbes that biochemically process certain toxic substances into harmless by-products.

A practical concern of developed nations is the economic implications of pollution management. Review of historical examples shows promise—the benefits often exceed the financial investment. In the early 1970s only a third of America's lakes were considered safe for swimming and fishing, and in less than three decades almost two-thirds were deemed safe—these improvements took place during a period of significant economic growth. By the end of the twentieth century, smog in the United States was reduced by nearly a third, while the number of automobiles nearly doubled. The benefits to health costs, as compared to antismog investment, were estimated at forty to one in a recent report from the EPA. Not all environmental investments are as clearly financially rewarding, with the Endangered Species Act an example often cited by critics.

Natural Resource Depletion

The Earth has limited natural resources, many of which took millions of years to produce. Consumption of these resources must be at a rate less than (or at most equivalent to) the rate at which they are replenished; otherwise the balance is one of resource depletion. Popularized by the media are such things as the decline of fossil fuel reservoirs, deforestation, and the slaughter of various creatures (from insects to whales). The danger of depleting these natural resources is not simply limited to their permanent loss but also includes a consequential imbalance in the biological hierarchy. The complex webs of creature and plant interactions in nature exist in delicate balance—rapid loss of principal resources or biomass can quickly disrupt the large-scale natural balance, with serious consequences.

A classic example of environmental exploitation is the case of fossil fuels. The fluctuations in the price of oil, strongly tied to both international politics and supply, repeatedly grab the attention of consumers in industrialized nations. Historically, the extraction of crude oil has been sufficient to exceed demand. Despite the optimistic reports of extensive oil reserves promoted by the oil industry, recent independent analyses suggest that by the end of the first decade of the twenty-first century supply will begin to fall behind a steadily increasing demand. The consequences will be the loss of this relatively cheap resource. In the interim, consumers will endure politically driven price fluctuations with eventual dramatic increases in price as the supply dwindles. The extraction of oil may continue through more advanced (and expensive) technology, but the inevitable depletion of this fossil fuel demands an alternative.

There has been mounting concern regarding the destruction of tropical environments, mostly in the pursuit of fossil fuels and arable land—approximately 42 million acres of rain forest are lost every year. Despite heroic planting efforts to replenish lost foliage, the coincident loss of established habitats requires many years to return (if ever). This loss of habitats brings with it the loss of indigenous plant and animal species. In addition, the potential benefit of the rich and complex ecosystems of the rain forests has been only recently recognized, particularly in the field of medicine. Just as the hidden benefits of rain forests to humanity are being discovered, there is also danger in encountering new pathogens (i.e., viral or bacterial) as humans are exposed to these previously isolated regions.

As the human population expands on land, the oceans are also affected. Development of coastal areas, overfishing, and chemically polluted agricultural runoff destroy marine life and natural habitats. The recently released UN report *Pilot Analysis of Global Ecosystems* points out that commercial fishing fleets are already harvesting about 40 percent beyond the capacity of the oceans. Although vast in extent, the oceans are also a limited resource and they cannot indefinitely sustain such exploitation.

Global Climate

The Earth's climate is not constant: although it has varied between extremes over periods measured in thousands of years, significant fluctuations may occur over much shorter intervals. Relatively brief alterations in the global climate can result from large volcanic eruptions. In 1991, the eruption of Mount Pinatubo in the Philippines sent plumes of volcanic ash and gases high into the atmosphere, which prevented some of the Sun's heat energy from reaching the Earth's surface and resulted in a transient mild global cooling in the weeks that followed. More dramatic climatic changes are exemplified by the ice ages.

During the last billion years there have been four or five major ice age periods. Theories of the origin of global ice ages point to several factors. The Earth's orbit gradually changes from a relatively circular path to a more elliptical one over a period of 100,000 years. The variation in the orbit alters the amount of sunlight reaching the Earth, which affects the surface temperature. The tilt of the Earth's axis varies over a 40,000-year interval as the axis simultaneously wobbles with a 20,000-year cycle. These variations in the Earth's orbit and rotation axis lead to cycles of warm and cold periods. Currently the Earth is in a cool period which began about 2.5 million years ago. During this period there have been about twenty glacial periods, with the most recent, peaking about twenty thousand years ago, representing the last ice age. Typically the glacial periods last about one hundred thousand years, with interglacial periods of ten thousand years—we are currently in an interglacial period. During the last interglacial period, about 120,000 years ago, the Earth was warmer (by about two degrees). During the ice ages, when global temperatures were relatively low, massive glaciers slowly flowed from the Earth's poles toward the equatorial regions and carved the land along their path. During periods of relative warmth, the glaciers receded and their melting ice gradually flowed to the seas. When the glaciers receded, they left large accumulations of rock and debris in their wake known as end moraines. In the case of the glaciers of North America, end moraine sediments formed Cape Cod, Martha's Vineyard, Nantucket, and Long Island. The cycles of global temperature change have been the results of "natural" phenomena, but the activities of man have raised concern of anthropogenic effects on the global climate.

The effects of greenhouse gases on the Earth's climate, which can ultimately lead to global warming, have received much attention in the media. The greenhouse gases (carbon monoxide, carbon dioxide, nitrous oxide, ozone, the chlorofluorocarbons, methane, and water vapor) and water droplets act as a transparent insulator to sunlight. Light easily penetrates the Earth's atmosphere, but when it is absorbed by the planet's surface, some of the energy is re-emitted at longer (infrared) wavelengths.

The greenhouse gases prevent escape of the longer wavelength energy, which leads to warming. The latest predictions suggest an increase in global temperature of about 0.01°C/year. In this context, global warming, as a consequence of the greenhouse effect, will further increase atmospheric moisture through evaporation of water from the oceans, plants, and soil by virtue of the increased temperatures. The increase in atmospheric water content will contribute to increased severity of storms in wet regions and the increased temperatures promote droughts in drier regions. However, some climatologists propose that the consequences of global warming may lead to a counterintuitive cooling. The elevated temperatures of global warming may cause increased freshwater flow into the North Atlantic (through melting of Canadian and Greenland glaciers), which, in turn, will alter the flow pattern of the gulf stream. The gulf stream brings warm water from southern latitudes—the theoretical rerouting of the gulf stream will diminish its warming effect and could plunge the Northern Hemisphere into an ice age.

There are two principal mechanisms responsible for the greenhouse effect: natural and anthropogenic warming. The presence of greenhouse gases and water vapor are necessary to support terrestrial life; they produce a vital amount of natural warming. In the absence of greenhouse gases, the Earth's surface equilibrium temperature would fall to -18°C. Under such conditions, the Earth's oceans would turn to ice and life may have never developed. Fortunately, the greenhouse gases raise this equilibrium temperature to 15°C, with the majority of the warming effect from the presence of atmospheric water vapor. *Anthropogenic warming* refers to the increases in global temperature from the activities of man. The majority of greenhouse gas production arises from the burning of fossil fuels by factories, power plants, and automobiles. Hundreds of millions of tons of these gases are released into our atmosphere each year.

Levels of the greenhouse gases have clearly shown increases during recent centuries. Carbon dioxide has undoubtedly increased since the industrial revolution, and levels of methane have increased by more than 140 percent during the same period. One hundred years ago, the atmospheric carbon dioxide concentration was about 270 parts per million (ppm), as evidenced from analysis of air bubbles preserved in samples of the Greenland ice sheet—the carbon dioxide concentration has since risen to over 360 ppm. There are estimates that the atmospheric carbon dioxide levels may be double the preindustrial levels by the end of the twenty-first century. There has been a trend over the years toward fuels that contain lower levels of carbon. The combustion of wood was largely replaced by coal (lower in carbon) in the late nineteenth century. In the mid–twentieth century, oil became available, which further decreased the atmospheric carbon load. The use of natural gas still further reduces the carbon level. Should hydrogen become a major fuel source in the future, levels of atmospheric carbon dioxide may further decrease, as the combustion of hydrogen does not produce carbon (just water and oxygen). Although the implementation of "clean fuels" (e.g., hydrogen or electricity) seems ideal from an emissions standpoint, production of these fuels generally entails combustion of fossil fuels, thereby overshadowing much of the promoted benefit. In the meantime, global economic and population growth places increasingly higher energy demands, which in turn raise the atmospheric carbon levels. Currently the United States is the leading greenhouse gas–producing nation, but during the early twenty-first century it may be replaced by China. Trends toward lower carbon fuels and population stabilization may reduce atmospheric carbon levels. However, more aggressive measures are necessary to actively avoid adverse climactic consequences. The duration of the greenhouse effect depends on how long the greenhouse gases remain in the atmosphere. Carbon dioxide has a long period of stability within the atmosphere (about a century), making its depletion a lengthy process. Despite the reality of the greenhouse effect, the trend in global temperature remains uncertain.

Climatologists have been using computer technology to model the dynamics of weather since the mid–twentieth century. The more recently developed coupled general circulation mathematical models (CGCMs) of climate divide the Earth's surface, oceans, and atmosphere into an imaginary three-dimensional grid of contiguous box volumes. Complex equations describe the weather-related properties within each volume and how each volume interacts with its neighbor volumes to create weather systems. Computer algorithms based on the general circulation model produce much of the data used for global warming predictions, but their accuracy relies on many factors and assumptions, leaving a significant margin of error in the results. Studies using models from NASA and the Goddard Institute for Space Studies on global temperature trends have shown both warming and cooling periods since the 1880s, with a net increase in temperature of 0.8°C. Contributions to global warming, other than greenhouse gases, include environmental factors such as sun spot activity, atmospheric dust and soot levels, various weather phenomena, and volcanic activity. In fact, these factors contribute substantial roles in the global temperature and probably overshadow the effects of greenhouse gases. However, the greenhouse effect is certainly a contributor to global climate, and in consideration of the rising levels of its component gases, contribution by the greenhouse effect will become increasingly significant.

Among the most worrisome consequences of global warming is the anticipated elevation of sea level. When water is heated, it expands and thereby occupies a larger volume. This same process affects the Earth's oceans; as global temperatures rise, the oceans undergo volumetric expansion, which leads to a rise in sea level. In fact, theory predicts that the amount of heat energy already stored in the oceans will cause continued expansion for centuries. Thermal expansion alone may cause up to thirty centimeters of sea-level rise over the next century. Precise determination of sea level, which is not trivial, has been estimated by various techniques, including decades of tidal gauge records and recently more sophisticated measurements from satellites. Study of geological and coral records has provided reliable sea-level information dating back thousands of years. Recent calculations indicate that the global sea level has been rising at about two millimeters each year, which has been fairly constant for at least the past few decades.

Conflicting data from models of global temperature effects on the Antarctic and Greenland ice sheets complicate climate predictions. Initial predictions were that elevations in global temperature would lead to melting of polar ice, with subsequent release of this water into the oceans producing an overall rise in sea level. About 120,000 years ago, the Earth was undergoing a period of rapid warming following an ice age, analogous to the situation of the past 20,000 years. At that time, sea levels rose to about five meters above levels of today, which coincides with the predicted elevation should the polar ice caps of today melt. Consequences of sea-level elevation include flooding and reduction in the amount of habitable land space, particularly along shoreline communities, which harbor much of the world's population. The two-millimeter annual rise in sea level is expected to continue, but uncertainty exists regarding acceleration of the rise by contribution from global warming. The international consortium of over two thousand climate scientists of the Intergovernmental Panel on Climate Change (IPCC) predicts a significant rise in sea level—between 20 and 100 centimeters—during the next century.

The effects of global warming may delay recovery of the protective ozone layer in the upper atmosphere over the Arctic. The ozone layer, which provides protection from damaging ultraviolet radiation, has been weakened by the unrestrained use of various agents, including perfluorocarbons and chlorine. An atmospheric computer model used by NASA indicates that unfavorable trends in global wind patterns secondary to global warming will impede the ozone recovery, despite significant reductions in the emissions of the injurious agents.

The initial approach to decreasing the levels of greenhouse gases is to limit their production, particularly limiting fossil fuel consumption—abrupt deprivation with strict imposed limits of fossil fuel usage is impractical. However, through improved efficiency of utilization, reduction of consumption without major change in lifestyle may be achieved. The United States economy actually reduced energy consumption by 32 percent per dollar of gross domestic product between 1975 and 1995, even as more Americans acquired the benefits of energy through appliances, air-conditioning, and travel during the same period. In fact, the amount of change necessary to recover from the greenhouse gas overproduction (based on recent proposals) is likely within the realm of future technology. Much of the technology already exists or is under investigation: energy efficient appliances, hybrid auto engines, and alternate energy sources, including oceanic thermal and tidal energy, and continued development of solar and wind power technology. Implementing improved efficiency measures often come with a high initial cost and a payoff benefit recovered in the future—a situation that fosters hesitation by investors. Another approach to improve energy efficiency is by reducing energy waste, which could save up to $300 billion for the United States economy. It has been proposed that taxation measures be instituted for this purpose. Application of this concept to greenhouse gas pollution may come in the form of a "carbon tax" to encourage consumers and corporations to more actively pursue improved efficiency. Finally, the reserves of fossil fuels are limited; experts predict oil production will peak (followed by decline) in the middle of the twenty-first century, which will likely lead to price escalation. In the face of rising fossil fuel prices, the economics of conservation and alternative sources will become much more attractive.

Climatologists estimate that the Earth's temperature increased by about 0.5°C during the twentieth century. The IPCC predicts the average global temperature will increase by about 2°C (with a range of 1.4 to 5.8°C) by the year 2100. Climatic fluctuations are not unknown during human history, including the 5°C drop during the last ice age and several interim transient episodes of climate change of up to 5°C (some of which

308

took place over periods measured in decades). In fact, between A.D. 950 and 1350 the average temperature was about 1°C higher than today and was regarded as a period of climactic stability. Although the consequences of global temperature fluctuations of one or two degrees historically have not necessarily been devastating, mild changes in today's climate could adversely affect the industry of high-yield agriculture, which is crucial in maintaining the human population.

In December 1997, a meeting of 161 nations was held in Kyoto, Japan, to outline a global effort to control the release of greenhouse gases. The result was the Kyoto Protocol. The guidelines include reduction of carbon dioxide emission (and five other gases) by about 5 percent (of 1990 levels) by 2012 by the industrialized countries. Some studies suggest it will take closer to a 60 percent reduction to have a discernible effect on the levels. The guidelines include the option of crediting emissions of several greenhouse gases against a CO_2-equivalent limit (based on a "global warming potential" index for each gas). In fact, the limitation of non-CO_2 greenhouse gases appears to have a beneficial effect on global warming according to climate models and may substantially reduce implementation costs by as much as 60 percent. Developing nations have the option of curbing their production of the dangerous emissions. The largely forested nations have less strict criteria because their trees absorb the carbon dioxide—industrialized nations should consider tree-planting campaigns! Interestingly, there is an option of "trading," under which a nation that will not meet its reduction quota may purchase unused percentages from another nation (that is ahead of its requirement). Critics argue that penalties for nonadherence to the protocol were not sufficiently addressed. Although the Kyoto policy has been signed, it is still not ratified by enough countries to be completely effective. The results of the Kyoto Protocol were mixed, and no great victory for a globally accepted solution was obtained. Although such protocols hold little political power, the recognition of the problem in global terms is a crucial first step on the way to a global solution.

The efforts to control emissions of greenhouse gases must be global in their scope; they involve a "cast of characters" including conservation, technology, economics, and political cooperation of the developed and developing nations. The potential dangers of global warming are alarming, but the rising of atmospheric levels of the greenhouse gases has been a gradual process, which suggests a similar gradual change may be adequate to compensate and perhaps reverse the effect. Thus the situation may not require immediate and drastic measures, but implementation of a well-designed campaign of reasonable controls over an extended period may be sufficient. Perhaps the advice of Epicurus, a third century B.C. Greek philosopher, is appropriate: "Moderation in all things."

Extraterrestrial Encounters

The third angel sounded his trumpet, and a great star, blazing like a torch, fell from the sky . . .

—Revelation 8:10

The boundaries of our environment can be extended to include that which lies outside the confines of our planet. Interplanetary material is constantly swept up as the Earth travels around the Sun. This debris varies in size from grains of dust to asteroids many kilometers in diameter. The atmosphere provides a safety net by preventing the smaller particles from reaching the surface as they burn up as harmless "shooting stars." Almost one thousand tons of such material is incinerated in the atmosphere daily. The celebrated annual Perseid meteor shower occurs when the Earth passes through the path of a comet whose evaporated ice and dust forms a tenuous stream of tiny particles in interplanetary space. Thus, the brilliant flashes of the Perseid meteors emanate from the same location in the sky over a brief period on an annual basis. Larger pieces of material may reach the Earth's surface in the form of meteorites. Each June the Earth passes through the particle stream of the Beta Taurid meteor shower; in 1908 the Earth may have crossed the path coincidentally with its parent comet (or comet fragment), resulting in a massive airburst explosion over Siberia.

July 1994 was an exciting month for astronomers, a time when attention was focused on the planet Jupiter. Orbiting Jupiter was a recently discovered comet, found to be on a collision course with the planet. The Shoemaker-Levy 9 comet, originally a little smaller than Halley's comet, was pulled apart into a long chain of many smaller fragments by the tug of Jupiter's gravity. Over several days, the impact plumes of the twenty discernible fragments of the comet were carefully followed as they marched one-by-one at sixty kilometers per second and plunged to their explosive finales in the clouds of Jupiter (Figure 15.2). As the individual fragments plummeted into the Jovian atmosphere, they exploded with multimegaton releases of energy, producing Earth-sized defects in the upper cloud decks

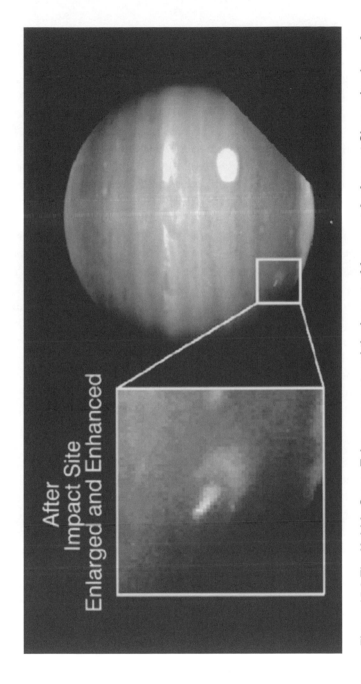

After
Impact Site
Enlarged and Enhanced

Figure 15.2 The Hubble Space Telescope captured the impact of fragment A of comet Shoemaker-Levy 9 with the planet Jupiter. The crescent-shaped feature may be a remnant of the plume ejected by the impact. *Credit: Heidi Hammel, Massachusetts Institute of Technology, and NASA.*

and spewing 50,000° C ejecta hundreds of kilometers into space. Jupiter survived the impact event without lasting effects, as the blemishes of the impacts faded from Jupiter in the weeks that followed. However, if a cometary impact event of similar proportions took place on the Earth, the consequences would be terminal.

The Earth's fossil and geological records suggest a cyclic nature to celestial impacts. There appears to be a 28-million-year cycle, with its most recent maximum about 11 million years ago; we are therefore approaching a minimum. The Oort Cloud represents a shell (about one light-year in diameter) of about 100 billion comets around our solar system, which is occasionally gravitationally disturbed—a time when some of its comets may be released from their orbits to fall into the inner solar system. The mechanism of disturbance is not entirely understood but may involve stars passing nearby or possibly the gravitational influence of a yet-undiscovered planet beyond Pluto. There has been recent indirect evidence of a massive object (between 1.5 and 6.0 times the mass of Jupiter) orbiting the solar system within the Oort Cloud. Some comets may fall into periodic orbits, such as Halley's comet, which passes through the inner solar system every seventy-six years on its trip from beyond the orbit of Neptune.

On the morning of June 30, 1908, a ten-million-ton comet entered the Earth's atmosphere and exploded with the force of a thousand atomic bombs 8,500 meters above the Tunguska River valley of Siberia. The night sky was lit for thousands of kilometers. About twenty thousand square kilometers of forest were leveled and burned, and herds of wildlife were incinerated. Fortunately, the area was sparsely populated and there was no loss of human life. The Tunguska object was about fifty meters in size—objects of this size have an expected Earth-impact frequency of about once every two centuries.

The Earth has many impact craters; there are about a hundred craters measuring over one kilometer in diameter alone. Statistically, a major impact that will produce a one-kilometer crater occurs less than once every 10,000 years. Larger impacts (those great enough to globally extinguish life) are much less common. An impact from a ten-kilometer comet (about the size of Halley's) would produce an explosion of several million megatons (the equivalent of tens of millions of atomic bombs) but happens about once every 100 million years. About once every 50 to 100 years, we can expect an encounter with a smaller object (under fifty meters in size), which would tend to explode above the ground as it enters the atmosphere and cause only local damage. Meteors of only a few meters in size, which will cause air burst explosions equivalent to an atomic bomb, occur almost on a yearly basis. It is estimated that the overall risk of death for an individual by an asteroid (of any size) during his lifetime is about 1 in 20,000, similar to the risk of death in a commercial airline round-trip or on a long-distance automobile trip. The majority of this risk comes from the assumption of a global-scale devastation; a global catastrophe is rare, but it would affect billions of people.

Scrutiny of our inner solar system has revealed the presence of many asteroids that have orbits that cross the Earth's path. The majority of these celestial trespassers are less than a few kilometers in diameter. In recent years alone there have been several "close" encounters in which the Earth and an asteroid have crossed paths—missing a collision by just days or even hours. The 50-million-ton meteor 1989 FC passed within 650,000 kilometers of the Earth and was unnoticed at the time as it approached from the direction of the Sun (making it difficult to detect with telescopes). In 1937, Hermes (a two-kilometer asteroid) crossed the Earth's orbital path, avoiding an impact by 900,000 kilometers; at the rate the Earth orbits the Sun, the miss was by only eight hours. The asteroid Hermes has since been lost to tracking efforts. Interestingly, about three hundred other asteroids that pose an Earth impact threat are known to exist but were subsequently "lost" due to lack of knowledge of their orbital dynamics.

The reality and dire consequences of a major Earth impact are well known as exemplified by the Chicxulub meteor impact 65 million years ago. More recently, another encounter that occurred in the American Southwest created a tourist attraction in the state of Arizona; 50,000 years ago a meteor about a hundred meters in diameter struck the surface, creating an impact crater about two kilometers in diameter. A much more recent ground impact occurred in 1965 when a meteorite fell in an uninhabited region of Canada with the energy of an atomic bomb.

Meteor and comet impacts have occurred throughout history but were more common during the earlier stages of Earth's history when the abundance of debris in the solar system was much more prevalent. In the modern era, impacts with smaller objects are far more common than those with larger ones. The sixteen-kilometer Chicxulub meteor, which left an impact crater about two hundred kilometers across, eliminated about 75 percent of life and brought fires that burned about 90 percent of the Earth's biomass, brought the end of the Cretaceous period. The largest recent impacts (since Chicxulub) occurred at nearly the same time, roughly 36 million years ago.

The craters formed were 100 kilometers (Popigai in Siberia) and 80 kilometers (Chesapeake Bay in Maryland) in diameter. Although these more recent impacts were immense, there is no clear evidence in the fossil record of major associated extinctions. Thus a more massive impact is probably required to affect species survival on a global level. An impact of sufficient magnitude to produce global biological devastation occurs only once every 100 million years or so.

In March 1998, apocalyptic predictions of a global doomsday on October 26, 2028, caught the attention of the public media. Discovered in December 1997, Asteroid 1997 XF11 was initially predicted to pass as close as 50,000 kilometers from the Earth. In astronomical terms, 50,000 kilometers (about an eighth the distance to the Moon) is a near miss and raises serious concern of possible impact. Although XF11 is a small fry (just under two kilometers in size) compared to Chicxulub, an impact with it would produce an explosion with a force about 20 million times that of the uranium core atomic bomb that destroyed Hiroshima in 1945. The impact crater would be up to 50 kilometers across, with at least a 1,000-kilometer incinerating blast wave. It would produce devastating effects for a land hit anywhere near populated areas, since its kill radius would extend for thousands of kilometers. Consequences of a ground impact include a dense dust cloud from the explosion and ejecta that would reduce global temperatures for weeks or months from its shielding effects on sunlight. Random statistical probability favors an ocean hit—the resulting 800-meter-high tsunami waves would circle the globe at least once and wreak havoc on coastal regions. The day following the initial press release about XF11 by the International Astronomical Union, independent calculations from NASA's Jet Propulsion Laboratory reestablished the closest encounter distance to approximately 950,000 kilometers (about two and a half times the distance to the Moon). Although it will be one of the closest encounters of such a large asteroid in recent times, the revised margin of safety is comforting. However, discovered in early 1999, the one-kilometer asteroid 1999 AN10 will pass the Earth as close as 30,000 kilometers in 2027 and may pass even closer seventeen years later. In fact it may remain a potential threat for the next 600 years. Presently, the highest known probability (1 in 300 chance) of Earth-impact with global consequence is on March 16, 2880 with the 1.1 kilometer asteroid 1950 DA.

In July 1999, the International Astronomical Union endorsed a scale that categorizes the probability and devastation of a celestial impact. The "Torino Scale" rates an impact likelihood and consequences within a range

314

from 0 (no impact) to a maximum of 10 (definite collision with global dev-astation). The scale was devised as a tool to express to the general public the very low probabilities and wide range of damage involved—similar to the Richter Scale as applied in the rating of earthquakes.

There has been some effort to organize an early-detection network to provide an early warning of potential Earth-impact objects. In 1973, as-tronomers Gene Shoemaker and Eleanor Helin began the "Palomar Earth Crossing Asteroid Survey" as the first dedicated search for asteroids with trajectories that bring them close to the Earth. Currently there are at least two groups actively involved in discovering and tracking these "near-Earth objects" (NEOs). Although destruction from an asteroid or comet is relatively remote by most standards of environmental catastro-phes, the threat has multiple undeniable historical precedents. The concern has encouraged some scientists to propose a $50 billion defense program in the hope of averting, or at least minimizing, the damage of such an event. The proposal tiers the impact problem where the most serious threats are given the highest priority. In the case of celestial impacts, the object size is of prime importance. The goal of the asteroid tracking effort is to detect 90 percent of the NEOs greater than one kilometer in size within a ten-year period. Impacts with meteors in the one kilometer (or larger) size range imply not just local but also global threats of damage. Estimates indicate there are between 2,000 and 2,500 asteroids (measuring larger than one kilometer) that pose a threat of terrestrial impact. Currently about 200 of these have been discovered and are being followed, leaving the whereabouts of the vast majority unknown. An impact with one of the far more numerous smaller asteroids would produce local damage. Meteor impacts with objects as small as twenty meters across could destroy a large city, and estimates imply there are up to a hundred million such asteroids out there. Although most asteroids have their orbits near the plane of our solar system (which narrows the region in the sky to search), many comets have their origin within the Oort Cloud, which lies far away and envelopes the solar system in all directions. Thus comets may approach the Earth from any direction. Detection and tracking of these objects allows an early warning of an approaching nemesis, providing valuable time to determine a course of action in such an event.

The basic first-line defense against an anticipated impact is to alter the course of the approaching object. If the celestial object is small enough or relatively far away, the trajectory may be altered sufficiently by "push-ing" it with a modest explosive deployed by a rocket. Larger objects, or ones dangerously close to the Earth, will require more drastic measures, perhaps involving destruction of the object with nuclear warheads. In one proposal, a large parabolic mirror in space would be used to focus the Sun's light energy on the threatening object in order to vaporize it. The consistency of the object is of importance, as a solid iron asteroid may be deflected by a nearby blast, while a primarily rocky object could be shattered into multiple (smaller) fragments that may continue on a collision course with the Earth. Recent investigations of the properties of asteroids suggest that many of them may be agglomerations of small rocks loosely held together by their collective gravity. The ability to conduct a controlled deflection of a loosely bound collection is far more difficult than a solid object, because the force of an explosion would be dispersed among the many fragments and may leave the bulk of the body intact. In the case of a solid object, rocket boosters could be sent and secured to the object's surface, where they would produce a thrust force to alter the trajectory. Such a method would require several months' warning to implement. In situations where an object is farther away from Earth impact, deployment of solar sails on the object's surface may catch the pressure of the solar wind and use that force to alter the object's course. At the beginning of the twentieth century, the Russian engineer I. O. Yarkovsky proposed that the warming of an asteroid by sunlight will cause its surface to emit thermal radiation which in turn produces a weak force that will affect the trajectory (Yarkovsky effect). Thus, an asteroid's trajectory may be altered by changing its surface color (i.e., "painting" it with a light or dark substance). In the event deflection is ineffective or not possible, terrestrial preparations include stockpiling of food supplies and large-scale regional evacuations (providing there is enough advance warning). Last-ditch efforts to save humanity in the event of an impending globally terminal impact (with sufficient time for preparation) include subterranean sanctuaries and even relocation to another planet. However, such plans have significant issues of practicality. The situation may be rendered academic: since the majority of the NEOs are undetected, the most likely scenario would be a detection too late to allow any substantial preparations prior to impact.

Extended human occupation of a fully self-sufficient environment has already been seriously investigated, in the interests of future interplanetary space travel and colonization. In 1991, the largest-scale example of such an experiment was carried out in Arizona at the Biosphere II lab ("Biosphere I" is the Earth itself). For two years, eight people lived within the environmentally self-contained lab, but ultimately the experiment failed for multiple reasons and further long-term occupancy work has been abandoned. Our current technology is capable of delivering a small group of pioneers to the planet Mars, but sustaining human life for an extended time in the Martian environment requires further development. Surveys of the cosmos suggest that planetary systems abound, but detection is currently limited to Jupiter-sized worlds (unlikely candidates for human inhabitation), leaving the prevalence of Earth-like worlds uncertain. In the event that such a candidate planet is found, the distance would be prohibitive by contemporary technology.

As we look beyond the confines of our solar system, there may be another source of danger lurking. Analysis of the interstellar gas in the vicinity of our solar system suggests that for the majority of the past 5 million years we have been drifting through a relatively "empty" region of interstellar space. Ahead of us, perhaps less than 2 trillion kilometers (about three hundred times as far away as Pluto), lies the leading edge of an interstellar cloud—the Sun is not likely to enter the cloud for at least another 2,500 years. The cloud of interstellar gas and dust, although mostly a thin vapor, may contain regions within it that are hundreds of thousands of times denser. These dense areas have the potential to affect our solar system as it drifts through the cloud. The Sun's solar wind provides a degree of protection from interstellar gases and sets up a "hydrogen wall" as its wake. However, as the density of interstellar gases increases, the hydrogen wall is compressed closer to the core of the solar system. Currently it is thought that the hydrogen wall lies about four to five times the distance to Pluto but could be brought to within the orbit of Saturn over a period of decades should a dense cloud be encountered. Scenarios of such an encounter range from difficulty with communications and increased levels of cosmic radiation to disastrous environmental effects including depletion of our atmospheric oxygen. A "killer cloud" scenario is not yet a completely understood phenomenon, leaving conclusions drawn from such analyses controversial. Less drastic encounters with interstellar clouds may have contributed to ice ages in our Earth's past—clouds of gas and dust could have passed through our system and filtered enough sunlight to produce a drop in temperature on Earth.

16

Extinction

It took several billion years of evolution to reorganize the organic molecules of the terrestrial prebiotic era into sentient creatures. The constant influence of natural selection has sculpted terrestrial life into a plethora of species. The dynamics of species long-term survival are complex, often with intermittent periods of large fluctuations in the population numbers. At the extreme are periods of massive loss of species diversity on a global scale—even placing the survival of all life in jeopardy. During terrestrial life's history, there have been five major extinctions (Figure 16.1) and although the fossil record is incomplete, certain generalizations can be surmised. The long-term survival of most vertebrate species falls between about two hundred thousand and a few million years. According to the fossil record, the average duration of mammalian species is between 2 and 3 million years. Extinction has its strongest associations with population size and geographical distribution. Small populations occupying a limited area are more susceptible to extinction, which often results from random effects of population variation rather than environmental or interspeciate competition. The major global extinctions are generally attributed to purely terrestrial environmental causes, exceptions being the extinction at the end of the Cretaceous and possibly also at the end of the Permian, which may have been the results of meteor impacts. The growing presence of man during the Pleistocene has been suggested as the cause for the last period extinction, which was mostly limited to large mammals. Theories implicate man's overhunting of these species around eleven thousand years ago as the primary cause of the extinctions. However, these conclusions are controversial because environmental changes were also taking place that may have contributed substantially to the extinctions.

The Five Major Phanerozoic Extinctions

Ordovician	438 million years ago
Devonian	354 million years ago
Permian	245 million years ago
Triassic-Jurassic	201 million years ago
Cretaceous-Tertiary	65 million years ago

Figure 16.1 The five major extinctions of Earth's life history. During the Permian extinction, 96 percent of all life was lost. The Triassic extinction led to the eventual domination by the dinosaurs, themselves succumbing to the Cretacous extinction, followed by the rise of mammals.

There is genetic evidence suggesting humans were near extinction at some point within the past million years. Comparison of 1,158 control regions of human mitochondrial DNA sequences with those of other primates demonstrates relatively fewer variations in the human DNA. The lack of variation may represent the consequence of a severe decline in the human population. The reduced population size probably occurred after the *Homo sapiens–neanderthalensis* divergence. Some experts place the human population bottleneck at about sixty-five thousand years ago, when, they propose, the total human population was reduced to as few as ten thousand. Furthermore, it has been proposed that a global cooling may have occurred at that time from ash sent into the atmosphere by a massive volcanic eruption in Indonesia, bringing difficult times for our ancestors.

The Earth has a rich biodiversity, including nearly 2 million documented living species, with experts claiming that there are from 10 to 100 million that have not yet been discovered—an overall consensus favors a conservative estimate of about 14 million living species. However, recent review of the categorization of fossilized mammals raises concern of significant duplications in the record (i.e., individual species accidentally listed more than once with different names); species diversity may be overestimated by up to 31 percent. The current biodiversity the Earth enjoys comes from the billions of years of species diversification, which are entering a new era of global extinctions. It is estimated that up to about thirty thousand species of plants and animals are lost annually from the direct influence of humans through alteration and destruction of ecosystems, poor agricultural techniques, and overhunting. Almost in desperation, there are ongoing efforts for long-term storage of the genetic material of endangered species before it is lost forever. Experts estimate that the rate of species extinction is now about 100 to 1,000 times greater than it was before the arrival of modern humans. The rise in the species extinction rate, coupled with the decrease in development of new species (through loss of natural habitats), contributes to the trend of a loss of biodiversity. Study of the fossil record suggests that on an annual basis about 2 to 2.5 species become extinct while about three new species are formed during the periods between massive extinctions. The decline in biodiversity has raised concern, as it has been estimated to rival the magnitude of previous mass extinction events. At the current rate of global species extinction, some experts argue that if the trend continues we may have only a few hundred "good years" left. In fact, 99 percent of all terrestrial creatures that have ever existed have gone on to extinction; among those are all hominid ancestors.

Clearly, if all the Earth's species are brought to extinction, man will perish as well. The biomass of our planet is not composed of millions of species and plants and animals living in their isolated ecosystems but is a complicated matrix of life where individual species are dependent on many others for survival. Consequently, the loss of one species can affect the survival of many others, and so on. There have been several historic global massive extinctions that have been attributed to various environmental events, ranging from dramatic climate changes to falling meteors. Ironically, the next major extinction may indeed be directly related to a single species: man.

To avoid such a grave ecological disaster, certain preventive steps are necessary. Much of the environmental exploitation is driven by the needs of an expanding human population. Stabilization of the population would reduce pressures on the environment. Developing nations contribute significantly to these environmental demands, which in turn requires their cooperation to avert disaster. The demands of industrialized nations on natural resources and the production of waste must also be tempered. In an attempt to maintain a semblance of the diversity of living species, natural ecosystems and habitats should be allocated and protected in the interest of permitting their continued survival. The success of these measures will require a global commitment.

However, despite all measures, extinction of all terrestrial life is inevitable in the distant future. Currently the Sun increases its heat output by about 10 percent every billion years. The Sun is about 30 to 40 percent brighter today than when it stabilized its nucleosynthesis reactions over four billion years ago. In another billion years the Earth will have been warmed enough such that the oceans will begin to evaporate. In 2 to 3 billion years there will be sufficient water vapor in the Earth's atmosphere to initiate a runaway greenhouse effect—Earth's surface temperature will soar to intolerable levels (approximately 300°C). A couple billion years later, the Sun will expand as a red giant and burn off the outer layers of the Earth, ending any surviving terrestrial life. In fact, as the Sun expands, the Earth's orbit may gradually enter the solar atmosphere where drag forces will slow the Earth until it spirals into the Sun and is incinerated. Although the future will bring an end to terrestrial-based life, other places in the solar system may become more hospitable. With the expansion of the Sun, the polar caps of Mars will thaw to provide liquid water. Jupiter's moon Europa and Saturn's moon Titan will also warm and possibly become more suitable to develop their own forms of life. Alternatively, these places may provide temporary refuge for distant human descendants as they venture out of the solar system in search of a new home.

17

Destiny of Infinitum

About 14 billion years ago a spark ignited the Universe, which through a series of fortuitous circumstances created many worlds, eventually including our Earth. After billions of years of contending with evolutionary forces, humans arose. We are at a point in history when our dominance over the Earth has serious consequences, which affect not only other life-forms but our destiny as well. It would be a tragedy for galactic anthropologists of the future studying the archaeological record of the Earth to find *Homo sapiens* an evolutionary dead end, the aftermath of a lethal mutation.

Nemesis

If humanity is to survive, it must be able to control those phenomena that can potentially bring its downfall. Throughout human history, many great civilizations have risen, prospered, and fell into oblivion. These cycles of fortune may come with a substantial cost—from the loss of civil liberties to genocidal extermination of large populations. Equally devastating can be the ravages of disease (AIDS, the Black Death, etc.) or the policies of a corrupt ruling party. Furthermore, technology has brought the ability for total human annihilation into the realm of possibility. Is it the destiny of mankind to survive these obstacles or fall victim?

The most plausible candidate for the extermination of mankind in the near future is technologically driven mutual destruction. The most likely scenarios include nuclear holocaust, global warfare, and biochemical weapons deployment through terrorism. The capability to eliminate the world's population has existed for only a short time. During the vast majority of human history, aggressive encounters between warring parties were generally brief and had relatively few casualties. There are innumerable accounts of genocide campaigns, particularly during times of warfare, but escalation to a global level was not feasible until recently. The development of weapons of mass destruction has made total extermination of life (perhaps sparing cockroaches) a plausible threat. It seems inconceivable that such apocalyptic decisions could be made, but it need be done only one time to be absolute.

The majority of the world's nuclear weapons are maintained by the United States and Russia. Smaller arsenals are held by several nations, including France, China, India, Pakistan, and the United Kingdom, while other countries are actively developing or acquiring the technology. The threat of a nuclear weapon accident is a possibility with any arsenal. It had been popular opinion that with the close of the Cold War the threat of nuclear weapons deployment would be vanquished. To the contrary, some of these arsenals continue to be operated as if a threat still exists. Although no purposeful breach has been proven, there have been several instances of missiles capable of carrying nuclear warheads accidentally flying over or crashing near or within other nations. Within the United States, thousands of employees of the nuclear operations group have been removed from duty for a variety of reasons, including drug abuse and psychiatric difficulties. Technical failures also occur, which is of particular concern regarding the aging Russian equipment. The most likely scenario of an accidental launch would result from a false alarm situation. Both human error and computer failures have resulted in serious false alarms for both the United States and Russia. An accidental launch carries an estimated risk of nearly 7 million deaths; retaliation would increase the toll. However, in the event of an intentional nuclear weapons deployment, the human death toll would be far greater. There are almost 32,000 nuclear warheads worldwide.

Although the enlightenment of philosophy has been instrumental in man's pursuit of knowledge and a congenial existence with nature and society, it can also forge a destructive path. Under the philosophy of relativism, where one considers ethical truths are relevant only to those who hold them, one can legitimately rationalize criminal acts against humanity. Furthermore, the notion that humanity does not serve a beneficial purpose could lead one to consider the elimination of it as a sovereign act. Perverted interpretation of religious doctrines is frequently used to rationalize heinous acts against humanity. Fortunately, the vast majority of philosophical and religious convictions are compassionate and cherish human life.

Universal Destiny

As we go about our daily activities, our Earth spins on its axis once each day, it orbits the Sun once a year, and our solar system travels around the Milky Way galaxy once every 250 million years. Within our galaxy, we are drifting through space at 250 kilometers each second in the direction of the star Altair in the northern constellation of Aquila. Although the average distance between the stars in our "neighborhood" is about 7 light-years, the closest star (Proxima Centauri) lies about 4.3 light-years away. Our Sun and the stars in our proximity have random motions through space, which gradually change the relative positions over time. In about twenty-five thousand years, the Alpha Centauri triple star system, which includes Proxima Centauri, will have come to within 3 light-years and shine twice as bright in the night sky. A closer encounter is anticipated with the nondescript red dwarf star Gliese 710, located 63 light-years away at present. In about a million years it may come to within a light-year, bringing it inside our system's Oort Cloud. In fact, the future flyby of Gliese 710 may be announced by a shower of comets to the inner solar system as they are gravitationally released from the Oort Cloud by the intruding star.

Our solar system is about 5 billion years old. In another 5 billion years, as the Sun exhausts its hydrogen fuel at its core, the core will begin to collapse. Simultaneous with the core collapse, the outer layers of the Sun will expand and cool, bringing its fiery surface ever closer to the Earth. As the Sun's outer layers continue to expand, its light will redden as it becomes a red giant. The disk of the Sun will continue to enlarge as it gradually fills the sky. The Earth's polar ice caps will melt, the oceans will boil, and the Earth's surface will be scorched—all terrestrial life will have been long since extinguished. The Sun will continue to expand and may eventually engulf the Earth. Following this era of incinerating heat, the Sun will shrink into a white dwarf, then cool for an eternity, leaving the Earth (if not entirely vaporized in the inferno) a frozen, lifeless rock orbiting the smoldering solar remnant.

The Milky Way galaxy has its own relative motion through the Universe, carrying us in the direction of the Hydra-Centaurus galactic cluster at about 620 km/sec. As we travel through the Universe, in about 3 billion years our journey may be interrupted by a galactic collision. Just over 2 million light-years away lies the Andromeda galaxy, which is currently approaching us at about 130 km/sec. As the two galaxies draw closer, the mutual gravitational attraction will increase, which will in turn accelerate the speed of approach. Most likely the galaxies will pass by each other at first, but the mass of the galactic systems will be sufficient to bring them back together. If the collision takes place, the two galaxies may eventually combine to form an elliptical galaxy. The collision process may take a couple billion years to complete, as many of the stars of the two galaxies are thrown into random orbits. The planets of our solar system may be pulled from the Sun by gravitational interactions brought by interloping stars, gas, and dust of the galactic collision. Thus, if the Earth should still exist at that time, it may find its twilight years spent witnessing the galactic merger. Alternatively, should the tidal forces send the Earth into intergalactic space, the final chapter would be increasingly cold and dark on an eternal path to oblivion.

The fate of the Universe is controversial—the path it will follow largely depends upon the amount of mass and energy it contains. If there is sufficient quantity to eventually halt and reverse the expansion, the final chapter will be the Big Crunch. In this scenario, the Universe eventually implodes upon itself in a fiery finale. If there is just enough mass and energy to balance expansion, the Universe will continue to expand but gradually slow, cool, and eventually come to a halt (theoretically after an infinite amount of time). The third scenario, which is favored by a slim margin, indicates the Universe will continue to expand indefinitely. Similar to balanced expansion, indefinite expansion will leave the Universe as a cold, desolate place in the distant future.

The mass density of the Universe is related to its geometry. Theory suggests that analysis of the pattern of the distribution of matter throughout the Universe may reveal its geometry—observational investigations have not yet been able to clearly support one geometry over another. Combining models based on Einstein's gravity and the latest observational findings suggests a nearly flat geometry with an accelerating rate of expansion that will continue indefinitely. In a Universe with an accelerating rate of expansion, distant galaxies that are visible in the present will disappear in the future as they recede ever faster. As the recession velocities approach the speed of light, their emitted light will become increasingly redshifted until it becomes undetectable. In fact, it is estimated that within about 2 trillion years everything beyond our own galactic cluster will disappear from sight.

Cosmologists have devised a dimensionless quantity that is intimately related to the future of the Universe, Ω (omega). The value of Ω is expressed as the ratio of the energy density of the Universe divided by the theoretical energy density that would be just enough to stop expansion after an infinite amount of time (a flat universe). In theory, if Ω was exactly unity (1) at the beginning of the Universe, it would remain that value for eternity, leaving a flat universe that would cease expanding after an infinite time. However, slight variations in Ω from unity would be rapidly amplified with the passage of time. There are a wide range of recent estimates of the value of Ω placing it somewhere between 0.1 and 2, which leaves the eventual fate in question. In the Superstring Theory model, Ω is less than 1, implying an open universe that will expand forever. The Inflationary Model of cosmology, which stretches and flattens space-time, suggests Ω has a value of 1. Proponents of the Inflationary Theory argue the mounting scientific evidence placing the value of Ω around unity as support for the model. In fact, for Ω to be so close to unity at this time in the Universe's history it would have had to be within one part in 10^{15} of unity about a second after the Universe's expansion began.

Some of the latest evidence suggests the Universe is accelerating in its rate of expansion (from supernovae observations) and has a flat geometry (suggested by the microwave background measurements). Should the expansion rate truly be increasing, the previously abandoned Cosmological Constant may get resuscitated as an "antigravity force" that exists throughout the Universe and accelerates the expansion. Adding up all of the matter in the Universe, the total comes to about a third of that required to establish a flat geometry. To make up this difference in energy density (suggested by the apparent shortage of matter), there may be a contribution by the cosmological constant through an as yet unknown energy, perhaps derived from the vacuum itself (Quantum Gravity models) or from the multidimensional geometry of space-time (Superstring theories).

Assuming continued expansion of the Universe, the galaxies will disperse throughout a growing cosmos. In about 1 trillion years, the stars within the aging galaxies will burn out, collapse, and settle into dense white dwarf stars while others form neutron stars or black holes. Black holes will become more common than they are now, as random gravitational interactions between the orbiting remnants of extinct stars eventually bring them in the proximity of their galactic center, where they will be swept up by large core black holes. According to the GUT, any matter left in the Universe will undergo proton decay into electrons, positrons, neutrinos, and photons, a process becoming prevalent by about 10^{32} years. In this

326

expanding dark Universe populated by gigantic black holes, energy may again be released as the black holes begin to disintegrate around 10^{67} years, creating a fireworks finale lasting billions of years, marked by occasional flashes scattered throughout a vast abyss. As the Universe expands further, its temperature will continually fall as it approaches absolute zero, leaving the Universe infinitely large, completely dark, and absolutely cold.

One of the fundamental principles of thermodynamics is the law of entropy. The law states that all systems, with time, favor progression to a condition of higher entropy (synonymous with increasing disorder). The progression from order to disorder is something we experience in everyday life. In the rushed preparation of a gourmet meal, the kitchen (ideally) begins in an orderly condition and after completion of the meal's preparation the kitchen is disorganized (messy); the entropy of the kitchen has increased. The remedy is to clean and reorganize. Thus the entropy may be reduced within the kitchen but only after energy is put into the system (the work of cleaning). As with the Universe, it began in an orderly state, an infinitesimal point at initial singularity. The subsequent expansion has allowed an overall element of randomness (loss of order) by spreading all the matter into a larger volume. Within this model of overall increasing entropy, there are regions where entropy is being reduced locally (as in the formation of planetary systems and life), which requires energy. If this entropy concept is applied to the cosmos, one would anticipate an infinite entropy at the end of time, an epitaph of complete disorder.

In an indefinitely expanding Universe, all life will eventually come to an end. As the Universe expands, matter and energy disperse and stars fizzle out; any remaining life-forms will find an increasingly difficult struggle to survive as their resources slip away—living organisms require matter and energy. In the interest of preserving humanity in the distant future, many intriguing theories have been proposed. As temperature is directly related to energy requirements, decreasing the human body temperature and metabolism has been entertained as a means to reduce the amount of energy necessary to sustain life—perhaps through genetic modifications. As with cold-blooded creatures, metabolic and biological function would be slowed, including cognition. However, there are limitations of practicality, as biochemical processes are temperature-dependent and tissues have freezing points. More drastic measures include transferring humans (consciousness) to alternative hosts, such as highly complex android machines. However, such measures would still be temporizing as the available energy to function would still become increasingly scarce until it is altogether insufficient. The future may seem bleak, but some consolation may be found, as the inevitable outcome is not expected for many billions of years to come.

Launched in 1972, the *Pioneer 10* space probe successfully completed its mission to study the planet Jupiter and now drifts through space over 12 billion kilometers from the Earth. It has not yet left the Sun's influence but will soon enter into interstellar space on a long and presumably peaceful journey. The probe is traveling in the direction of the star Aldebaran, a giant red star marking the eye of the bull in the constellation Taurus, which lies sixty-eight light-years away. The probe will reach the star's vicinity in about 2 million years and will likely continue far beyond. The probe carries a plaque fastened to its side depicting its home and creators—our first ambassador to the cosmos. Should there be sentient life-forms that someday encounter the probe in the distant future, what will they find has become of humanity?

Part V

Manifest Destiny

I am awake.

—Buddha, sixth century B.C.

18

Final Reality

Philosophers, theologians, sorcerers, and scientists have devoted many lifetimes over the millennia of human history in pursuit of defining man's origin, place, and destiny in the Universe. Improved understanding of the physical sciences has brought new tools to lift some of the layers of uncertainty. Despite our inability to fully depict every aspect of humanity's legacy, much of the journey has been brought to light—enabling investigators to at least partly explain our existence and permit rational speculation regarding the arcane.

It has become increasingly clear that space, matter, energy, time, and the forces of nature are interrelated, and their properties (perhaps including their very existence) were determined with the unfolding of the Universe during its earliest moments. Different approaches to the mystery—offered by diverse scientific disciplines including cosmology, astronomy, and particle and theoretical physics—appear to converge on certain fundamental concepts. Beginning with the early theories of electromagnetism, followed by Special Relativity, General Relativity, quantum mechanics, and the more recent theories of Quantum Cosmology and String Theory, the intimate relationships between matter, energy, and space-time are coming into focus.

The evidence clearly points to a Universe that began in a hot, dense form. We don't fully understand the properties in effect at the beginning—our current concepts of physical laws break down at the extreme conditions presumably existing at the beginning, precluding our ability to describe those earliest moments. Thus we must assume that the Universe was in a completely different form at that time, which defies our ability to explain it. Several theories point to a linkage among matter, energy, space, and time, suggesting these ordinarily discrete properties may have (at the beginning of the Universe) been combined in a single form. In fact, the distortions of matter, space, and time were so extreme that time may have lost its identity—blurring the meaning of a beginning (in the conventional sense of time). Furthermore, the surprising complexity of the vacuum of space, once considered empty and dull, may in fact hold the key to unifying these properties. This vacuum, which surrounds us, permeates us, and occupies the vast majority of the Universe, is the realm of quantum particles and their forces—it acts as a medium that transmits the forces of gravity and electromagnetism. Hidden within this vacuum may be the infinitesimal remains of multiple dimensions of space, curled upon themselves since the genesis of our Universe. Those infinitesimal dimensions may define the laws of nature, which govern everything that exists.

One might conclude that much of what transpired in our history was the product of chance occurrences. In fact, the laws that describe the most fundamental properties of nature, those of quantum mechanics, claim that uncertainty is the rule. The arrival of the Universe, its rate of expansion, the formation of our planet, the development of terrestrial life, the series of massive extinctions eventually leaving mammals to proliferate, and the final complex path to modern humans are all events that could have followed a different course. If events had not occurred as they did, one may propose that man wouldn't be here to ponder them; thus is it valid to consider otherwise? It can be argued that it is human egocentricity to assume that the Universe is simply a vehicle to produce *Homo sapiens* on one of its planets! In fact, many terrestrial species have existed much longer than humans—some of those species still survive, while others have succumbed to extinction.

Is the existence of human life a series of random events that just happened, or is it influenced by a supreme intelligence? We are familiar with the chain of events that set the stage on which organic life was able to prosper. Theoretical models suggest that the physical properties of a universe that can sustain life fall within a narrow range, from the rate of spatial and temporal expansion to the charges and masses of elementary particles. If any of these parameters were changed only slightly, organic life would not have developed in our Universe. Perhaps we are simply lucky in the manner in which our Universe unfolded this time around: if one throws dice for an eternity, there comes a time for any particular combination to be cast. The weak anthropic principle argues that in order for intelligent life (e.g., humans) to exist, a universe must have the appropriate characteristics to support development of intelligent life in the first place. Taking it further, one can argue that the properties of the Universe have occurred with such precision that a form of divine intervention must be involved: the strong

332

anthropic principle assumes the Universe was designed for intelligent life. However, the anthropic principle is a philosophical fallacy: the argument is redundant as the outcome is assumed a priori.

What will the future bring for man? A first approximation may be surmised by reviewing man's history. If we look back a thousand years or even just a hundred, society consisted of urban centers where social and political activities were concentrated. There has been a trend over the centuries for urban centers to become more densely populated, with simultaneous territorial expansion and increase in the global population. These circumstances will likely continue but will eventually reach a balance when the global population stabilizes.

As for individuals living in distant future civilizations, lifestyles will likely be similar to those of today, as certain qualities will remain stable—human nature, in particular. Hair styles, dress codes, and household commodities will differ from today, but the essence of humanity will persist. Likewise, the human anatomic form will remain genetically stable for many millions of years to come. Although genetic manipulations may introduce some alterations of the human genome, the modifications will probably be relatively minor, directed to improvement of health and elimination of genetically based diseases. However, modification of the human genome alone will not cure all of today's ailments, as most afflictions are not genetic in origin. One must keep in mind (as nature has taught us), that most (random) changes to an individual's genetic constitution are not beneficial. Therefore, the practice of genetic interventions must be carefully conducted and tightly controlled. As human nature remains constant, so will the concerns and ambitions of the populace, which will foster societies similar to those of today and of those that existed in antiquity.

However, the constancy of human nature conjoined with the expanding base of technology invites some concern. The conflicts of man in the past, brought on by aggression, suspicion, paranoia, or greed, will undoubtedly continue far into the future. Even though the boundaries between people (be they political, cultural, or geographical) seem less significant than in times past, there will always be a core of distrust that must be held in check to ensure congenial relations and the long-term endurance of civilization. The capability to effectively subjugate one's enemy with technological innovations enables rapid escalation to domination, devastation, or even complete extermination. Thus control of aggression will become ever more crucial in maintaining a peaceful civilized society.

333

One of the great legacies of the human adventurous spirit is the drive to venture into new territory. Ancient man covered much of the globe with little more than walking and using simple boats. Twentieth-century technology brought the Moon into reach. Some physicists have proposed designs of much faster spaceships that are able to reach the nearest stars in a human lifetime. Are there absolute limits to our exploration? Will man ever venture beyond the boundaries of our galaxy? or our Universe? Although the immense size of the Universe alone presents a major obstacle, one can deduce an even more substantial barrier that ultimately confines humanity. If we return to the beginning of our Universe, we find an extremely compact mass representing all matter in one infinitesimal package, perhaps a singularity. The early Universe, with its tightly concentrated mass, was subjected to extensive distortion of gravitational fields. In the smaller-scale situation of a black hole, where matter is compacted to lie within its Schwarzschild radius, its gravitational fields are distorted to such an extreme degree that matter and light are unable to escape its grasp. In fact, its isolation is so complete that it is no longer part of the Universe from which it originated. A star becomes a black hole if its mass is concentrated into a volume with a radius less than the Schwarzschild limit (about three kilometers for one solar mass). The early Universe was analogous to a black hole, as its mass was packed into an infinitesimally small volume. The Schwarzschild radius of our Universe (which has a margin of error from the uncertainty of the total mass in the Universe) is estimated to lie somewhere between 10 and 100 billion light-years. Thus the evidence suggests our Universe lies within its Schwarzschild radius—we are essentially inside a vast black hole, precluding any opportunity to explore what lies outside.

The grand philosophies of a "great order in the Universe" and even the notion of predetermined fate are not clearly supported by the emerging picture of our Universe—both on vast cosmological scales and in its microcosm. Although one may argue that there is an elegant order, citing the existence of galaxies, stars, planets, and even life itself as evidence, the underlying processes are actually chaotic. Although in theory born from nothing (i.e., the spontaneous formation of a virtual particle), the Universe may have originated in an orderly state (a singularity), many cosmologists argue that the origin may have actually been in a chaotic state prior to the onset of the primordial expansion. Nonetheless, the overall trend is clearly one of dispersion through cosmic expansion (increasing entropy). At the other extreme, life itself, which may be the most complicated and ordered natural phenomenon, is fundamentally driven by biochemical processes, which themselves rely on the random interactions of molecules often immersed in radiant energy. Furthermore, the underlying statistical nature of the fundamental processes of quantum mechanics removes the certainty of future events. In fact, stochastic properties were dominant in the original quantum Universe from which our Universe arose. Although the Universe and its laws may seem heartless, through philosophical reflection one may consider the uncertainty of future events as an opportunity for humanity to pursue productive motivations and initiatives for a better future. Although the legacy of humanity, a period measured in thousands of years, is an infinitesimal fraction of the Universe's existence, the achievements thus far are staggering.

To many, the human mind is the crowning achievement of the evolutionary path of *Homo sapiens*. Although still controversial, the qualities of sentience and self-awareness were certainly in effect by the time of the arrival of modern man and may have been present much earlier. Sentience, a quality derived from the brain, is primarily related to the complex configuration of the cortical association areas and their interaction with memory centers. The notion of sentience as a consequence of complex neurobiology and biochemistry seems reasonable, but explanation of this intriguing phenomenon by metaphysical or other unscientific premises continues.

As we are sentient creatures, our comprehension of reality is shaped by our senses. It is disconcerting that so much of reality is, in fact, counterintuitive. Things thought to be solid are actually mostly empty—composed merely of infinitesimal particles held in a rigid structure by electromagnetic forces. Objects felt to be heavy actually distort the space around them, creating the sensation of a force (gravity). The passage of time is really unidirectional travel along a single dimension, which may be one dimension of a multidimensional space-time reality. In fact, space-time may be composed of up to ten or eleven dimensions, rather than the concept of three dimensions of space coupled with time. Furthermore, regarding the spatial dimensions, every location in the Universe is theoretically identical—assuming the Universe began from a singularity. Finally, sentience may be ultimately reduced to the biochemistry and electrical impulses that collectively form a construct within the brain; an organ composed mostly of water. Although it may seem defeatist, the laws of nature are not necessarily required to fall within the pale of human comprehension. Despite these limitations, we accept our interpretation of reality and function quite well within it. In the early days of the Roman Empire, the freed slave Phaedrus elegantly summarized the situation across the millennia with his famous statement: "Things are not always what they seem."

In an attempt to formulate an understanding of ourselves and the Universe in which we exist, the destination is somewhere in the realm of unbounded imagination. If we turn to one of the great ancient philosophers, Aristotle, who contemplated such questions, we find that he concluded such intellectual journeys with his doctrine of "teleology of nature." Aristotle reasoned that a purpose lies behind the actions of man and those of his environment. Although identifying an underlying purpose may be intangible, humanity's course is determined by a higher being, which is called *nature*.

Appendix

Time Compressed

The scales of time to describe the processes of this book are immense. The Universe has existed for billions of years, life for a few billion, modern man for a hundred thousand, and civilization for a few thousand years. To put these large periods of time into perspective, we can compress them into a more familiar period, one year, where the Universe began on January 1 at 12:00 A.M. and the present corresponds to midnight on December 31.

Big Bang	January 1
Our galaxy begins forming	late January
Our Sun ignites	mid-September
The solid Earth forms	late September
Life begins	
First cells	October 7
First animal life	December 7
Primitive fish thriving	December 19
First land vertebrates	December 22
First dinosaurs	December 25
Primates	December 29, afternoon
Modern man arrives	December 31, 11:56:30 P.M.
Civilization	
First tools	December 31, 10:28:54 P.M.
First civilization	December 31, 11:59:48 P.M.
Roman Empire begins	December 31, 11:59:54.7 P.M.
Roman Empire ends	December 31, 11:59:56.8 P.M.
Europe discovers New World	December 31, 11:59:58.9 P.M.
American independence	December 31, 11:59:59.5 P.M.
First Moon landing	December 31, 11:59:59.9 P.M.
Half life of matter	700,000,000,000,000,000,000 years

On our compressed scale, the average life span of a person is 0.15 second, which may help to gain a perspective on the vastness of time in the story of the birth of our Universe and ourselves. There is certainly no need for concern regarding the possible decay of matter, which will not have a significant effect for many years to come.

Matter and Its Properties

Matter is composed of atoms and collectively has mass. Atoms represent the basic building blocks of matter and are composed of a nucleus (positively charged) with electrons (negatively charged) around it. The nucleus is composed of neutrons (neutrally charged) and protons (positively charged). The electrons, which are much smaller than the protons and neutrons (which are roughly equal in size), have a negative charge (equal in magnitude but opposite in polarity to the positive charge of the proton).

A force of repulsion is produced between like charges and one of attraction between opposite charges. The force increases strongly as the charges are brought closer together. In the nucleus of an atom, the positively charged protons are closely packed (intermixed with neutrons). Why don't the protons fly apart? The strong nuclear force, which is effective over only very short distances (as those within an atom), holds the protons together.

The electrons that surround the nucleus will balance the overall charge of the atom (equal positive charge from the protons in the nucleus to equal negative charge from the surrounding electrons). Since the magnitudes of the proton and electron charges are the same (differing only in polarity), in a neutral atom the number of protons equals the number of electrons. If an atom is ionized, at least one electron has been taken away, so that there is at least one less negative charge and the atom will have an overall positive charge (from the unbalanced positive charge in the nucleus).

Since matter has mass, which is the sum of the masses of its component parts, so, too, must the atoms have mass. The protons and neutrons have nearly equal (but still very small) masses, while the electron has far less mass. One can further subdivide the atomic particles into quarks, but that is beyond the scope of this example.

Atoms (and molecules) are extremely small. The simplest atom, hydrogen (a proton with its orbiting electron), is roughly one angstrom (1 X 10^{-10} meters) in diameter. A molecule is a combination of atoms which are linked together. Common examples are the oxygen we breathe (O_2, two

338

oxygen atoms) and water (H_2O, an oxygen and two hydrogen atoms). A gram of water contains approximately 10,000,000,000,000,000,000,000,000 (or 1×10^{25}) molecules of H_2O. Atoms and molecules are small indeed!

Density is a property of matter often used in science. Mathematically it is the mass of something divided by the volume that it occupies. If we use the example of water, we would find it has a density of 1 gram/cubic centimeter. That is, if one collects a cubic centimeter (cc), a volume that is about one-fifth of a teaspoon, and put it on a scale to determine its mass, it would be 1 gram. Thus the density is 1 gram per cc. If 5 cc's were collected, it would weigh 5 grams, but the density would remain unchanged; 5 grams/5 cc's is equal to 1 gram/cc.

Density may be increased by applying pressure. If one takes a gram of water and somehow squeezes it so that it is only one-half a cc in volume, the density has been doubled (1 gram/½ cc is equal to 2 gram/cc). It is not easy to compress matter. In doing so, the atoms are brought closer together and they start "bumping" into each other. These interactions produce heat. Thus when something is compressed it will heat up. Conversely, as matter expands it will cool (like our expanding Universe).

Light

Radiant energy (which includes visible light) is part of a continuous (electromagnetic) spectrum. It may be measured in terms of energy, wavelength or frequency and travels at 3×10^8 meters/second (speed of light in a vacuum). Frequency and wavelength are inversely proportional, so that an increase in frequency represents a decrease (shortening) of wavelength. High-energy radiant energy (high frequency or short wavelength) include gamma rays (X rays), and lower-energy waves (low frequency or long wavelength) include microwaves or radio waves. In between these is visible light, where higher-energy light is blue and lower-energy light is red.

Light may be produced by a variety of energy-emitting reactions of atoms. Interestingly, light has properties of both particles (matter) and waves, which seems contradictory. Customarily, light is thought of as a stream of individual energy packets (photons) traveling at the speed of light. Each photon has a characteristic energy (i e , frequency and wave length) but no mass.

Since light may be thought of as a wave phenomenon, similarities ex-

339

ist with sound (also a wave phenomenon). In particular, higher frequencies of light and sound represent higher energies and lower frequencies lower energy. With sound (as with light), as one moves toward a source emitting a tone (or color of light), the pitch (frequency) increases. An increase in frequency represents an increase in energy, and in the case of light that represents a change (shift) toward the blue end of the spectrum. Conversely, if one moves away from a source, the pitch lowers (the waves are of lower energy) for sound; the analogy with light is that the frequency has been moved toward the red end of the spectrum (redshifted). The faster one moves away (or toward) a source, the greater the amount (magnitude) of the shift. The magnitude of the shift may be used to determine the velocity (of recession or approach) between the observer and source. This property is known as the Doppler Effect.

Powers of Ten

A convenient tool in mathematics is the expression of very large (or small) numbers by a power of ten. The *power* refers to the exponent. Let us use the following example:

$$A \times 10^B$$

A is the argument of the expression. B is the power (exponent). If we try to use this technique to express a large number, say 20,000, A = 2 and B = 4 ($10 \times 10 \times 10 \times 10 = 10,000$). A special case is where B = 0 so that $10^B = 1$ and our expression reduces to A ($A \times 10^0 = A \times 1 = A$). For numbers less than 1, the power is negative. As an example, if we try to express two ten-thousandths (2/10,000), it would be 2×10^{-4}. Other examples:

Speed of light in a vacuum, 300,000,000 meters/second, becomes 3×10^8 m/s.

A light-year, 9,500,000,000,000 kilometers, becomes 9.5×10^{12} km.

The mass of an electron, 0.00000000000000000000000000091 grams, becomes 9.1×10^{-28} gm.

Glossary

A

abiotic in the absence of life

acceleration increase in velocity with time

acidophile a form of archaebacteria accustomed to living in acidic environments

adenosine triphosphate (ATP) an organic molecule which is an immediate source of chemical energy for biological systems

aerobic utilizing oxygen

algorithm a protocol designed to solve a problem

anaerobic in the absence of oxygen

A.D. anno Domini (Medieval Latin: "in the year of the Lord")

allopatric occupying a different region or existing in isolation

amino acid an organic acid that is an essential structural component of proteins

Andromeda galaxy spiral galaxy in the Local Group of galaxies, of which our Milky Way galaxy is a member

angular momentum the product of mass and angular velocity of an object that is rotating

anisotropy variation with direction

anthropic principle the theory that presence of life and other qualities of the cosmos constrain the fundamental properties of the Universe

anthropogenic resulting from the activities of mankind

anthropomorphic having human form or qualities

antimatter matter composed of particles that have identical properties to ordinary matter but have an opposite charge

archaebacteria ancient bacteria, of which there are living species, including thermophiles, halophiles, and acidophiles

artificial intelligence technological approximation of intelligent function

association cortices large expanses of the cerebral cortex involved in the processing of sensory and other information

asteroid rocky object representing debris from the formation of the solar system

astronomical unit the mean distance between the Earth and Sun, 150 million kilometers

asymmetry lack of symmetric qualities, which has particular importance in the early Universe

atavism reappearance of characteristics from an ancestral form

atom constituent of all matter consisting of a positively charged nucleus surrounded by a cloud of negatively charged electrons

atomic bomb nuclear warhead that utilizes nuclear fission (splitting plutonium or uranium nuclei) to supply the energy of the explosion

auditory cortex region of cerebral cortex, located in the temporal lobes, which is involved with the processing of auditory information

autotrophs cells capable of manufacturing their own biomolecules through metabolism

B

baryons massive particles, including protons and neutrons, that respond to the strong nuclear force

basal minimal

basal ganglia paired gray matter structures deep within the cerebral hemispheres involved with regulation of voluntary movements

B.C. before Christ

behavior pattern of response made spontaneously or elicited from an organism

Big Bang theory the hypothesis that following an extremely hot and dense initial condition the Universe is expanding from an infinitesimal origin

Big Crunch theory that the expansion of the Universe will eventually reverse and end with a cosmic collapse

biodiversity the collection of all different species

biomass the entire substance of living matter

biosphere the region to which living material is confined

Black Death the bubonic plague epidemic of 1348–49

black hole an object with an intense gravitational field that has an escape velocity beyond the speed of light

brain the primary processing center of the central nervous system

brainstem portion of the central nervous system residing between the brain and spinal cord

Broca's Area region of the cerebral motor cortex involved with initiation of speech, which is usually located in the left frontal lobe of right-handed individuals

C

c speed of light in a vacumm, defined as 299,792,458 m/sec (symbol *c* from Latin *celeritas*)

Calabi-Yau space the infinitesimal space in which the extra dimensions of String Theory are curled

carbohydrate an organic molecule representing a complex of sugars for storage as a chemical energy source

carnivore flesh-eating animal

Cephied variable a variable star with a period of brightness fluctuation that is determined by the star's mass

cerebral cortex the outer cell layers of the cerebral hemiopheres, the gray matter

cerebrum the highest-developed portion of the brain, divided into two cerebral hemispheres

chemotroph cells that obtain energy from oxidation of food materials

Chicxulub site of the meteor impact in the Yucatán Peninsula that ended the Cretaceous period 65 million years ago

chordates members of the phylum Chordata, comprising of animal species that develop a notochord, dorsally located central nervous system, and gill clefts

chromodynamics quantum theory of a force that binds quarks to form nuclear particles

chromosome structure within the nucleus of a cell composed of DNA and binding proteins (twenty-three pairs in humans)

civilization a relatively high level of cultural and technological development

climate the long-term general pattern of weather over a large region

closed universe cosmological model of a geometrically spherical universe that will collapse following an initial period of expansion

codon the basic unit of the genetic code designating a single amino acid

cognition state of awareness and comprehension

color force fundamental force that binds quarks and stabilizes nucleons (strong nuclear force)

comet icy debris left from the formation of the solar system

conscious state of awareness of one's self and environment

Cosmic Microwave Background Radiation the microwave background detected in all directions, which represents a lingering cooling remnant of the early universe

Cosmological Constant (Ω_Λ) a theoretical antigravity force

cosmology science of the structure and properties of the universe

cranium part of the skull that contains the brain

critical density cosmic matter density, symbolized by the Greek letter omega (Ω), where the collective gravitational influence of all matter will balance the expansion of the Universe

Cro-Magnon early *Homo sapiens* that inhabited Europe 40,000 years ago

culture customary beliefs, social aspects, and material artifacts associated with a racial or social group

cytoplasm the substance of a cell, excluding the nucleus, which contains various organelles within the protoplasm

D

dark energy a proposed energy that exists everywhere and affects cosmological expansion

dark matter unseen matter that is thought to represent the vast majority of the mass in the Universe

decoupling isolation of particles from normal interactions under certain conditions

determinism the belief that all events are predictable

deuterium hydrogen isotope with a nucleus containing two neutrons and one proton; heavy hydrogen

DNA Deoxyribonucleic acid, nucleic acid organized as a mirror-imaged double helix located within the nucleus of cells and containing the genetic code in its sequence of bases

Doppler Effect the change (shift) in frequency of light from a source that occurs with relative velocity between the source and observer

dorsal located near or on the back

Drake equation calculation to determine prevalence of extraterrestrial civilizations

duality situation where different theories yield identical results

dust disk transient remnant of stellar formation thought to contain material that may aggregate into planets

dwarf star small main-sequence star

E

ecosystem environment in which an organism lives

electroencephalogram (EEG) pattern of brain electrical activity

electromagnetic force force between charged particles

electromagnetic wave wavelike alteration of an electromagnetic field, which travels at the speed of light, with examples including visible light, microwaves, and X rays

electron negatively charged elementary particle with 1/1836 the mass of a proton

electronuclear force unification of the three nongravitational forces; grand unified force

electroweak theory unification of the electromagnetic and weak nuclear forces

embryo early developmental stage of a living organism

endogenous originating within an organism

endorphins naturally occurring opioid peptides that are found in particularly high concentration in brain tissue and can affect pain tolerance, mood, and behavior

entanglement ability of quantum particles to instantaneously influence each other over any distance

entropy property measuring the degree of a system's disorder

escape velocity the speed required for an object to break the binding force of gravity from another object

eugenics the improvement, by control of breeding, of hereditary qualities

eukaryote a cellular organism with a nucleus that harbors genetic material

event horizon the "surface" around the singularity of a black hole marking the boundary between the Universe and the realm of the black hole, from which there is no escape

evolution theory that life-forms have their origin in preexisting forms, with genetic differences creating changes in successive generations

exogenous originating from outside an organism

extinction the termination of a species lineage

extraterrestrial originating beyond the Earth's influence

F

fermion matter particles with a half-integer spin number, including electrons, protons, and neutrons

fission reaction whereby splitting the nuclear particles of an atom releases energy

flatness problem the failure of the classic Big Bang theory to account for the Universe being neither clearly open or closed but nearly balanced in its rate of expansion

frequency oscillations per second measured in Hertz (1 Hertz = 1 oscillation per second)

functional magnetic resonance imaging (fMRI) application of magnetic resonance imaging techniques to demonstrate and localize metabolic activity of the brain

fusion reaction whereby combining nuclear particles to form a new atom releases energy

G

Gaia controversial theory where the diverse species of the Earth are analogous to a single life-form

galactic halo spherical concentration of stars and gas that is centered on the galactic nucleus but extends radially beyond the reaches of matter in the galactic disk or spiral arms

gamete mature germ cell

gamma ray short wavelength photon

gauge boson force-carrying subatomic particle

gene a sequence of nucleotides in a segment (or segments) of DNA that code for a protein

gene pool collective of all genes in an interbreeding population

General Theory of Relativity Einstein's theory that describes the relationship between matter, space, and time through gravity; also known as Einstein's Theory of Gravity

genetic code triplet code adapted by all terrestrial life-forms to convey hereditary information stored in nucleic acids to future generations

genetic drift tendency for variations to occur in the genotype of a species over generations

genetics the science of heredity as determined by the genetic code

genocide systematic destruction of a racial, political, or cultural group

genotype the genetic constitution of an organism

global warming elevation of the Earth's temperature as a consequence of the greenhouse effect

gluon carrier particle of the strong nuclear force

Gondwanaland large transient southern landmass present during the Jurassic period

Grand Unification theories that unite the three nongravitational forces

graviton carrier particle of gravitational force; gravity between two bodies is an exchange of gravitons (virtual particles) between the particles of each body, which would produce gravitational waves by classical physics

gravity force of attraction between masses

gray matter the outer cellular layer of the brain

greenhouse effect trapping of solar energy by the presence of carbon dioxide and other molecules in the atmosphere

H

hadrons class of elementary particles that respond to the strong nuclear force

346

halophile a form of archaebacteria accustomed to living in high-salinity environments

Heisenberg Uncertainty Principle quantum principle that holds particle position and velocity cannot be simultaneously determined with precision

herbivore plant-eating animal

Hertz (Hz) unit of measuring frequency of periodic signals in cycles per second

hieroglyphics pictorial written language

Higgs fields fields that exist everywhere in the vacuum and generate the property of mass of a particle through interaction between the particle and the fields

hominids bipedal primates, including man, his immediate ancestors, and related forms

homogeneous uniform composition throughout

horizon problem the failure of the classic Big Bang theory to account for the homogeneity of the Universe

Hubble's Constant (h) the constant of proportionality between a galaxy's distance and its velocity of recession, which is related to the rate of expansion of the Universe

Hubble's Law rule that increasing galactic distance correlates with increased velocity of recession

hydrogen bomb nuclear warhead that utilizes nuclear fusion (of hydrogen into helium) to supply the energy of the explosion with approximately one thousand times the force of an atomic bomb

hypothalamus area deep within the brain that is active in a variety of functions, including body temperature maintenance, drives for thirst and hunger, and sexual function

I

indigenous native to a particular location or environment

inertia the property of mass whereby it remains at rest or in uniform motion unless a force acts upon it

Inflationary theories cosmological models that propose an early brief period of exponential (inflationary) expansion

instinct pattern of innately determined behavior

intelligence possessing the capability of reason

intelligence quotient (IQ) a relative index of intellectual ability

interferometry the study of interference phenomena of light to determine spectral patterns and wavelength

ion an atom that has a net charge from either a loss or gain of electrons

iridium a silver-white metallic element found in relatively high concentration in meteorites

isotope an atom possessing the appropriate number of protons of an element but containing a different number of neutrons

isotropic uniform appearance in all directions

J

Jeans mass theoretical minimum mass of a gas cloud that will contract under its own gravitational force to form a proto-galaxy, approximately 10^5 solar masses

K

Kelvin temperature scale where 0 represents the absolute minimum temperature

kinetic energy energy associated with particle motion and related to temperature

Kuiper belt ring of cometary material that lies beyond the orbit of Neptune

L

Lamarckism theory of inheritance of acquired traits

larynx organ in the throat that participates in breathing and speech

lateral relating to the side

Laurasia large transient northern landmass present during the Jurassic period

leptons elementary particles that are not affected by the strong nuclear force, including electrons, muons, and neutrinos

life an entity characterized by metabolism, growth, interaction with the environment, and reproduction

light electromagnetic radiation at, or near, visible wavelengths

light-year distance light travels in one year: 9.5×10^{12} kilometers

local group association of galaxies of which the Milky Way is a member

Lorentz transformation alterations of physical qualities of an object traveling at relativistic speeds as seen by an observer

M

macromolecules complex organic molecules involved in biochemistry

magmetism volcanic activity

magnetic resonance imaging (MRI) technique to create anatomical images by application of magnetic and radio frequency fields

main sequence astrophysical categorization of typical stars such as the Sun

mammal member of class Mammalia comprising of higher vertebrate species that nourish their young with milk from mammary glands and have skin containing hair

mass amount of matter in an object

mass density (Ω_m) average amount of mass per unit volume in the Universe

megaparsec 1 million parsecs; 3.26 million light-years

Mesolithic period between the Paleolithic and Neolithic stone age cultures, which was most notable in Europe

messenger particle quantum bundle of force

metabolism the chemistry in living organisms by which energy is made available for the vital processes of life

meteor flash of light seen when a dust grain or small fragment of interplanetary rock burns up in the Earth's atmosphere (e.g., shooting star)

meteorite a meteor that reaches the Earth's surface

microwave photon energies between infrared and radio wavelengths

Milky Way the spiral galaxy in which our Sun exists

mitochondrion cellular organelle involved in energy metabolism

mitosis the process of cell division

molecule a combination of atoms producing the smallest unit of a chemical compound

morphology structural characteristics

M-theory eleven-dimensional Superstring Theory associating five ten-dimensional Superstring theories using two-dimensional membranes

muon short-lived negatively charged elementary particle

mutation an alteration in genetic material not caused by normal genetic processes

N

natural selection the tendency for fit individuals to survive and produce offspring whereby after many generations a species' genetic constitution will be affected and may ultimately lead to origins of new species

Neanderthal extinct subspecies of *Homo sapiens* that populated Europe and western Asia between 200,000 and 35,000 years ago

near Earth objects (NEOs) asteroids and comets that have orbits that cross the Earth's orbit and present a risk of an Earth-impact

nebula interstellar gas cloud

Neolithic latest period of Stone Age culture, characterized by the use of polished quality stone tools

neuroanatomy the anatomy of the nervous system

neuron specialized cell of the nervous system used in signal transmission and processing

neurophysiology the science of the biochemistry and function of the nervous system

neurotransmitter chemical substance used to convey signals between neurons

neutrino an uncharged particle, previously thought to be massless, which comes in three known varieties: electron neutrino, muon neutrino, and tau neutrino

neutron neutral particle of the atomic nucleus with equal mass to a proton

neutron star dense star core remnant composed of neutrons

niche a habitat occupied by a species

nucleon a nuclear particle, proton or neutron

nucleoplasm the protoplasm of the nucleus of a cell

nucleosynthesis fusion of nucleons to produce new atoms

nucleotide basic structural unit of the DNA and RNA molecules

nucleus, atomic the core of an atom, which contains protons and neutrons

O

Occam's Razor precept that entities should not be multiplied unnecessarily; the least complicated of competing theories are preferred or that explanations of unknown phenomena be pursued first by known quantities

Olbers' paradox theory that infinite composition and age of the Universe are contradicted by a night sky that is predominantly dark

omega (Ω) the ratio of the energy density of the Universe to the theoretical energy density required to halt expansion in an infinite time

Oort Cloud shell of comets that surrounds our solar system

organelle part of a cell involved in specific functions, such as the mitochondrion

organic relating to the carbon-based chemistry of living organisms

ozone a triatomic type of oxygen formed in the upper atmosphere by a photo-chemical reaction with solar ultraviolet light

P

Paleolithic earliest period of Stone Age culture, characterized by the use of crude stone tools

Paleozoic geological period between 590 and 245 million years ago

Pangea single landmass that comprised all the Earth's continents during the Permian period

panspermia theory where space-borne "seed" life-forms fall to planets where, if conditions are right, they flourish

parsec distance from the Earth where an object would have a parallax of one arc second, approximately 3.26 light-years

pathogen an entity, such as a bacterium or virus, that causes disease

phenotype the observable characteristics of an organism expressed from its genetic constitution

photon quantum of the electromagnetic force that has energy, frequency, and wavelength but no mass

photosynthesis formation of biochemical compounds utilizing light energy

phototroph cells that obtain energy from trapping light energy

phrenology unfounded doctrine that parts of the cerebral cortex vary in size relative to specific mental faculties, which in turn affect the external configuration of the skull

Planck Epoch the earliest moment of the expanding Universe when quantum effects were dominant

Planck time the time interval of the Planck Epoch, 10^{-43} second

plasma a state of matter composed of ions and electrons

plasticity capacity for being altered

plate tectonics the geological theory of continental drift whereby the landmasses of the lithosphere glide over an underlying deformable region of upper mantle, the asthenosphere, propelled by new crust formation

plebeian member of the common people

polygenic involving multiple genes

pongid apelike

positron antimatter particle partner of the electron

predestinarian one who believes all events are predictable

prenatal in a stage before birth

primate member of order Primata, comprising of humans, apes, monkeys, and related forms

prokaryote a cellular organism lacking a distinguishable nucleus, such as bacteria and blue-green algae

prosthetic relating to an artificial origin

protein biological factor composed of amino acids that may carry out a variety of cellular processes

proteinoid aggregations of macromolecules that spontaneously form tiny spheres in aqueous solution

proto-cell ancient precursor form of living cells

proto-galaxy a galaxy in the midst of formation

proton positively charged atomic nuclear particle with mass equal to the neutron

proton decay the disintegration of protons as predicted by Grand Unification theories

protoplasm the living matter of which plant and animal cells are formed

proto-star a star in the midst of formation

pulsar spinning neutron star that periodically emits radio frequency radiation

Q

Quantum Chromodynamics quantum theory of the strong nuclear force where quarks are bound by gluons to form matter particles

quantum computer computer that utilizes quantum mechanical principles for calculation

quantum cosmology cosmological approach combining quantum theory and relativistic principles

quantum electrodynamics quantum theory of the electromagnetic force where photons are the force carriers

quantum mechanics branch of physics that describes the properties of matter and radiation on small (atomic and subatomic) scales

quarks subatomic particles comprised of six types (up, down, strange, charmed, bottom, and top), with each coming in three "colors" (red, green, and blue); protons and neutrons are composed of three quarks, one of each color. A proton contains two up quarks and one down; a neutron contains two down and one up. Particles made of the other quarks are much heavier and quickly decay into protons and neutrons.

quasars Quasi-Stellar Radio Source; active galactic core black holes of distant galaxies

R

radiation energy radiated in the form of waves or particles

red giant large star with a relatively cool surface temperature such that it has a reddish color; a phase late in the stellar evolution of main sequence stars

redshift shift of spectral lines of light to longer wavelengths indicating a relative recession between the observer and the light source

relativistic relating to properties encountered at or near the speed of light

replicator agent under the influence of natural selection such as a gene complex

reptile member of class Reptilia, comprising of vertebrates most noted for a body covered with scales or bony plates

RNA Ribonucleic Acid, nucleic acid occurring in the nucleus and cytoplasm of cells that is involved with synthesis of proteins; RNA stores the genetic material in some viruses.

ribosome protein involved in the synthesis of proteins

S

savanna subtropical grassland with scattered trees

scavenger animal that feeds on refuse or dead carcasses

Schwarzschild radius distance from a nonrotating black hole's center to its event horizon

senescence process of aging

sentience the quality of conscious awareness and interaction derived from processing of information from the senses

SETI Search for Extraterrestrial Intelligence, program that utilizes radio telescopes to detect signals from the cosmos originating from extraterrestrial civilizations

singularity a dimensionless point with an associated infinite distortion of the fabric of space-time

solar wind stream of electrically charged particles (electrons and protons) from the Sun

space-time four-dimensional continuum of the General Theory of Relativity

space-time foam description of space-time on quantum scales where space and time are discontinuous and virtual particles spontaneously form and recombine

speciation act of forming a new species

species the smallest unit in classifying living organisms where individuals generally bear close resemblance and produce fertile offspring

spectrum the pattern of photon energies of a light source according to its wavelengths (colors)

spin an elementary particle's intrinsic angular momentum

spontaneous generation the doctrine of living organisms originating from nonliving precursor

Standard Model combination of the electroweak and quantum chromodynamics theories to form the particle physics theory of the three nongravitational forces and their interactions with matter

String Theory theory of hyperdimensional space-time where particles are formed by the characteristics and vibrational modes of infinitesimal one-dimensional strings

strong nuclear force fundamental force that binds quarks and stabilizes nucleons (color force)

supernova a final catastrophic explosion of a massive star where the core may become a neutron star or black hole

Superstring Theory String Theory under a condition of supersymmetry that represents a unified theory of the cosmos

sympatric occupying the same region

synapse junction between neurons

T

telescope instrument used to collect energy (visible light, radio waves, etc.) from the cosmos to create an image or other forms of analysis

telomere region at the tip of chromosome arms that is related to stability of the chromosome and life-span of the cell

terrestrial pertaining to the Earth

thalamus paired gray matter structures deep within the cerebral hemispheres involved with processing of sensory information and probably creating conscious awareness of external environment

thermodynamics the characteristics of energy and entropy in dynamic physical systems

thermophile a form of archaebacteria accustomed to living in high-temperature environments

time period during which a process occurs distinguishing past, present, and future; a principal dimension of space-time

Torino Scale scale that rates the probabilities of meteor impact scenarios

trephination act of cutting through the skull

U

Unified Field Theory hypothesis that nature's four forces (gravity, electromagnetism, nuclear strong, and weak forces) become a single force at extremely high temperatures or densities

universe the entire extent of the cosmos

V

vacuum the fabric of space-time, emptied of macroscopic matter, that is classically considered empty but under quantum principles is represented by a space-time foam

vacuum energy theoretical energy analogous to an antigravity force (see *cosmological constant*) that may contribute to an acceleration of the expansion of the Universe; zero-point energy

variable star a star exhibiting periodic variations in brightness

vertebrates members of division Verbrata, comprising animal species with a segmented spinal column or backbone

vestigium rudimentary structure

virtual particles particles that spontaneously form and recombine in the quantum environment of the vacuum

visual cortex region of cerebral cortex, located in the occipital lobes, that is involved with the processing of visual information

W

Wallace's paradox theory that human intellectual capacity is beyond that explained by evolutionary principles alone

wavelength the distance between two crests of a wave

weak gauge boson carrier particle of the weak nuclear force; W or Z boson

weak nuclear force fundamental force that stabilizes nucleons and mediates certain forms of radioactive decay; also involved in reactions of neutrinos

white dwarf star end-stage of a main sequence star, such as the Sun, characterized by a slowly cooling stellar remnant that no longer undergoes nucleosynthesis

white matter nerve fiber networks in the brain that lie beneath the gray matter layer of the brain's surface

WIMP acronym for Weakly Interacting Massive Particles; proposed heavy particles, not yet detected, that satisfy conditions in the Unified Field Theory

wormhole interdimensional conduit connecting two different points of space-time (Einstein-Rosen bridge)

X

xenophobia fear of foreign customs or people

X ray high-energy photon between that of ultraviolet light and gamma ray energies

Z

zero-point energy energy contained within the construct of the vacuum

ziggurat temple structure of ancient Mesopotamia

zygote fertilized egg

Bibliography

The Beginning

Alcock, C. "The Dark Halo of the Milky Way." *Science* January 7, 2000, Vol. 287, No. 5451, pp. 74–79.

Barrow, J. D. *The Origin of the Universe.* New York: Basic Books, 1994.

Basri, G. "The Discovery of Brown Dwarfs." *Scientific American.* April 2000, Vol. 282, No. 4, pp. 76–83.

Branchini, E., Teodoro, L., Frenk, C. S., Schmoldt, I., Efstathiou, G., White, S. D. M., Saunders, W., Sutherland, W., Rowan-Robinson, M., Keeble, O., Tadros, H., Maddox, S., and Oliver, S. "A non-parametric model for the cosmic velocity field." *Monthly Notices of the Royal Astronomical Society.* September 1, 1999, Vol. 308, No. 1, pp. 1–28.

Bucher, M. A., and Spergel, D. N. "Inflation in a Low-Density Universe." *Scientific American.* January 1999, Vol. 280, No. 1, pp. 62–69.

Buser, R. "The Formation and Early Evolution of the Milky Way Galaxy." *Science.* January 7, 2000, Vol. 287, No. 5451, pp. 69–74.

Caldwell, R. R., and Kamionkowski, M. "Echoes from the Big Bang." *Scientific American.* January 2001, Vol. 284, No. 1, pp. 38–43.

Carr, B. J. "On the Origin, Evolution and Purpose of the Physical Universe." *The Irish Astronomical Journal.* March 1982, Vol. 15, No. 3, pp. 237–253.

Cowen, R. "Computer model captures missing matter." *Science News.* June 20, 1998, Vol. 153, No. 25, p. 390.

Cowen, R. "Kuiper belt may hold fragments of Pluto." *Science News.* October 16, 1999, Vol. 156, No. 16, p. 245.

Davies, P. *The Edge of Infinity.* New York: Simon and Schuster, 1981.

Davies, P. "That Mysterious Flow." *Scientifice American.* September 2002, Vol. 287, No. 3, pp. 40–47.

De Bernardis, P., Ade, P. A. R., Bock, J. J., Bond, J. R., Borrill, J., Boscaleri, A., Coble, K., Crill, B. P., DeGasperis, G., Farese, P. C., Ferreira, P. G., Ganga, K., Giacometti, M., Hivon, E., Hristov, V. V., Iacoangeli, A., Jaffe, A. H., Lange, A. E., Martinis, L., Masi, S., Mason, P. V., Mauskopf, P. D., Melchiorri, A., Miglio, L., Montroy, T., Netterfield, C. B., Pascale, E., Piacentini, F., Pogosyan, D., Prunet, S., Rao, S., Romeo, G., Ruhl, J. E.,

Scaramuzzi, F., Sforna, D. and Vittorio, N. "A flat Universe from high-resolution maps of the cosmic microwave background radiation." *Nature.* April 27, 2000, Vol. 404, No. 6781, pp. 955–959.

Deutsch, D. *The Fabric of Reality.* New York: Penguin Press, 1997.

Disney, M. "A New Look at Quasars." *Scientific American.* June 1998, Vol. 278, No. 6, pp. 52–57.

Duff, M. J. "The Theory Formerly Known as Strings." *Scientific American.* February 1998, Vol. 278, No. 2, pp. 64–69.

Evrard, A. E. "Real or virtual large-scale structure?" *Proceedings National Academy of Sciences of the USA.* April 27, 1999, Vol. 96, pp. 4228–4231.

Ferris, T. *The Whole Shebang.* New York: Simon & Schuster, 1997.

Folger, T., "The Real Big Bang." *Discover,* December 2002, Vol. 23, No. 12, pp. 40–47.

Geller, M. J., and Huchra, J. P. "Mapping the Universe." *Science.* November 17, 1989, Vol. 246, No. 4932, pp. 897–903.

Ghez, A. M., Morris, M., Becklin, E. E., Tanner, A., and Kremenek, T. "The accelerations of stars orbiting the Milky Way's central black hole." *Nature.* September 21, 2000, Vol. 407, No. 6802, pp. 349–351.

Greene, B. *The Elegant Universe: Superstrings, Hidden Dimensions, and the Quest for the Ultimate Theory.* New York: W. W. Norton & Company, 1999.

Gribbon, J. "The genesis of the Earth." *UNESCO. The Courier.* July 1986, pp. 4–9.

Guth, A. H., and Steinhardt, P. "The Inflationary Universe." *Scientific American.* May 1984, Vol. 250, No. 5, pp. 116–120.

Habing, H. J., Dominik, C., Jourdain de Muizon, M., Kessler, M. F., Laureijs, R. J., Leech, K., Metcalfe, L., Salama, A., Siebenmorgen, R., and Trams, N. "Disappearance of stellar debris disks around main-sequence stars after 400 million years." *Nature.* September 30, 1999, Vol. 401, No. 6752, pp. 456–458.

Hartmann, W. K. *The History of Earth.* New York: Workman Publishing, 1991.

Hawking, S. W. *A Brief History of Time: From the Big Bang to Black Holes.* New York: Bantam, 1988.

Hawking, S. W. *The Universe in a Nutshell.* New York: Bantam, 2001.

Hawking, S. W., and Penrose, R. *The Nature of Space and Time.* Princeton, New Jersey: Princeton University Press, 1996.

Hawking, S W., Thorne, K. S., Novikov, I., Ferris, T., Lightman, A., and Price, R. *The Future of Spacetime.* New York, NY: W. W. Norton & Company, 2002.

Helmi, A., White, S. D. M., de Zeeuw, P. T., and Zhao, H.-S. "Debris streams in the solar neighborhood as relics from the formation of the Milky Way." *Nature.* November 4, 1999, Vol. 402, No. 6757, pp. 53–55.

Hodge, P. W. *Galaxies.* Cambridge, MA: Harvard University Press, 1986.

Hoffman, P. F., and Schrag, D. P. "Snowball Earth." *Scientific American*. January 2000, Vol. 282, No. 1, pp. 68–75.

Hogan, C. J., Kirschner, R. P., and Suntzeff, N. B. "Surveying Space-time with Supernovae." *Scientific American*. January 1999, Vol. 280, No. 1, pp. 46–51.

Ibata, R. A., Gilmore, G., and Irwin, M. J. "A dwarf satellite galaxy in Sagittarius." *Nature*. July 21, 1994, Vol. 370, No. 6486, pp. 194–196.

Icke, V. *The force of symmetry*. Cambridge, Great Britian: Cambridge University Press, 1995.

Irion, R. "Lunar Prospector Probes Moon's Core Mysteries." *Science*. September 4, 1998, Vol. 281, No. 5382, pp. 1423–1425.

Irion, R. "Hubble Sees All the Light There Is." *Science*. January 16, 1998, Vol. 279, No. 5349, p. 322.

Jolie, J., "Uncovering Supersymmetry." *Scientific American*. July 2002, Vol. 287, No. 1, pp. 70–77.

Kaku, M. "What Happened before the Big Bang?" *Astronomy*. May 1996, Vol. 24, No. 5, pp. 34–41.

Kaku, M. *Beyond Einstein: The Cosmic Quest for the Theory of the Universe*. New York: Anchor Books, 1995.

Kirshner, R. P. "Supernovae, an accelerating universe and the cosmological constant." *Proceedings National Academy of Sciences of the USA*. April 27, 1999, Vol. 96, pp. 4224–4227.

Konopliv, A. S., Binder, A. B., Hood, L. L., Kucinskas, A. B., Sjogren, W. L., and Williams, J. G. "Improved Gravity Field of the Moon from Lunar Prospector." *Science*. September 4, 1998, Vol. 281, No. 5382, pp. 1476–1480.

Krauss, L. M. "Cosmological Antigravity." *Scientific American*. January 1999, Vol. 280, No. 1, pp. 52–59.

Lee, Y.-W., Joo, J.-M., Sohn, Y.-J., Rey, S.-C., Lee, H.-c., and Walker, A. R. "Multiple stellar populations in the globular cluster ω Centauri as tracers of a merger event." *Nature*. November 4, 1999, Vol. 402, No. 6757, pp. 55–57.

Lineweaver, C. H. "A Younger Age for the Universe." *Science*. May 28, 1999, Vol. 284, No. 5419, pp. 1503–1507.

Lubowich, D. A., Pasachoff, J. M., Balonek, T. J., Millar, T. J., Tremonti, C., Roberts, H., and Galloway, R. P. "Deuterium in the Galactic Centre as a result of recent infall of low-metallicity gas." *Nature*. June 29, 2000, Vol. 405, No. 6790, pp. 1025–1027.

Macchetto, F. D., and Dickinson, M. "Galaxies in the Young Universe." *Scientific American*. May 1997, Vol. 276, No. 5, pp. 92–99.

Mather, J. C., Cheng, E. S., Cottingham, D. A., Eplee, Jr., R. E., Fixsen, D. J., Hewagama, T., Isaacman, R. B., Jensen, K. A., Meyer, S. S., Noerdlinger, P. D., Read, S. M., Rosen, L. P., Shafer, R. A., Wright, E. L., Bennett, C. L., Boggess, N. W., Hauser, M. G., Kelsall, T., Moseley, S. H., Silverberg, R. F., Smoot, G. F., Weiss, R., and Wilkinson, D. T. "Measurement of the Cosmic Microwave Background spectrum by the COBE FIRAS instrument." *The Astrophysical Journal. Part I.* January, 1994, Vol. 420, No. 2, pp. 450–456.

Metz, A., Jolie, J., Graw, G., Hertenberger, R., Gröger, J., Günther, C., Warr, N., Eisermann, Y., "Evidence for the Existence of Supersymmetry in Atomic Nuclei." *Physical Review Letters*, August 23, 1999, Vol. 83, Issue 8, pp. 1542–1545.

Milgrom, M., "Does Dark Matter Really Exist?" *Scientific American.* August 2002, Vol. 287, No. 2, pp. 42–52.

Milgrom, M., "A Modification of the Newtonian Dynamics as a Possible Alternative to the Hidden Mass Hypothesis." *Astrophysical Journal, Part 1.* July 15, 1983, Vol. 270, pp. 365–370.

Murray, J. B. "Arguments for the presence of a distant large undiscovered Solar system planet." *Monthly Notices of the Royal Astronomical Society.* October 11, 1999, Vol. 309, No. 1, pp. 31–34.

Musser, G. "The Flip Side of the Universe." *Scientific American.* September 1998, Vol. 279, No. 3, p. 22.

Musser, G. "Inflation is dead; long live inflation." *Scientific American.* July 1998, Vol. 279, No. 1, pp. 19–20.

Narlikar, J. "Was there a Big Bang?" *New Scientist.* July 2, 1981, Vol. 91, No. 1260, pp. 19–21.

O'Neill, L., Murphy, M., and Gallagher, R. B. "What are we? Where did we come from? Where are we going?" *Science.* January 14, 1994, Vol. 263, No. 5144, pp. 181–183.

Ostriker, J. P., and Steinhardt, P. J. "The Quintessential Universe." *Scientific American.* January 2001, Vol. 284, No. 1, pp. 47–53.

Peterson, I. "Evading quantum barrier to time travel." *Science News.* April 11, 1998, Vol. 153, No. 15, p. 231.

Peterson, I. "Loops of Gravity; calculating a foamy quantum space-time." *Science News.* June 13, 1998, Vol. 153, No. 24, pp. 376–377.

Rees, M. J., "Piecing Together the Biggest Puzzle of All." *Science.* December 8, 2000, Vol. 290, No. 5498, pp. 1919–1925.

Rovelli, C., and Smolin, L. "Discreteness of area and volume in quantum gravity." *Nuclear Physics B.* March 1995, Vol. 442, No. 3, pp. 593–619.

Schilling, G. "Spying on Solar Systems in the Making." *Science.* April 24, 1998, Vol. 280, No. 5363, pp. 523–524.

Silk, J. *The Big Bang.* New York: W. H. Freeman and Company, 1989.

Silk, J. *A Short History of the Universe.* New York: Scientific American Library, 1994.

Svitil, K. A. "When Earth Tumbled." *Discover.* November 1997, Vol 18, No. 11, p. 48.

Tarlé, G., and Swordy, S. P. "Cosmic Antimatter." *Scientific American.* April 1998, Vol. 278, No. 4, pp. 36–41.

Tripp, T. M., Savage, B. D., and Jenkins, E. B., "Intervening O IV Quasar Absorption Systems at Low Redshift: A Significant Baryon Reservoir." *The Astrophysical Journal Letters.* May 1, 2000, Vol. 534, No. 1, Part 1, pp. L1–L5.

Tryon, E. P., "Is the Universe a Vacuum Fluctuation?" *Nature.* December 14, 1973, Vol. 246, No. 5433, pp. 396–397.

Wagoner, R. V., and Goldsmith, D. W. *Cosmic Horizons.* San Francisco, CA: W.H. Freeman and Company, 1982.

Weinberg, S. "A Unified Physics by 2050?" *Scientific American.* December 1999, Vol. 281, No. 6, pp. 68–75.

Weissman, P. R., "The Oort Cloud " *Scientific American.* September 1998, Vol. 279, No. 3, pp. 84–89.

Witten, E. "Duality, Spacetime and Quantum Mechanics." *Physics Today.* May 1997, Vol. 50, No. 5, pp. 28–33.

Witten, E. "Reflections on the Fate of Spacetime." *Physics Today.* April 1996, Vol. 49, No. 4, pp. 24–30.

Yam, P. "Exploiting Zero-Point Energy." *Scientific American.* December 1997, Vol. 277, No. 6, pp. 82–85.

Zehavi, I., and Dekel, A. "Evidence for a positive cosmological constant from flows of galaxies and distant supernovae." *Nature.* September 16, 1999, Vol. 401, No. 6750, pp. 252–254.

Zimmer, C. "Ancient Continent Opens Window on the Early Earth." *Science.* December 17, 1999, Vol. 286, No. 5448, pp. 2254–2256.

Zimmer, C. "In Times of Ur." *Discover.* January 1997, Vol. 18, No. 1, pp. 18–19.

Origin of Life

Alberts, B., Bray, D., Johnson, A., Lewis, J., Raff, M., Roberts, K., and Walter, P. *Essential Cell Biology.* New York: Garland Publishing Inc., 1998.

Asfaw, B., White, T., Lovejoy, O., Latimer, B., Simpson, S., and Swa, G. "*Australopithecus garhi*: A New Species of Early Hominid from Ethiopia." *Science.* April 23, 1999, Vol. 284, No. 5414, pp. 629–635.

Ayala, F. J., Rzhetsky, A., and Ayala, F. J. "Origin of the metazoan phyla: Molecular clocks confirm palentological estimates." *Proceedings National Academy of Sciences of the USA.* January 1998, Vol. 95, pp. 606–611.

Balavoine, G., and Adoutte, A. "One or Three Cambrian Radiatons?" *Science.* April 17, 1998, Vol. 280, No. 5362, pp. 397–398.

Banfield, J. F., and Marshall, C. R. "Genomics and the Geosciences." *Science.* January 28, 2000, Vol. 287, No. 5453, pp. 605–606.

Beck, W. S., Liem, K. F., and Simpson, G. G. *Life: An Introduction to Biology*. *3rd Edition*. New York: Harper Collins Publishers, 1991.

Becker, L., Poreda, R. J., Hunt, A. G., Bunch, T. E., and Rampino, M. "Impact Event at the Permian-Triassic Boundary: Evidence from Extraterrestrial Noble Gases in Fullerenes." *Science*. February 23, 2001, Vol. 291, No. 5508, pp. 1530–1533.

Bernstein, M. P., Sandford, S. A., and Allamandola, L. J. "Life's Far-Flung Raw Materials." *Scientific American*. July 1999, Vol. 281, No. 1, pp. 42–49.

Bower, B. "Cutting-edge pursuits in Stone Age." *Science News*. April 11, 1998, Vol. 153, No. 15, p. 238.

Brocks, J. J., Logan, G. A., Buick, R., and Summons, R. E. "Archean Molecular Fossils and the Early Rise of Eukaryotes." *Science*. August 13, 1999, Vol. 285, No. 5430, pp. 1033–1036.

Canfield, D. E., Habicht, K. S., and Thamdrup, B. "The Archean Sulfur Cycle and the Early History of Atmospheric Oxygenation." *Nature*. April 28, 2000, Vol. 288, No. 5466, pp. 658–661.

Caroll, R. L. *Vertebrate Paleontology and Evolution*. New York: W.H. Freeman and Company, 1988.

Childe, V. G. "The Urban Revolution." *Town Planning Review*. April 1950, Vol. 21, No. 1, pp. 3–17.

Crick, F., and Orgel, L. "Directed Panspermia." *Icarus*. July 1973, 19: 341–346.

DeRoberts, E., and DeRoberts, E. M. *Cell and Molecular Biology. 8th Edition*. Philadelphia, PA: Lea and Febiger, 1987.

Fox, S. W., and Dose, K. *Molecular Evolution and the Origin of Life*. San Francisco, CA: W.H. Freeman and Company, 1972.

Frazer, J. G. *The Golden Bough*. New York: Macmillan Publishing Company, 1922.

Gamble, C. "Time for Boxgrove man." *Nature*. May 26, 1994, Vol. 369, No. 6478, pp. 275–276.

Gibson, E. K., Jr., McKay, D. S., Thomas-Keprta, K., and Romanek, C. S. "The Case for Relic Life on Mars." *Scientific American*. December 1997, Vol. 277, No. 6, pp. 58–65.

Gilbert, W. "The RNA world." *Nature*. February 20, 1986, Vol. 319, No. 6055, p. 618.

Gonzalez, G., Brownlee, D. and Ward, P. D., "Refuges for Life in a Hostile Universe," *Scientific American*, October 2001, Vol. 285, No. 4, pp. 60–67.

Gould, S. J., ed. *The Book of Life, an Illustrated History of the Evolution of Life on Earth*. New York: W. W. Norton & Company, Inc, 1993.

Greaves, R. L., Zaller, R., Cammistraro, P. V., and Murphy, R. *The Civilizations of the World: The Human Adventure. Third Edition*. Reading, MA: Addison-Wesley Educational Publishers, Inc., 1997.

Haile-Selassie, Y. "Late Miocene hominids from the Middle Awash, Ethiopia." *Nature.* July 12, 2001, Vol. 412, No. 6843, pp. 178–181.

Hoffman, P. F., Kaufman, A. J., Halverson, G. P., and Schrag, D. P. "A Neoproterozoic Snowball Earth." *Science.* August 28, 1998, Vol. 281, No. 5381, pp. 1342–1346.

Jaroff, L. "Life on Mars." *Time.* August 19, 1996, Vol. 148, No. 9, pp. 58–64.

Jarvik, E. *Basic Structure and Evolution of Vertebrates. Vol. II.* New York: Academic Press, 1980.

Jull, A. J. T., Courtney, C., Jeffrey, D. A., and Beck, J. W. "Isotopic Evidence for a Terrestrial Source of Organic Compounds Found in Martian Meteorites Allan Hills 84001 and Elephant Moraine 79001." *Science.* January 16, 1998, Vol. 279, No. 5349, pp. 366–369.

Kerr, R. A. "Pushing Back the Origins of Animals." *Science.* February 6, 1998, Vol. 279, No. 5352, pp. 803–804.

Kivelson, M. G., Khurana, K. K., Russell, C. T., Volwerk, M., Walker, R. J., and Zimmer, C. "Galileo Magnetometer Measurements: A Stronger Case for a Subsurface Ocean at Europa." *Science.* August 25, 2000, Vol. 289, No. 5483, pp. 1340–1343.

Klein, R. G. *The Human Career, Human Biological and Cultural Origins.* Chicago: The University of Chicago Press, 1989.

Krings, M., Stone, A., Schmitz, R. W., Krainitzki, H., Stoneking, M., and Svante, P. "Neandertal DNA Sequences and the Origin of Modern Humans." *Cell.* July 11, 1997, Vol. 90, pp. 19–30.

Kyte, F. T. "A meteorite from the Cretaceous/Tertiary Boundary." *Nature.* November 19, 1998, Vol. 396, No. 6708, pp. 237–239.

Luo, Z.-X, Crompton, A.W., and Sun, A.-L. "A New Mammaliaform from the Early Jurassic and Evolution of Mammalian Characteristics." *Science.* May 25, 2001, Vol. 292, No. 5521, pp. 1535–1540.

Marzoli, A., Renne, P. R., Piccirillo, E. M., Ernesto, M., Bellieni, G., and De Min, A. "Extensive 200-million-year-old Continental Flood Basalts of the Central Atlantic Magmatic Province." *Science.* April 23, 1999, Vol. 284, No. 5414, pp. 616–618.

Menon, S. "Clovis R.I.P." *Discover.* January 1998, Vol. 18, No. 1, pp. 100–101.

Miller, S. L. "Production of Some Organic Compounds under Possible Primitive Earth Conditions." *Journal of the American Chemical Society.* May 12, 1955, Vol. 77, No.9, pp. 2351–2361.

Miller, S. L., and Urey, H. C. "Organic Compound Synthesis on the Primitive Earth." *Science.* July 31, 1959, Vol. 130, No. 3370, pp. 245–251.

Mojzsis, S. J., Arrhenius, G., McKeegan, K. D., Harrison, T. M., Nutman, A. P., and Friend, C. R. L. "Evidence for life on Earth before 3,800 million years ago." *Nature.* November 11, 1996, Vol. 384, No. 6604, pp. 55–59.

Nash, M. J. "Was the Cosmos Seeded with Life?" *Time.* August 19, 1996, Vol. 148, No. 9, p. 62.

Nemecek, S. "Who Were the First Americans?" *Scientific American.* September 2000, Vol. 283, No. 3, pp. 80–87.

O'Brien, S., Menotti-Raymond, M., Murphy, W. J., Nash, W. G., Wienberg, J., Stanyon, R., Copeland, N. G., Jenkins, N. A., Womack, J. E., and Graves, J. A. M. "The Promise of Comparative Genomics in Mammals." *Science.* October 15, 1999, Vol. 286, No. 5439, pp. 458–481.

Owens, K., and King, M.-C. "Genomic Views of Human History." *Science.* October 15, 1999, Vol. 286, No. 5439, pp. 451–453.

Paton, R. L., Smithson, T. R., and Clack, J. A. "An amniote-like skeleton from the Early Carboniferous of Scotland." *Nature.* April 8, 1999, Vol. 398, No. 6727, pp. 508–513.

Pennisi, E., and Roush, W. "Developing a New View of Evolution." *Science.* July 4, 1997, Vol. 277, No. 5322, pp. 34–37.

Qiang, J., Zhexi, L., and Shu-an, J. "A Chinese triconodont mammal and mosaic evolution of the mammalian skeleton." *Nature.* March 25, 1999, Vol. 398, No. 6725, pp. 326–330.

Rampino, M. R., Prokoph, A., and Adler, A. "Tempo of the end-Permian event: High resolution cyclostratigraphy at the Permian-Triassic boundary." *Geology.* July 2000, Vol. 28, No. 7, pp. 643–646.

Rasmussen, B. "Filamentous microfossils in a 3,235-million-year-old volcanogenic massive sulphide deposit." *Nature.* June 8, 2000, Vol. 405, No. 6787, pp. 676–679.

Roberts, M. B., Stringer, C. B., and Parfitt, S. A. "A hominid tibia from Middle Pleistocene sediments at Boxgrove, UK." *Nature.* May 26, 1994, Vol. 369, No. 6478, pp. 311–313.

Robinson, R. "The Origins of Petroleum." *Nature.* December 17, 1966, Vol. 212, No. 5068, pp. 1291–1295.

Rowe, T. "At the roots of the mammalian family tree." *Nature.* March 25, 1999, Vol. 398, No. 6725, pp. 283–284.

Ruff, C. B., Trinkaus, E., and Holliday, T. W. "Body mass and encephalization in Pleistocene *Homo.*" *Nature.* May 8, 1997, Vol. 387, No. 6629, pp. 173–176.

Rutter, N. W. "Late Pleistocene history of the Western Canadian ice-free corridor." *Canadian Journal of Anthropology.* 1980, Vol. 1, pp. 1–8.

Saito, H., Kourouklis, D., and Suga, H. "An *in vitro* evolved precursor tRNA with aminoacylation activity." *The EMBO Journal.* April 2, 2001, Vol. 20, No. 7, pp. 1797–1806.

Scholz, M., Lutz, B., Nicholson, G. J., Bachmann, J., Giddings, I., Rüschoff-Thale, B., Czarnetzki, A., and Pusch, C. M. "Genomic Differentiation of Neanderthals and Anatomically Modern Man Allows a Fossil-DNA-Based Classification of Morphologically Indistinguishable

Hominid Bones." *American Journal of Human Genetics.* June 2000, Vol. 66, No. 6, pp. 1927–1932.

Smith, H. *The Illustrated World's Religions.* New York: Harper San Francisco, 1994.

Steinbock, R. T. *Paleopathological Diagnosis and Interpretation; Bone Diseases in Ancient Human Populations.* Springfield, IL: Charles C. Thomas, 1976.

Strait, D. S., and Wood, B. A. "Early hominid biogeography." *Proceedings National Academy of Sciences of the USA.* August 1999 Vol. 96, No. 16, pp. 9196–9200.

Stryer, L. *Biochemistry, 3rd Ed.* New York: W.H. Freeman and Company, 1988.

Tattersall, I. "Once we were not alone." *Scientific American.* January 2000, Vol. 282, No. 1, pp. 56–62.

Wickramasinghe, C. "Life from space; Does life on Earth have its origins in comets?" *UNESCO. The Courier.* July 1986, pp. 33–34.

Wray, G. A., Levinton, J. S., and Shapiro, L. H. "Molecular evidence for deep precambrian divergances among metazoan phyla." *Science.* October 25, 1996, Vol. 274, No. 5287, pp. 568–573.

Wright, K. "Empires in the Dust." *Discover.* March 1998, Vol. 19, No. 3, pp. 94–99.

Sentient Ascension

Editorial. "The Mirror of your soul." *The Economist.* January 3, 1998, Vol. 345, pp. 74–76.

Allman, J. "The origin of the neocortex." *Seminars in the Neurosciences.* August 1990, Vol. 2, No. 4, pp. 257–262.

Arey, L. B. *Developmental Anatomy, 7th Edition.* Philadelphia, PA: W.B. Saunders Company, 1974.

Baily, J. *Genetics and Evolution: The Molecules of Inheritance.* New York: Oxford Press, 1995.

Beardsley, T. "The Machinery of Thought." *Scientific American.* August 1997, Vol. 277, No. 2, pp. 78–83.

Begley, S. "Aping Language." *Newsweek.* January 19, 1998, Vol. 131, No. 3, pp. 56–58.

Benson, H., and Friedman, R. "Harnessing the power of the placebo effect and renaming it 'remembered wellness.'" *Annual Reviews of Medicine.* 1996, Vol. 47, pp. 193–199.

Bodnar, A. G., Ouellette, M., Frolkis, M., Holt, S. E., Chiu, C.-P., Morin, G. B., Harley, C. B., Shay, J. W., Lichtsteiner, S., and Wright, W. E. "Extension of

Life-Span by Introduction of Telomerase into Normal Human Cells." *Science.* January 16, 1998, Vol. 279, No. 5349, pp. 349–352.

Brewer, J. B., Zhao, Z., Desmond, J. E., Glover, G. H., and Gabrieli, J. D. E. "Making Memories: Brain Activity that Predicts How Well Visual Experience Will Be Remembered." *Science.* August 21, 1998, Vol. 281, No. 5380, pp. 1185–1187.

Churchland, P. M., and Churchland, P. S. "Intertheoretic reduction: a neuroscientist's field guide." *Seminars in the Neurosciences.* August 1990, Vol. 2., No. 4, pp. 249–256.

Conroy, G. C., Weber, G. W., Seidler, H., Tobias, P. V., Kane, A., and Brunsden, B. "Endocranial Capacity in an Early Hominid Cranium from Sterkfontein, South Africa." *Science.* June 12, 1998, Vol. 280, No. 5370, pp. 1730–1731.

Cooke, B. M., Tabibnia, G., and Breedlove, S. M. "A brain sexual dimorphism controlled by adult circulating androgens." *Proceedings National Academy of Sciences of the USA.* June 22, 1999, Vol. 96, No. 13, pp. 7538–7540.

Crick, F., and Koch, C. "Towards a neurobiological theory of consciousness." *Seminars in the Neurosciences.* August 1990, Vol. 2, No. 4, pp. 263–275.

Damasio, H., and Damasio, A. R. "The neural basis of memory, language and behavorial guidance: advances with the lesion method in humans." *Seminars in the Neurosciences.* August 1990, Vol. 2, No. 4, pp. 277–286.

Dawkins, R. *The Selfish Gene.* New York: Oxford University Press, 1976, rev. 1989.

DeGusta, D., Gilbert, W. H., and Turner, S. P. "Hypoglossal canal size and hominid speech." *Proceedings National Academy of Sciences of the USA.* February 16, 1999, Vol. 96, No. 4, pp. 1800–1804.

de Lange, T. "Telomeres and Senescence: Ending the Debate." *Science.* January 16, 1998, Vol. 279, No. 5349, pp. 334–335.

Dieckmann, U., and Doebeli, M. "On the origin of species by sympatric speciation." *Nature.* July 22, 1999, Vol. 400, No. 6742, pp. 354–357.

Ehrenstein, D. "Immortality Gene Discovered." *Science.* January 9, 1998, Vol. 279, No. 5348, p. 177.

Flowers, C. *A Science Odyssey.* New York: William Morrow and Company, Inc., 1998.

Gannon, P. J., Holloway, R. L., Broadfield, D. C., and Braun, A. R. "Asymmetry of Chimpanzee Planum Temporale: Humanlike Pattern of Wernicke's Brain Language Area Homolog." *Science.* January 9, 1998, Vol 279, No. 5348, pp. 220–222.

Giedd, J. N., Vaituzis, C., Hamburger, S. D., Lange, N., Rajapakse, J. C., Kaysen, D., Vauss, Y. C., and Rapoport, J. L. "Quantitative MRI of the Temporal Lobe, Amygdala, and Hippocampus in Normal Human Development: Ages 4-18 Years." *Journal of Comparative Neurology.* March 4, 1996, Vol. 366, No. 2, pp. 223–230.

Greenough, W. T., and Bailey, C. H. "The anatomy of a memory: convergence of results across a diversity of tests." *Trends in the Neurosciences.* April 1988, Vol. 11, No. 4, pp. 142–147.

Greyson, B., "Dissociation in people who have near-death experiences: out of their bodies or out of their minds?" *The Lancet,* February 5, 2000, Vol. 355, No. 9202, pp. 460–463.

Gur, R. C., Turetsky, B. I., Matsui, M., Yan, M., Bilker, W., Hughett, P., and Gur, R. E. "Sex Differences in Brain Gray and White Matter in Healthy Young Adults: Correlations with Cognitive Performance." *The Journal of Neuroscience.* May 15, 1999, Vol. 19, No. 10, pp. 4065–4072.

Hutchison III, C. A., Peterson, S. N., Gill, S. R., Cline, R. T., White, O., Fraser, C. M., Smith, H. O., and Venter, J. C. "Global Transposon Mutagenesis and a Minimal Mycoplasma Genome." *Science.* December 10, 1999, Vol. 286, No. 5447, pp. 2165–2090.

Inglehart, R., Basañez, M., and Moreno, A. *Human Values and Beliefs: A Cross-Cultural Sourcebook.* Ann Arbor, MI: The University of Michigan Press, 2001.

Ishai, A., Ungerleider, L. G., Martin, A., Schouten, J. L., and Haxby, J. V. "Distributed representation of objects in the human ventral visual pathway." *Proceedings National Academy of Sciences of the USA.* August 1999, Vol. 96, No. 16, pp. 9379–9384.

Judis, J. B. "Taboo you." *The New Republic.* October 31, 1994, Vol. 211, No. 18, p. 18.

Kandel, E. R., and Squire, L. R. "Neuroscience: Breaking Down Scientific Barriers to the Study of Brain and Mind." *Science.* November 10, 2000, Vol. 290, No. 5494, pp. 1113–1120.

Kay, R. F., Cartmill, M., and Balow, M. "The hypoglossal canal and the origin of human vocal behavior." *Proceedings National Academy of Sciences of the USA.* April 28, 1998, Vol. 95, No. 9, pp. 5417–5419.

Kirsch, I., and Sapirstein, G. "Listening to Prozac but Hearing Placebo: A Meta-Analysis of Antidepressant Medication." *Prevention & Treatment* (American Psychological Association online journal). June 26, 1998, Vol. 1, Article 2a.

Knight, R. D., and Landweber, L. F. "The Early Evolution of the Genetic Code." *Cell.* June 9, 2000, Vol. 101, No. 6, pp. 569–572.

Kosslyn, S. M., and Osherson, D. N. (Editors). *An Invitation to Cognitive Science. Vol 2: Visual Cognition, 2nd Ed.* Cambridge, MA: MIT Press, 1995.

Kurian, G. T. *DATAPEDIA of the United States 1790–2000, America Year by Year.* Lanham, MD: Bernan Press, 1994.

Lande, R. "Quantitative genetic analysis of multivariate evolution, applied to brain body size allometry." *Evolution.* March 1979, Vol. 33, No. 1, part 2, pp. 402–416.

Laubach, M., Wessberg, J., and Nicolelis, M. A. L. "Cortical ensemble activity increasingly predicts behaviour outcomes during learning of a motor task." *Nature.* June 1, 2000, Vol. 405, No. 6786, pp. 525–571.

Llinás, R. R., and Paré, D. "Commentary: of Dreaming and Wakefulness." *Neuroscience.* 1991, Vol. 44, No. 3, pp. 521–535.

Mattei, J.-F. "Prenatal diagnosis." *World Health.* September-October 1996, Vol. 49, No. 5, pp. 22–23.

McEwen, B. S., "Permanence of brain sex differences and structural plasticity of the adult brain." *Proceedings National Academy of Sciences of the USA.* June 22, 1999, Vol. 96, No. 13, pp. 7128–7130.

Melov, S., Ravenscroft, J., Malik, S., Gill, M. S., Walker, D. W., Clayton, P. E., Wallace, D. C., Malfroy, B., Doctrow, R. R., and Lithgow, G. J. "Extension of Life-Span with Superoxide Dismutase/Catalase Mimetics." *Science.* September 1, 2000, Vol. 289, No. 5484, pp. 1567–1569.

Moravec, H. "Rise of the Robots." *Scientific American.* December 1999, Vol. 281, No. 6, pp. 124–135.

Ochert, A. "Transposons." *Discover.* December 1999, Vol. 20, No. 12, pp. 59–66.

Ojemann, G. A. "Organization of language cortex derived from investigations during neurosurgery." *Seminars in the Neurosciences.* August 1990, Vol. 2, No. 4, pp. 297–305.

Pakkenberg, B., and Gundersen, H. J. G. "Neocortical Neuron Number in Humans: Effect of Sex and Age." *Journal of Comparative Neurology.* July 28, 1997, Vol. 384, No. 2, pp. 312–320.

Pinker, S. *How the Mind Works.* New York: W.W. Norton & Company, 1997.

Puca, A. A., Daly, M. J., Brewster, S. J., Matise, T. C., Barrett, J., Shea-Drinkwater, M., Kang, S., Joyce, E., Nicoli, J., Benson, E., Kunkel, L. M., and Perls, T., "A genome-wide scan for linkage to human exceptional longevity identifies a locus on chromosome 4," *Proceedings National Academy of Sciences of the U.S.A.,* August 28, 2001, Vol. 98, No. 18, pp. 10505–10508.

Schmucker, D., Clemens, J. C., Shu, H., Worby, C. A., Xiao, J., Muda, M., Dixon, J. E., and Zipursky, S. L. "*Drosophila* Dscam Is an Axon Guidance Receptor Exhibiting Extraordinary Molecular Diversity." *Cell.* June 9, 2000, Vol. 101, No. 6, pp. 671–684.

Serruya, M. D., Hatsopoulos, N. G., Paninsky, L., Fellows, M. R., and Donaghue, J. P., "Brain-machine interface: Instant neural control of a movement signal." *Nature.* March 14, 2002, Vol. 416, No. 6877, pp. 141–142.

Service, E. R. *Origins of the State and Civilization.* New York: W.W. Norton & Company, 1975.

Sher, L. "The placebo effect on mood and behavior: the role of the endogenous opioid system." *Medical Hypotheses.* April 1997, Vol. 48, No. 4, pp. 347–349.

Shu, D.-G., Luo, H.-L., Morris, S. C., Zhang, X.-L., Hu, S.-X., Chen, L., Han, J., Zhu, M., Li, Y., and Chen, L.-Z. "Lower Cambrian vertebrates from south China." *Nature.* November 4, 1999, Vol. 402, No. 6757, pp. 42–46.

St George-Hyslop, P. H. "Piecing Together Alzheimer's." *Scientific American.* December 2000, Vol. 283, No. 6, pp. 76–83.

Swaab, D. F., and Fliers, E. "A Sexually Dimorphic Nucleus in the Human Brain." *Science.* May 31, 1985, Vol. 228, No. 4703, pp. 1112–1115.

Tang, Y.-P., Shimizu, E., Dube, G. R., Rampon, C., Kerchner, G. A., Zhou, M., Liu, G., and Tsien, J. Z. "Genetic enhancement of learning and memory in mice." *Nature.* September 2, 1999, Vol. 401, No. 6748, pp. 63–69.

Thompson, M. *Philosophy.* Chicago, IL: NTC Publishing Group, 1995

Tregenza, T., and Butlin, R. K. "Speciation without isolation." *Nature.* July 22, 1999, Vol. 400, No. 6742, pp. 311–312.

van Lommel, P., van Wees, R., Meyers, V., and Elfferich, I., "Near Death experience in survivors of cardiac arrest: a prospective study in the Netherlands," *The Lancet,* December 15, 2001, Vol. 358, No. 9298, pp. 2039–2045.

Vasser, R., Bennett, B. D., Babu-Khan, S., Kahn, S., Mendiaz, E. A., Denis, P., Teplow, D. B., Ross, S., Amarante, P., Loeloff, R., Luo, Y., Fisher, S., Fuller, J., Edenson, S., Lile, J., Jarosinski, M. A., Biere, A. L., Curran, E., Burgess, T., Louis, J.-C., Collins, F., Treanor, J., Rogers, G., and Citron, M. "ß-Secretase Cleavage of Alzheimer's Amyloid Precursor Protein by the Transmembrane Aspartic Protease BACE." *Science.* October 22, 1999, Vol. 286, No. 5440, pp. 735–741.

Wagner, A. D., Schacter, D. L., Rotte, M., Koutstaal, W., Maril, A., Dale, A. M., Rosen, B. R., and Buckner, R. L. "Building Memories: Remembering and Forgetting of Verbal Experiences as Predicted by Brain Activity." *Science.* August 21, 1998, Vol. 281, No. 5380, pp. 1188–1191.

Weihrauch, T. R., and Gauler, T. C. "Placebo—efficacy and adverse effects in controlled clinical trials." *Arzneimittelforschung.* May 1999, Vol. 49, No. 5, pp. 385–393.

Weiss, H., and Bradley, R. S., "What Drives Societal Collapse?" *Science.* January 26, 2001, Vol. 291, No. 5504, pp. 609–610.

Wilkins, W. K., and Wakefield, J. "Brain evolution and neurolinguistic preconditions." *Behavioral and Brain Sciences.* 1995, Vol. 18, No. 1, pp. 161–226.

Witelson, S. F., Glezer, I. I., and Kigar, D. L. "Women Have Greater Density of Neurons in Posterior Temporal Cortex." *Journal of Neuroscience.* May 1995, Vol. 15, No. 5, Part 1, pp. 3418–3428.

Zimmer, C. "In Search of Vertebrate Origins. Beyond Brain and Bone." *Science.* March 3, 2000, Vol. 287, No. 5458, pp. 1576–1579.

Ad Infinitum

Adams, F., and Laughlin, G. *The Five Ages of the Universe: Inside the Physics of Eternity.* Monroe, LA: Free Press, 1999.

Basalla, G. *The Evolution of Technology.* New York: Cambridge University Press, 1989.

Benioff, P. "The Computer as a Physical System: A Microscopic Quantum Mechanical Hamiltonian Model of Computers as Represented by Turing Machines." *Journal of Statistical Physics.* 1980, Vol. 22, No. 5, pp. 563–591.

Brown, P., Spalding, R.E., Revelle, D. O., Tagliaferri, E., Worden, S. P., "The flux of small near-Earth objects colliding with the Earth." *Nature*, November 21, 2002, Vol. 420, No. 6913, pp. 294–296.

Campbell, C. J., and Laherrère, J. H. "The End of Cheap Oil." *Scientific American.* March 1998, Vol. 278, No. 3, pp. 78–83.

Caroll, R. L. *Patterns and Processes of Vertebrate Evolution.* New York: Cambridge University Press, 1997.

Cirac, J. I., and Zoller, P. "A scalable quantum computer with ions in an array of microchips." *Nature.* April 6, 2000, Vol. 404, No. 6778, pp. 579–581.

Close, F. *Apocalypse When?* New York: William Morrow and Company, Inc., 1988.

Cohen, J. E. *How Many People Can The Earth Support?* New York: W.W. Norton and Company, 1995.

Cohen, J. E. "Population Growth and Earth's Human Carrying Capacity." *Science.* July 21, 1995, Vol. 269, pp. 341–346.

Deevey, E. "The Human Population." *Scientific American.* September 1960, Vol. 203, pp. 194–204.

DiVincenzo, D. P. "Quantum Computation." *Science.* October 13, 1995, Vol. 270, No. 5234, pp. 255–261.

Drancourt, M., Aboudharam, G., Signoli, M., Dutour, O., and Raoult, D. "Detection of 400-year-old *Yersinia pestis* DNA in human dental pulp: An approach to the diagnosis of ancient septicemia." *Proceedings National Academy of Sciences of the USA.* October 13, 1998, Vol. 95, No. 20, pp. 12637–12640.

Dyson, F. "Time without end: Physics and Biology in an open universe." *Reviews of Modern Physics.* July 1979, Vol. 51, No. 3, p. 447–460.

Easterbrook, G. "Greenhouse common sense." *U.S. News & World Report.* December 1, 1997, Vol. 123, No. 21, pp. 58–62.

Eldredge, N. *Life in the balance: Humanity and the Biodiversity Crisis.* Princeton, NJ: Princeton University Press, 1998.

Forrow, L., Blair, B. G., Helfand, I., Lewis, G., Postol, T., Sidel, V., Levy, B. S., Abrams, H., and Cassel, C. "Accidental Nuclear War: A Post Cold-War Assessment." *The New England Journal of Medicine.* April 30, 1998, Vol. 338, No. 18, pp. 1326–1331.

Gagneux, P., Wills, C., Gerloff, U., Tautz, D., Morin, P. A., Boesch, C., Fruth, B., Hohmann, G., Ryder, O. A., and Woodruff, D. S. "Mitochondrial sequences show diverse evolutionary histories of African hominids." *Proceedings National Academy of Sciences of the USA.* April 27, 1999, Vol. 96, No. 9, pp. 5077–5082.

Gehrels, T. ed., *Hazards Due to Comets and Asteroids.* Tucson, Arizona: The University of Arizona Press, 1995.

Gershenfeld, N., and Chuang, I. L. "Quantum Computing with Molecules." *Scientific American.* June 1998, Vol. 278, No. 6, pp. 66–71.

Gibbs, W. W. "Trailing a Virus." *Scientific American.* August 1999, Vol. 281, No. 2, pp. 80–87.

Giorgini, J. D., Ostro, S. J., Benner, L. A. M., Chodas, P. W., Chesley, S. R., Hudson, R. S., Nolan, M. C., Klemola, A. R., Standish, E. M., Jurgens, R. F., Rose, R., Chamberlin, A. B., Yeomans, D. K., and Margot, J.-L. "Asteroid 1950 DA's Encounter with Earth in 2880: Physical Limits of Collision Probability Prediction." *Science.* April 5, 2002, Vol. 296, No. 5565, pp. 132–136.

Grassl, H. "Status and Improvement of Coupled General Circulation Models." *Science.* June 16, 2000, Vol. 288, No. 5473, pp. 1991–1997.

Hansen, J. E., Sato, M., Ruedy, R., Lacis, A., and Glascoe, J. "Global Climate Data and Models: A Reconciliation." *Science.* August 14, 1998, Vol. 281, No. 5379, pp. 930–932.

Jensen, A. R. "The Limited Plasticity of Human Intelligence." *Eugenics Bulletin.* Fall, 1982.

Karl, T. R., and Trenberth, K. E. "The Human Impact on Climate." *Scientific American.* December 1999, Vol. 281, No. 6, pp. 100–105.

Kates, R. W. "Population, Technology, and the Human Environment: A Thread Through Time." *DAEDALUS.* Summer 1996, Vol. 125, No. 3, pp. 43–71.

Krauss, L. M., and Starkman, G. D. "The Fate of Life in the Universe." *Scientific American.* November 1999, Vol. 281, No. 5, pp. 58–65.

Leakey, R. "A race to extinction." *U.S. News & World Report.* August 5, 1996, Vol. 121, No. 5, pp. 46–47.

Leslie, J. *The End of the World: The Science and Ethics of Human Extinction.* New York: Routledge, 1998.

Lyons, S. A., and Petrucelli, R. J. *Medicine: An Illustrated History.* New York: Abradale Press, 1987.

May, R. M. "Kyoto and Beyond." *Science.* December 5, 1997, Vol. 278, p. 1691.

Monastersky, R. "Greenhouse Warming Hurts Arctic Ozone." *Science News.* April 11, 1998, Vol. 153, No. 15, p. 228.

O'Neill, G. K. "The colonization of space." *Physics Today.* September 1974, Vol. 27, No. 9, pp. 32–40.

Packan, P. A. "Pushing the limits." *Science.* September 24, 1999, Vol. 285, No. 5436, pp. 2079–2081.

Reilly, J., Prinn, R., Harnisch, J., Fitzmaurice, J., Jacoby, H., Kicklighter, D., Melillo, J., Stone, P., Sokolov, A., and Wang, C. "Multi-gas assessment of the Kyoto Protocol." *Nature.* October 7, 1999, Vol. 401, No. 6753, pp. 549–555.

Russell, G. J., Brooks, T. M., McKinney, M. M., and Anderson, C. G. "Present and future taxonomic selectivity in bird and mammal extinctions." *Conservation Biology.* December 1998, Vol. 12, pp. 1365–1376.

Schilling, G. "And Now, the Asteroid Forecast . . ." *Science.* July 30, 1999, Vol. 285, No. 5428, p. 655.

Schneider, D. "The Rising Seas." *Scientific American.* March 1997, Vol. 276, No. 3, pp. 112–117.

Smil, V. " How many billions to go?" *Nature.* September 30, 1999, Vol. 401, No. 6752, p. 429.

Spitale, J. N., "Asteroid Hazard Mitigation Using the Yarkovsky Effect." *Science.* April 5, 2002, Vol. 296, No. 5565, p. 77.

Vesilind, P. J. "In Search of Vikings." *National Geographic.* May 2000, Vol. 197, No. 5, pp. 2–27.

Westbrook, G. T. "More data needed to support or disprove global warming theory." *The Oil and Gas Journal.* May 26, 1997, Vol. 95, No. 21, pp. 75–76.

Wilson, E. O. *The Diversity of Life.* New York: Norton, 1992.

Winters, J. "A Brief Tour of a Bad Cosmic Neighborhood." *Discover.* April 1998, Vol. 19, No. 4, pp. 54–61.

Zimmermann, T. "The mystery of Lake Nakuru." *U.S. News & World Report.* August 5, 1996, Vol. 121, No. 5, pp. 46–47.

Index

Page numbers in *italics* refer to figures and tables.

374

blueshift, 13, 340
redshift, 13–15, 26, 36, 325, 340
black holes and, 61, 62, 72
Doushantou formation, 122
Down's Syndrome, 242
Drake equation, 111, 112
Drake, Frank, 111
dust disk (accretion disk):
black hole, 56, 62
stellar, 75, *76, 77,* 78, 80, 82
dwarf:
galaxy, 56, 58
star, 28, 58, 64, 71, 72, 79, 324, 326

E

Earth, 11, 12, 74, 79–84, 86, 87
age, 74
atmosphere, 82, 92, 96, 119, 304, 321
carrying capacity, 294, 296
climate, 86, 87, 289, 303–310
composition, 71, 80, 81, 83
core, 81
cosmology and, 10, 11
development of life and, 79, 80, 86,
87, 91, 92, 94–96, 98, 106, 114,
115, 118
end of, 259, 321, 324
escape velocity, 61
formation, 4, 71, 74, 80, 117
formation of moon from, 81, 82
natural resources, 290, 302, 303
orbit/rotation dynamics, 82, 87, 304
plate tectonics, 83–86, *85,* 125, 130
snowball Earth, 86
earthquake, 83, 292, 315
Easter Island, 182, 288, 290
Ebola virus, 299
echinoderms, 204
ecosystem, 132, 138, 142, 303, 319, 320
Egypt, 160–163, *162,* 258, 259
Alexandria, 167
ancient civilization, 158, 160–165,
162, 250, 252, 286
decline of, 163, 164, 288
early hominids, 136

mummies, 163, 297
pyramids, 163, 176
Einstein, Albert, 12, 13, 19, 48, 272
Einstein-Rosen bridge (wormhole), 62
electromagnetism, 40, 44, 47, 48, 202, 331,
332, 335, 339
field, 22
force, 17, 23, 24, 38, *40,* 41, 47, 48,
49, 332, 335
radiation, 5, 6, 39, 339
electron(s), 30, 38, 39, *40,* 42, 48, 50, 59,
326, 338
antiparticle (positron), 39, 326
covalent bond, 97, 98
degenerate electron pressure, 71, 72
early universe and, 25, 26
mass, 29, 39, 338, 340
electron neutrinos, 29, *40*
elements, 71, 74, 80–83
cosmic abundance, 25, 26, 94
created in stars (nucleosynthesis), 70,
71, 92, 93
life and, 94, 187
radioactive, 80
energy, 12, 15–17, 20, 23, 24, 30, 35, 38,
43, 48, 50, 56, 327, 331
basal, 205
biological, 91, 96, 99, 101, 115, 119,
120, 202, 234, 327, 335
black holes and, 62, 63, 327
chemical bond, 99, 101
civilization and, 288, 293, 301, 306,
308
conservation of, 21, 28
dark energy, 32, 36, 37
density, 15, 36, 326
energy and matter equivalence, 17,
38, 43, 70
fossil fuel, 126, 288, 301
Heisenberg Uncertainty Principle
and, 17, 38
kinetic, 15, 25, 51, 58
nuclear fusion, 58, 70, 71, 283
particle-antiparticle annihilation, 24,
39
rotational, gas cloud, 55

378

Grand Unified Theory (GUT), 41, 44, 45, *49*, 284, 326
gravity, 12, 13, 15, 17–19, 23–25, 30, 36, 37, *40*, 41, 43, 47, 48, *49*, 51, 53, 55, 58, 59, 70, 93, 112, 332, 335
    artificial, 290
    black hole and, 59, 61, 62
    Einstein theory of gravity, *see* relativity, general theory of
    escape velocity and, 61
    negative energy and, 21
    neutron star and, 59, 72
    nucleosynthesis and, 71
    quantum, 23, 42, 46, 326
    string theory and, 44
    supergravity, 44–46
gravitational lensing, *7*
graviton, 23, *40*, 44, 47
Great Attractor, 65, 67
Great Wall, 171, 172
greenhouse effect, 86, 128, 292, 305–307, 321
greenhouse gases, 92, 301, 304–309
Greenland ice sheet, 86, 245, 305, 307

H
Hadrian, 233
hadrons, 39
    Large Hadron Collider, 45
Halley's comet, 107, *108*, 310, 312
halophiles, 118
Hammurabi, 159
    Code of Hammurabi, 159, 297, 298
Han Dynasty, 172
Hanta virus, 299
Hawking, Stephen, 18
HD141569, *76*
Hebb's rule, 223
Heisenberg Uncertainty Principle, 17, 38, 47, 272
Heisenberg, Werner, 272
Herculenium, 292
Hermes (asteroid), 313
heterotrophs, 115
hierarchy problem, 41

hieroglyphics,
    Egyptian, 160, 161
    Mayan, 176
Higgs field, 41
Hinduism, 160, 257
Hippocrates, 298
Homer, 165
hominids, 135–147, *140, 141, 143,* 320
    bipedalism, 137–139, *140, 141*
    brain, 137, *141,* 142, 144, 145, 193, 205, 224–227
    fossils, 136–139, 142, 205, 226
    intelligence, 139, 142, 144, 151, 152, 274
    language, 152, 206, 224–227
    predators of, 138, 148
    sentience, 218
    speciation of, 139, 142, *143*
    tools, 138–142, *140, 141,* 144–146, 148, 153, 154
Homo erectus, 143, 144–147, 151, 152
Homo habilis, 141, 143, 142, 144, 145, 150, 151, 208, 218, 226
Homo heidelbergensis, 144
Homo sapiens, 130, 135, 139, *143,* 145–147, 151, 152, 274, 275, 277, 319, 322, 332, 335
Homo sapiens neanderthalensis, 143, 145, 319
homosexuality, 221, 230
hormones, 210, 221, 229, 230, 234–236
HR4796A, 75, *77*
Hsia Dynasty, 171
Hubble Deep Field, 53, *54*
Hubble, Edwin, 13, 14
Hubble's Constant, 14, 15
Hubble's Law, 14
Hubble Space Telescope, 9, 28, 53, 56, 113
human genome project, 198
humanist movement, 253, 254
human races, 147, 148, 190, 275
    intelligence and, 241
humors, four, 201
hydrothermal vents, 118, 119
hydrogen wall, 317
hypoglossal canal, 225, 226

383

Middle Ages, 94, 154, 170, 200, 255, 287
migration, hominid, 142, 146, 147, 154,
    155, 156, 290
        Australian, *155,* 181
        Beringian, *155,* 156, 173, 174
        Egyptian, 161
        genetic analysis and, 147, 154, 156
        language and, 152
        Mesopotamian, 158, 159
        Oceana, *155,* 182, 183
        transatlantic, 175
Milky Way galaxy, 13, 27, 51, 55, 56, 58,
    65, 72, 324
        collision, 324, 325
        core black hole, 56, 64
        extraterrestrial civilizations in, 111,
            112
        formation of, 63
millennium man, 138
Ming Dynasty, 172, 173
Minoan, 165
mirror symmetry, 24
Mithraism, 168, 259
mitochondria, 120
        DNA, 120, 146, 152, 319
mitotic clock, 235
MOND (Modified Newtonian Dynamics),
    30
monoamine oxidase (MAO), 210, 211
        violent behavior and, 211
Monte Verde, 174
Moon, 115, 290, 314, 334
        Apollo missions and, 5, 32, 81
        colonization of, 290
        formation of, 81, 82
        stability of Earth's rotation, 82, 87
Moore's Law, 278
mortality, 237, 244, 274–276, 299
        Ch'in Shi Huang Ti and, 172
        female, childbirth related, 165
        infant, 242
Moses, 258
Mount Pinatubo, 304
Mount Tambora, 292
Mount Vesuvius, 292
M-theory, 45

Muhammad, 260
muon neutrinos, 29, *40*
mutation, 190–192, 267, 322

N
Nakuru, 300
Napoléon, 161
natural selection, 122, 188, 189, 192–194,
    222, 226, 227, 237, 243, 267, 273–275,
    318
Neanderthals, 139, 142–146, *143,* 153, 319
        culture, 153, 237, 256
        DNA, 146
        language and, 152
Nebuchadnezzar, 159
nebula, 13
        Eagle (M16), *73*
        solar, 72, 74, 75, 82
Neolithic, 150, 153, 154, 156, 158, 165,
    170, 247, 256
Neoplatonism, 253, 254
neuron(s), 199, 202, *203,* 206, 208–216,
    209, 219, 223, 230, 236, 237, 265–267
neurotransmitter, 202, *203,* 210, 218
        acetylcholine, 224, 231
        dopamine, 215, 230–232
        GABA, 202
        glutamate, 202
        metabolism of, 202, *203,* 210
        serotonin, 210, 231, 232
neutrino(s), 25, 28–30, 48, 70, 71, 326
        dark matter candidate, 28, 29
        detection of, 29
        electron, 29, *40*
        mass, 28, 29, 41
        muon, 29, *40*
        sterile, 29
        superheavy, 29
        tau, 29, *40*
neutron(s), 25, 39, 40, 72, 98, 338
        neutron spectrometer, 290
neutron stars, 28, 59, 64, 72, 326
Newton, Sir Isaac, 11, 272
nuclear fission, 283, 301
nuclear fusion:

Mercury, 59
orbits, 11, 27
Planet X, 79, 312
Pluto, 78, 79, 111
Saturn, 111, 132
planetary systems, detection of, 75, 78, 112, 113
planetesimals, 74, 81
plasma, 26, 50, 70
quark-gluon, 25
plasticity:
human developmental, 200
neuron, 212, 219, 221, 223, 265
organic macromolecules, 97
plate tectonics, 83–86, *85,* 125, 130
Plato, 166, 200, 237, 252
Plutinos, 79
Pluto, 78, 79, 111
Polo, Marco, 172
Polybus, 201
Polynesians, 182, 183, 288
Pompeii, 292
population:
carrying capacity, 294, 296
exponential growth of, 293
historic hominid, 144, 145, 147
human global, 291, 293–297, *295*
positron, 39, 326
Pot, Pol (Saloth Sar), 240
predestinism, 3, 272
Principia, 11
Proconsul africanus, 136
prokaryotes, 118–120, 122
Propliopithecus, 136
protein, 94–96, 101, 102, 120, 190, 191
aging and, 236
amyloid, 199
behavior and, 228, 230
DNA binding, 104
enzymes, 99
glucose cross-links with, 234
synaptic receptor, 202
synthesis of, 101, 102, 104, *105,* 195–198, *196*
tau, 199
proteinoid, 98, 99, 101

proto-cell, 98, 99, 101
proto-city, 156
proto-continent, 84
proto-galaxies, 6, 26, 51, 53, 56, 64
proton(s), 23, 25, 39, 40, 48, 98, 338
decay of, 41, 326
neutron star formation and, 72
nucleosynthesis and, 25
proto-planet, 80
proto-star, 58
proto-sun, 74
Proxima Centauri, 6, 324
Pterodactyls, 129
Ptolemaeus, Claudius (Ptolemy), 11
Ptolemy V, 161
pulsars, 59
Punic Wars, 167

Q
Quantum Chromodynamics, 40
quantum computer, 278–280
quantum cosmology, 47, 50, 331
Quantum Field Theory, 37, 42, 43, 45
quantum foam, 42
quantum gravity, 20, 23, 42, 46, 326
quantum physics/effects, 3, 10, 17, 18, 20–23, 30, 38, 41–44, 46, 47, 62, 71, 272, 278–280, 331, 332, 335
consciousness and, 219, 266
quantum space-time, 18, 20, 38, 42
quantum universe, 17–23, 25, 33, 35, 47, 335
quark(s), 24, 25, 30, *40,* 40, 48, 72
quark star, 72
quasars, 53, 55, 56, 63
quintessence, 37

R
races, 147, 148, 190, 275
intelligence and, 241
radiation, 5, 6, 26, 28, 29, 39, 70, 339
black holes and, 56, 62
cosmic microwave background, 6, 9, 15, 16, 25, 26, 35–37, 47, 50
galactic core, 56

life and, 96, 111, 112, 118, 190
radiation era, 26
Ramses, 163
rapid eye movement (REM), 214
redshift, 13, 14, 340
cosmic microwave background radiation and, 15, 26
of galaxies, 14, 325
gravitational, 61, 62, 72
of type Ia supernovae, 36
Reformation, 170, 254
Relativity, 41
General Theory of, 12, 13, 15, 19, 22, 23, 45, 331
Cosmological Constant and, 13, 15
gravity and, 43, 44
simplification, of 12, 42
space-time continuum of, 18, 19, 21, 42, 43
Special Theory of, 331
religion, 94, 233, 237, 246, 247, 249, 255–262
Chinese, 254, 261
Christianity, 164, 168, 170, 253, 259, 260, 262, 289
Hinduism, 160, 257
Islam, 164, 170, 260–262
Judaism, 258–260
Mesoamerican, 176, 178, 258
Mesopotamian, 257, 258
primal, 256, 257
Roman, ancient, 168, 259
Renaissance, 170, 181, 188, 201, 253
replicator, 189
reptilian brain, 204
ribonucleic acid (RNA), 94, 101, 102
origin of, 102, 104
protein synthesis and, 104, *105,* 197, 198
structure of, 102
viruses and, 104
RNA world, 102, 195
RNP world, 102
Rodinia, 84, *85,* 86, 122, 125
Roman Empire, 146, 163, 164, 167–171,

169, 176, 233, 253, 259, 260, 287, 288, 294, 336
early history, 167
engineering, 168, 283
Roy Hill Shale, 120

S
Sade, de, Marquis, 222
Sargon the Great, 158
Saturn, 111, 132
schizophrenia, 215, 228, 230
dopamine and, 215, 230
Schwarzschild radius, 61, 334
Search for Extraterrestrial Intelligence (SETI), 113
senescence, 233–238
senescent cells, 235, 236
sentience, 212–219, 266, 271, 276, 335
shaman, 165
Shang Dynasty, 171, 261
sickle cell disease, 228
singularities, 17–21, 43
black holes, 18, 61, 62
initial singularity, 10, 17–23, 25, 33, 46, 327, 334, 335
Sloan Digital Sky Survey, 67
society:
early, 135, 150, 153, 160, 187, 256
future, 282, 285–291
lifetime, 111, 112
Middle Ages, 255, 287
Neanderthal, 153
religion and, 256
stages, 149, 245–252
technology and, 262, 276, 280–282
utopian, 262, 291
Socrates, 166, 253
Solutreans, 175
Sons of Light, 258
space, 3, 12, 17, 20, 38, 42, 46, 272, 331, 332
cosmological expansion, 14, 15, 21, 22, 33, 35, 36, 47
geometry (curvature) of, 18, 30, 35, 36, 43, 44, 325, 326

AIDS, 197, 243, 274, 296, 299, 322
computer, 280
Ebola, 299
genetic engineering and, 242, 243
Hanta, 299
influenza, 299

W

Wallace, Alfred, 222
Wallace's paradox, 222
Watson, James, 195
wave function, 18, 19
Weakly Interacting Massive Particles (WIMPs), 28, 50
weak nuclear force, 17, 23, 24, *40,* 41, 47, 48, *49*
Wernicke's area, 226, 228

white dwarf stars, 58, 71, 324, 326
World Wide Web, 264
wormhole (Einstein–Rosen bridge), 62

X

X rays, 71, 339

Y

Y2K, 280
Yarkovsky effect, 316
Yersinia pestis, 299
Yunnanozoon lividum, 123

Z

zero-point energy, 39
ziggurats, 158, 159
zygote, 6, 9